D0931394

The Ethics of
Geometry

The Ethics of Geometry

A GENEALOGY OF MODERNITY

David Rapport Lachterman

Routledge

New York and London

Published in 1989 by

Routledge
an imprint of Routledge, Chapman and Hall, Inc.
29 West 35th Street
New York, NY 10001

Published in Great Britain by

Routledge
11 New Fetter Lane
London EC4P 4EE

Library of Congress Cataloging in Publication Data

Lachterman, David R., 1944–
    The ethics of geometry : a genealogy of modern thought / David R.
Lachterman.
        p.   cm.
    Bibliography: p.
    Includes index.
    ISBN 0-415-90053-0; ISBN 0-415-90141-3 (pb)
    1. Geometry—Philosophy.   I. Title.
QA447.L33 1989
516'.001—dc 19   88-26518
                                                                      CIP

British Library Cataloguing in Publication Data also available

Contents

Preface

To write a book is to be schooled in the abridgment of ambition. By venerable conventions, a preface allows an author to state his aims more amply, while also acknowledging the lacunae in their execution. Since the subtitle of this volume [and its planned sequel (*The Sovereignty of Construction*)] bears the term the "genealogy," I shall try to do justice to these conventions by saying something about how this work itself came into being.

It had its remote origin in a graduate seminar in the philosophy of mathematics I conducted some years ago, when I was pressed to explicate the meaning and philosophical roles of Kant's phrase "the construction of a concept" as it appears in his theory of mathematical knowledge. Before long I was able to convince myself that Kant's phrase and the notion it is meant to convey were neither novel nor afterthoughts loosely bonded to the main body of his thinking (especially as transcribed into the first *Critique*); on the contrary, they give expression both to a key historical legacy on which Kant drew and to the central import of that "revolution in [our] style of thinking" toward which all of his efforts were bent. The historical legacy in question turned out to be, in the broadest terms, Cartesian, transmitted to him with many modifications by Leibniz and what is often called the Leibnizian "scholasticism" of Christian Wolff and his disciples. A fairly direct line runs, so I discovered, from the "construction of a problem" (Descartes), through the "construction of an equation" (Leibniz), to Kant's "construction of a concept." Kant's "revolution in [our] style of thinking" showed up in this light as an attempt to expend the entire capital of this constructivist legacy within every philosophical domain he sought to occupy or to transform, including epistemology, the theory of natural science, ethics, and his oblique, truncated, but nonetheless seminal philosophy of history.

Not long afterwards, these philological-cum-philosophical inquiries into Kant's work began to intersect with, and finally to become absorbed into, a much larger set of questions with which I was preoccupied in my teaching and research. Put baldly, What are the most salient features by which philosophical modernity is distinguished from philosophical antiquity (or, more precisely, from the philosophical thinking of Plato and Aristotle and their heirs)? In even more abbreviated terms: What is it to be a "modern"? Several clarifications and brief illustrations may be helpful at this point.

If we take Kant at his revolutionary word and try to grasp how this (immensely successful) word announced a critical future at the same time that it echoed and re-echoed the original, founding voices of modernity in the seventeenth century, then an understanding of the interplay between the history of mathematics and the history of philosophy becomes crucial. (Gassendi's claim "Whatever we do know, we know in virtue of mathematics" might well serve as the aphoristic slogan on the banner of these revolutionaries, for whom establishing credentials as authentic moderns was *not* an academic or historiographic issue.) Thus, to analyze this interplay in its specifically modern form becomes a matter, first of all, of uncovering what is novel in the practice as well as in the self-understanding of modern mathematics and hence allows it to act as a catalyst for the wider shapes and designs of philosophical theory.

It was in the course of trying to uncover just this radical or unique feature that I found my antecedent investigations of Kant's employment of the notion of mathematical construction or constructibility starting to bear unexpected fruit. As I point out in chapter 1, it is nowadays a commonplace to read that concepts, theories, systems, indeed "worlds," even "the world," are all constructs, that is to say, fabrications, figments, or projective fictions put together, produced by the human intellect, the human will, or some tantalizing mélange of the two. The primary thesis I defend in the following pages is that this now commonplace usage is not only the index of Kant's philosophical triumph but also, and principally, the outcome of a signal alteration in the way mathematics itself is practiced and understood in the early modern, pre-Kantian period. This alteration can be captured in two interconnected expressions, both to be found at the deepest stratum of the "Cartesian" soul in which the seeds of that triumph and its aftermath were originally planted: one, that mathematics is essentially occupied with the solution of problems, not with the proof of theorems; two, that mathematics is most fertilely pursued as the "construction of problems or equations"—that is, as the transposition of mathematical intelligibility and certainty from the algebraic to the geometrical domain, or from the interior forum of the mind to the external forum of space and body. (Symbolization and formalization, while also obvious hallmarks of modern mathematics, need to be understood as a supplements to, or variations on, the root idea of construction. Moreover, it is geometrical *construction,* and not the axiomatic method of geometrical demonstration, the *mos geometricus,* that serves to distinguish modern from ancient mathematics.)

This "alteration," as I have thus far circumspectly called it, is not simply an isolated or separable piece of technical mathematics or metamathematics narrowly construed. In the manifold philosophical appropriations and extensions of mathematics the substitution of problem-solving for theorem-proving and, in tandem therewith, the desire to move from conceptual inwardness to outward embodiment are emblematic of the modern project at its most forthright. Kant seizes upon this point when he says, recollecting Descartes' *Discourse*, that the method of mathematical construction seems to make man "the master of nature." One of Kant's most thoughtful readers, Solomon Maimon, makes the same point with even less reserve: "In this [mathematical construction] we are therefore similar to God."

On the reading of modernity proposed in the body of this work, to be "modern" in the most exacting and exalting sense is to be carried along this trajectory from mathematical construction (in its precise technical sense) to self-deification. The mind is not nature's mirror; it is nature's generative or creative source.

A third, subsidiary strand of reflection was taken up into the genealogical web I was weaving at a somewhat later time, when the insistent voices of postmodernism made the question of the true identity of modernity even more pressing. (It should already be clear that by "modernity" I intend something of ampler scope than what has come to be called "aesthetic modernism" or the "high modernism" regnant in the arts of the first decades of this century. From the perspective of this study, it is nonetheless illuminating to find the young Eliot, in his dissertation on Bradley, writing: "We have the right to say that the world is a construction. Not to say that it is *my* construction, for in that way 'I' am as much 'my' construction as the world is.") What is at stake here is the nature of the tie between the radical modernity inaugurated by Descartes and others and its termination, repudiation, or supersession by the "postmodernists."

The currently ubiquitous rubric "deconstruction" already offers some clues to this tie, despite the frequent disclaimers that a single term signifies unerringly an unchanging set of interpretive practices and sensibilities. We need to reflect on the fashion in which deconstructive post-modernism disassembles, takes asunder, what modernity strove constructively to connect, and how it does so first by taking modernity "at its word" and then by turning this word (or the texts in which it finds expression) against itself. Deconstruction, *on this reading,* is an exposé of the only thinly disguised "secret" of modernity itself—namely, the willed or willful coincidence of human making with truth or intelligibility. If modernity is, as I said earlier, a trajectory from mathematical construction to self-deification, then deconstruction is principally the discovery that the trajectory described is a *finite* parabola, such as Galileo's missiles follow when they explode upon impact. Self-deification via construction is replaced by the shards of Emerson's "god in ruins," also known as the end, or erasure, or disappearance of "man." If the radical modernists saw in the fusion of making and truth an endless "functionality" (allowing the one-way inference from certainty to utility), postmodernists expose the other side of that coin, stamped with the visage of self-abnegating, but inexorable, fictionality.

In another formulation: for the program of Cartesian *mathesis universalis* to work, an infinite will, if not also an infinite intellect, is required, assuring the passage of problem-solving *à l'infini*. The postmodernists, having become persuaded of the essential finitude of human knowing and speaking, to say nothing of writing, somehow leave intact the infinitude of a will embedded or embodied in a finite *and* inescapably self-deluding intellect. The result is a bizarre, for some admittedly beguiling, version of the Enlightenment's promise of endless progress. Like war vis-à-vis diplomacy, postmodernity is a continuation of modernity by other means.

However, for the post-modernist, there is no fundamental difference between antiquity and modernity, the latter being in one fashion or another the destined outcome of what the former began. The indeterminate dyad they comprise gives us a single figure, the "history of metaphysics," "phono- or logo-phallocentrism," or, most simply, "Platonism," in Nietzsche's deliberately vulgarizing or superficial formulations. Accordingly, the self-styled "moderns," whether in the seventeenth or in later centuries, were quite mistaken both in their self-estimation and in their polemics against the ancients (and medievals), unaware as they were, as they had to be, of the ineluctable destiny they shared with their imagined opponents. The "quarrel between the ancients and the moderns," transposed from the arena of stylistic assessment to that of thinking and writing at large, was no more than a shadow-fight, since the texts once thought to be caught up in a radical "battle of the books" turn out to belong to that one, common epoch of "metaphysics" now facing its multivalent closure.

Although I say or imply a great deal about such matters in the body of this book, I should remark here that in my genealogical account the history of mathematics is meant to serve as the critical counter-example to this embracing figure (the history of metaphysics or the era of logocentrism) and thus to revive the claims to radical originality made by the "moderns" on their own behalf and given initial legitimacy by the new styles of mathematical construction.

Authors such as Gaston Bachelard, Thomas Kuhn, and Paul Feyerabend have for some time made the notion of epistemological breaks or "paradigm shifts" in the history of physics paramount themes of debate in contemporary philosophy of science as well as elsewhere. In these debates, with very few exceptions, the history of mathematics is left unscathed by threatened ruptures and discontinuities. Rather, it is believed to serve as the exemplar of what continuous progress in cognitive achievement signifies.

Although I do not delve into the present-day question of the meaning(s) of incommensurability, coreferentiality, translation versus interpretation, and the like, my "genealogical" account is intended to be an illustrative challenge to this long-dominant view of a continuous, conceptually homogeneous history embracing ancient and modern mathematics. Cast partially in the idioms of the day, the position I extract from the texts investigated here may be stated as follows: An Apollonian locus is *not* the same as a Cartesian locus, even though sentences or propositions containing terms designating them may have the same truth-value. (The same would also hold, *mutatis mutandis,* for a Euclidean and a Hobbesian circle or for a Archimedean and a Leibnizian spiral.) "Not to be the same" is, of course, the salient phrase, and its import will, I hope, become increasingly clear from this investigation itself. Briefly, it is not a matter of reference or meaning; rather, the difference concerns the source of the intelligibility of the figure (or statement) at issue: in the one, the ancient case, this source is the nature of the figure in its own right, while in the other, modern, case, it is to bring the figure into visibility, clarity, being. A distinction in the manner of knowing *entails* a difference in the

I have tried to capture what strikes me as the most subtle at work in differences by entitling this volume *An Ethics of Geometry*, which is not merely a play on the title of Spinoza's most celebrated work. "Ethics" must be understood here in the Aristotelian sense of *ta ēthē,* as the settled or characteristic ways human beings have of acting in the world or of comporting themselves toward one another or toward themselves (for example, as teachers and students). The sense of "ethics" intended here has its archaic roots in Heraclitus' adage "A human being's *ēthos* is his *daimōn*" (Frag. 114). Accordingly, in speaking of a radical difference between the ethos of Euclid and the ethos of Descartes I am *not* suggesting anything like a moral discrimination between persons; rather, I have in view the disparate ways (*mores*) and styles in which the Euclidean and the Cartesian geometer *do* geometry, comport themselves as mathematicians both toward their students and toward the very nature of those learnable items (*ta mathēmata*) from which their disciplined deeds take their name. Hence, the difference in the source of intelligibility is itself an expression of this *ethical* difference. Viewed under this light the present work may be read as an attempt to rebut Aristotle's remark, made in passing in the *Rhetoric,* that "mathematical speeches have no *ēthē,* since they do not involve any resolute choice (*prohairesis*) either. For they do not have 'that for the sake of which' " (1417a19–20).

In the exemplary cases studied here, Euclid and Descartes, I shall be striving to bring to light the resolute and deliberate choices by which their geometrical mores are determined. Geometry has origins, not an origin (transcendental or historical), and these distinct origins are at bottom *ethical*. (In this respect, the foil to my study is Husserl's essay "The Origin of Geometry." The institution of geometry among the Greeks is not, in my account, retained in the intentions of Descartes and his successors.)

This completes my recollections of the three lines of investigation converging in this book: an initial attempt to make sense of Kant's phrase "the construction of a concept"; the pursuit of the notion of construction as a possible identifying signature of modernity generally; and the subsidiary effort to suggest the ways in which "postmodernism" is a coda, an envoi to radical modernity. It is now time to rehearse very briefly the sequence of my actual discussion, before singling out those lapses and lacunae already apparent to me.

In chapter 1, after delineating the contours of the question "What is it to be modern?" I turn first to Vico and then to Kant in order to throw initial light on the complicity between constructive mathematics and modern philosophy. I am especially concerned to show how the paradigm of construction pervades and beguiles Kant's thinking even when he seems to be at the greatest remove from mathematics. The legacy left by Kant's "revolution" is appropriated not only by Fichte, but also by Nietzsche, who raises the original idea of knowing as construction to a still higher power, until he can draw the exhilarating or dispiriting inference: "Our salvation lies not in knowing, but in creating!" At the end of this chapter I suggest, in very compressed terms, how much of contemporary philosophy may be interpreted as a family quarrel among Nietzsche's descendants.

In chapter 2 I furnish a detailed account of some prominent features of Euclidean geometry as it is transmitted to us by Euclid. After studying the relationship between pedagogic prudence and technical virtuosity in the theory of ratio and proportion, I turn to the place of "construction" in Euclid (and other ancient geometers). This section is pivotal to my claim that early modern mathematics breaks with ancient mathematics by promoting *constructions* to a paramount position in the body of mathematics. Hence, it is necessary for me to undermine the orthodox view, first established by Kant and his disciples and now enshrined in Heath's translation and notes to the *Elements,* that for Euclid constructibility gives proof of the "existence" of the mathematical entities at issue. A long excursus into the history and contexts of the concept of "existence" is crucial to the success of this attempt; it is meant to bring home how theory-laden, how far from innocent, that concept is when it appears in theses concerning Euclidean constructions. A sketch of an alternative understanding of those constructions (and of the postulates) concludes this chapter. (Readers of this second chapter will want to compare two recent works which have been both provocative and useful to me during my final revisions: Wilbur R. Knorr, *The Ancient Tradition of Geometrical Problems* [Boston, 1986], and David H. Fowler, *The Mathematics of Plato's Academy: A New Reconstruction* [Oxford, 1987].)

Chapter 3 brings me to the heart of the matter: Descartes' qualifications to be named, as he often is, "the father of the modern mind." Before turning to the *Geometry* itself, I study the themes and acts of self-origination or self-making as they appear in the *Discourse on Method* and other writings. In the next two sections of this chapter I explore the strategies and commitments of Descartes' "geometry," first in the setting of the treatise of 1637 and then in that of the later *Meditations*. Here "construction" in name and in practice, receives its first decisive expression, for reasons that go to the core of the Cartesian revolutionary spirit, especially as this gives his programmatic *mathesis universalis* its *élan*.

In the sequel to the present volume, *The Sovereignty of Construction*, I intend to consider Hobbes, Leibniz, and Kant, respectively, in light of the genealogy begun here. Hobbes will be important by virtue of his emphasis—to the exclusion of symbolization—on the sovereign role of construction (and genetic definition) in mathematics and, indeed, in every field amenable to science. In contrast to Hobbes, but likewise ornamenting and modulating the Cartesian theme of construction, Leibniz will be examined in terms of his critique of the unsatisfied will to completeness or comprehensiveness marking Descartes' version of *mathesis universalis;* the significance of *analysis situs* as both a complement to Cartesian algebra and, on his interpretation, a restoration of figural geometry; and the symbolic underpinnings of his "universal characteristic." At stake in these discussions will be not how the mind can act as a faithful mirror of nature or rectify its own refractive impediments so that its ideas might resemble nature, but, in a phrase from the young Leibniz, how *mens facit phenomena*—how "the mind makes phenomena." This theme will carry us into the subsequent history of modern philosophy and especially to Kant and his continuation of the Cartesian geometrical and algebraic legacy. A question of central concern will be: How does the mind phenomenalize or exhibit itself, make its own appearance in the "external" world? My discussion of Kants' theory of mathematical construction will thus complete the circle begun in the first chapter of *The Ethics of Geometry.*

A final word on genealogies. Many do tend to be defensive or offensive (compare Nietzsche's *pudenda origo,*) offering palliatives or serving polemics. Several people kind enough to read earlier versions of *The Ethics of Geometry* have concluded that its genealogical narrative is an unambiguous defense of the ancients, even a recommendation for some sort of return to the Greeks. This is certainly not the aim of the present study. To understand why and in what sense the self-styled moderns understood themselves as radically modern, modern from the roots up, it is of course essential to grasp the shape of the antecedent tradition those moderns deracinate and discard. Only with this shape in our grasp can we begin to see how the moderns try to put an end to the "long-standing contention" between philosophy and poetry by making mathematics and its intellectual extensions into a kind of poetry, that is, of *poiēsis* in the sense of construction and production. Hence my Kantian epigraph, freed from its immediate context in the *Opus Postumum*, "Mathematics is pure poetry."

One "ethical difference" does, however, merit comment, if not assessment, in its own right. For the ancients who appear in my narrative, circumspect speech, about geometry or about the soul, seems to mark out the field in which learning and teaching take place. *Mania*, which may be the highest possibility of speech, somehow remains tied to pedagogical circumspection. The epigraph to chapter 3, from Pope's *Dunciad*, is an initial indication of how the radical moderns break this tie: "Mad mathesis alone was unconfined."

Acknowledgments

Early versions of some of the material in this volume were presented in invited lectures delivered to the Department of Mathematics, Swarthmore College and to the Departments of Philosophy, Vassar College, Syracuse University, The New School for Social Research, The Pennsylvania State University and San José State University. Chapter 3.II is adapted from my essay "Descartes and the Philosophy of History," *Independent Journal of Philosophy* 4 (1983); Chapter 3.IV is a revised version of " 'Objectum Purae Matheseos'. Mathematical Construction and the Passage from Essence to Existence," published in *Essays on Descartes' Meditations*, edited by Amélie O. Rorty (Berkeley: University of California Press, 1986). I am thankful to both sources for permission to reprint.

I am indebted to Mildred Tubby and Bonnie Schaedel for their help in the preparation of the manuscript. I am also grateful for the support given to me towards the completion of this volume by Dean Hart Nelsen, College of the Liberal Arts, and by Professor Carl G. Vaught, Head, Department of Philosophy, The Pennsylvania State University.

The assistance provided by Stanley Rosen, in this and in other undertakings, has been both exceptional in variety and magnanimous in depth.

Finally, I want to thank Brett Singer-Lachterman for her aid and my son Samuel Lachterman for inspiring the theme of a genealogy.

"Mathematik ist reine Dichtung." (Kant)

1

Construction as the Mark of the Modern

Aber, woher bin ich denn?
Kant

I Projection, Construction, and the Idea of Modernity

"Il nous faut être absolument modernes," Rimbaud writes toward the end of *Une Saison en enfer*—"We must be absolutely modern." All of us have felt the force of Rimbaud's imperative, since all of us, even those with the glint or mote of "postmodernism" in their eyes, find ourselves in a world for which we seem to have no designation more apt than "modern." But what does Rimbaud's imperative demand of us? What does this familiar designation *mean*?

The Latin word *modernus* and its later European derivatives—already veterans of many semantic and historiographical campaigns, beginning with Cassiodorus in the sixth century—have all along displayed resistance to univocal and lasting definition. In the new era on which the word "modernity" was to bestow a collective name, this aloofness from invariant denotation soon turned up as a readiness to embrace the most disparate possibilities of meaning. Hence, reflection upon the possible unity of "modernity" as well as frustration with its elusiveness seem somehow part and parcel of the phenomenon of modernity itself.[1]

We are dealing with a phenomenon far more important than the semantic vagaries of a simple word, for it is with this word (and its intended powers of reference) that the age it was chosen to designate signifies and epitomizes its historical self-consciousness and, indeed, its historical uniqueness. "Moderns," from the early seventeenth century on, are aware of being deliberately "modern" in a fashion and to a degree unparalleled in premodern epochs, even when thinkers and writers of those earlier ages also linked together chronological periodization and self-understanding or self-identification. (Compare, for example, the theme of the *aetas nova* in the Renaissance.)

Linkages of this kind—whether modern or premodern in form— seem to point to a deeply rooted desire not simply to locate oneself or one's coevals *in* time but, more fundamentally, to domesticate time, to make of one's age or epoch a dwelling. The dwelling might be hospitable or might prove uncongenial, but in any event it furnishes a sense of orientation, of one's whereabouts, as though having a sense of place were always a matter of timing. Periodization, then, is not a coincidental affair of counting the passage of mathematically equable time but a mode of reckoning with the possibilities of rootedness and vagrancy, homecoming and nostalgia, conviviality and exile. Historical periodization, with the self-knowledge meant to accompany or follow from it is, one could say, the temporary replacement or surrogate for the persistence of eternity.

"Modernity," even though it participates in this generic enterprise of periodization, does not regard itself as one specific instance among many, but as *unique* by virtue of being the consummate historical "period." From its inception, it views itself as consummate in a double sense: as the "final" period which brings to a permanent close all prior epochs by exposing their imperfections and as the one period without a temporal finale. Its immanent end or telos is endlessness. Hence, the image of growth from historical infancy to historical adulthood, often figuring in early modern rhetoric, had to be abandoned as soon as those using it recalled that adulthood is naturally followed by senescence and death. Pascal had already captured the true sense of modern endlessness more faithfully in the preface to his *Traité du vide* of 1647 when he asserted: "Man is produced only for the sake of infinity." Less confidently, perhaps, but still with recognizable empathy with the Pascalian sentiment, Karl Gutzkow in his work *Die Mode und das Moderne* (1836) writes: " 'How and whereby [*wodurch*] are we?' asked the Ancients; 'What are we?' the Medievals; 'Whereto [*wozu*] are we?' the Moderns."[2]

With these considerations something like a common shape begins to take form within what otherwise seems a chaotic manifold of the meanings of "modernity." To be modern now means at least this: to have a share in or to be caught up by a *project,* an enterprise *projecting* itself by anticipation into an unbounded future. It means, therewith, to gauge any present by its bearing on the future, however dark the details or design of the latter might now be.

This project does not get underway spontaneously, however willful it will turn out to be in its essence. Rather, as the modern exercise in periodization has already indicated, the momentum initiating this *projection* is drawn from the simultaneous *rejection* or thoroughgoing critique of the premodern, or, more specifically and fundamentally, the "ancients." The "quarrel of the ancients and the moderns" was much more than a literary parlor game, a bookish battle of shadow-contestants. Indeed, the literary versions of the quarrel often recapitulate in illuminating ways prior engagements on the field of philosophy and science; thus,

Perrault, spokesman for modern writers toward the end of the seventeenth century, invokes the examples of Copernicus and Descartes as proof of modern superiority.[3]

Consequently, the "shape" we initially discerned has a more complex inner structure. "Modernity" along with its projection *ad infinitum* makes sense to itself also by reason of its unabating antithesis to antiquity. Modernity, instead of being simply open-ended in the direction of the future, is perennially bound to, and partially determined by, those past habits and expectancies of thinking, speaking, and living it aims to uproot and decisively to surpass. Modernity, then, is "radically novel" by design and, at the same time, by the inextirpable reminiscence of what it rejects. Its relation to antiquity is one of "determinate negation" in Hegel's sense, not one of sheer incommensurability, discontinuity, or abrupt oblivion. The moderns, or at least the most self-conscious among them, must *silence* just those voices whose language they best understand. Herewith, we run up against the engima of a beginning which is, on the one hand, intended to be "absolute" and radical (presuppositionless) while, on the other, it can make sense and carry the day *only* insofar as it is a *new* beginning: The antistrophic radicalism of the moderns in fact presupposes the antecedent strophe of the ancients.

Projection and rejection, anticipation and reminiscence are themselves simply formats or horizons of significance still admitting a wide variety of more determinate shapes and contents. Thus, in the early nineteenth century, when the "quarrel of the ancients and the moderns" was expressly revived, most prominently in connection with the history of art, Friedrich Schlegel counterposed "the Beautiful" as the principle of antiquity to "the Interesting" as the principle of modernity, while Schelling, in his lectures on the philosophy of art, wrote:

> We can call the modern world generally the world of individuals; the ancient, the world of genera [*Gattungen*]. . . . In the latter, everything is eternal, lasting, imperishable, number has as it were no power, since the universal concept of the genus and that of the individual coincide; here—in the modern world—the ruling law is variation and change. Here, everything finite passes away, since it is not in its own right, but is only to signify the infinite.[4]

And Schleiermacher tried to capture the essential difference between ancient and modern thinking in the terse formulation: "The old philosophy is predominantly the becoming-conscious [*Bewusstwerden*] of Reason in the form of the Idea; the new is predominantly the becoming-conscious of Reason in the form of the Will."[5]

Each such formulation of the polarity "ancient-modern" belongs to its own wider domain of commitments and preoccupations and cannot be divorced from it without losing its particular tenor and allure. So, once again, it appears that a protean character is inherent in the form of "modernity," a character only temporarily fixed by acts of *force majeure*. The quandary of intrinsic versus extrinsic unity is inescapable: If there is no single "look" to the polarity "ancient-modern" and thus no true form we can persuade protean modernity to disclose, we seem to be left to our own devices for making a one out of an irrepressible many.

I want to claim that the "idea" giving significant shape to the "constellation" of themes ingredient in modernity, *in both its revolutionary and projective modes,* is the "idea" of construction or, more broadly, the "idea" of the *mind* as essentially the power of making, fashioning, crafting, producing, in short, the mind as first and last *poiëtic* and only secondarily or subsidiarily *practical* and *theoretical.* Let me emphasize from the outset this altogether basic claim which the body of my book is meant to elaborate and illustrate: For the radical moderns and their heirs, making—understood as (trans)formative or "creative" technique—is neither an occasional nor an indispensable "feat" performed by rational souls, the latter when it serves the indisputable needs of the body, the former when it puts artfully on display distinctively human activities; instead, making is *definitive* of the mind's "nature" or better of its comportment in and toward the "world." We could parody Aristotle's account of *nous* by saying of the modern mind, "Its actuality is making." It will also follow from the strongest version of my claim that the contest with the ancients intrinsic to the idea of the modern is not one between competing theories or conceptions of the "mind," as though this term named a philosophically neutral agency with ancient and modern renditions. "Mind," as Richard Rorty has recently suggested, is itself a modern "invention";[6] it is, one needs to add, tailor-made to fit the specifications required for competence in making and constructing. *Nous,* as seen from the vantage point of "mind," is congenitally deficient or incompetent in this vital respect, as Bacon insists in *The Great Instauration.* The ancients' failures or insurmountable limitations in other respects (for instance, in their ethical and political thinking) ultimately stem from this incompetence; just as, conversely, both the achieved and the anticipated excellences of the moderns result from their virtuosity as makers. Accordingly, this "idea" of the modern constellation, if it proves illuminating, should allow us to make coherent sense of a wide range of phenomena, all of which we have good reason to associate with modernity even when they stand at considerable distances from one another—phenomena such as individualism *and* the mechanical organization of the "state," a belief in human perfectibility *and* in the ineluctable need or desire for self-preservation, the inwardness of selfhood *and* the program of conquering external "Nature." Furthermore, we would have arrived at the "idea" most apt to shed light on the otherwise puzzling relationship between form and content in the signification of "modernity." Under the sign of making or construction, modernity *is* an empty form, indeed, the form of endlessly iterable projection, but it is so because it is receptive to all those possible contents which carry the seal of human "fabrication" in its most liberal sense. Projection, then, belongs to the idea of modernity, as the programmatic anticipation of an endless sequence of human feats, while productive virtuosity is the touchstone discriminating genuine from bogus feats, determining when a content did or did not result from a deliberately crafted project. Similarly, the polar antithesis between ancient and modern no longer simply marks off chronologically distinct periods, but turns on an ontological axis; the genuine and the bogus are distinct styles of being. The ancient, necessarily retained in memory as what modernity had to negate in order to secure its own identity, now names the inauthentic itself, that is, the recollected absence of the projected infinitude of human making. (In Greek, the *autarchēs*, the "authentic one," is the one who makes something with his own hand.)

I have several reasons for choosing the term "construction" to stand for the entire family of performances such as making, production, fabrication, and so on. First, the idioms of "constructing," "construction," and "construct" are pervasive, not to say ubiquitous, in contemporary parlance, both ordinary and technical, especially, but not only, when it is a matter of singling out the results of human ingenuity. "Concepts," "theories," "systems," "worldviews," "worlds," and even "reality" are alike labelled "constructs." "Constructing," on however vast a scale, is both child's play (as in Piaget's *The Child's Construction of Reality*) and a collective social deed (as in Berger and Luckmann's *The Social Construction of Reality*); its instances, major and minor, are either deliberately fictional or determinedly serious or both at once. It is used honorifically ("someone's marvellously impressive construction," said, for example, of a philosophical theory) and depreciatingly ("a mere construct"). But the very pervasiveness of the idiom muffles its original resonance, as I shall try to show; for the most part, the forceful claims inseparable from its inaugural setting remain alive in its now-colloquial inflections only in blunted or palliated form. Hence, to restore "construction" to something like its pristine context is already to be in search of the birthright passed down to the present-day descendants of the "founding fathers" of modernity.[7]

Secondly, "construction" in the strict and technical sense in which it is used in late sixteenth- and seventeenth-century *mathematics* (see chapter 3) stands patently at the center of the modern constellation in this inaugural, self-formative period. Let me try to clarify the resulting state of affairs. Over the history of its usage "construction" has had both strict and wide applications. It should be noted, however, that the wide applications (as in "construction of reality") remain tributaries to the strict (as in "construction of an algebraic equation"); in return, the former enjoy the latter's protective authority.[8] Put differently, "construction" can name both one paradigmatic element in the fully delineated constellation of modernity as well as the whole constellation (for which we also have the equivalent designations "making," "production," "creativity," or *poiēsis*.) This relation of part to whole is perhaps best captured by an image drawn from medieval heraldry, the *mise en abîme*, in which the design of an entire blazon is repeated in one of its quarters, and so on, without limit.

This image, despite its aptness for expressing the emblematic relations between the narrow and the wide uses of "construction," nonetheless remains too static to do full justice to the dynamic situation I shall spell out in some detail in later chapters. The successes thoughtful mathematician-philosophers enjoyed thanks to the technique of construction, *sensu stricto* (see especially chapter 3) transmitted a momentous, motivating power to analogous endeavors in other regions which, at first glance, might seem remote from "mathematics" as traditionally construed (namely, in "pre-modernity"). This transmitted power sustains the confidence of the early moderns that extending the same technique (or family of techniques) to ever wider domains will prove equally fruitful. We begin to witness in this process the workings of what I am tempted to call "promissory induction": the belief that one has the resources to make the future like the past or, more accurately, to ensure that nothing will check (or even retard, for long) the progressive expansion of power, since this campaign can always be backed up by appeal to the evident or unchallenged results obtained in the primary instance. As a result, projection is itself sustained by the workings of construction.

Consider the dramatic role played by the etymology of *mathematica* in Descartes' *Regulae* at just the moment when the expression *mathesis universalis* makes its one and only appearance in his discourse: The privilege given to a "part" (that is, only some disciplines and sciences) of bearing the name of the whole class can only be explained by understanding how that part fulfills to the supreme degree the pledge carried by the root-name *mathesis*, "learning." Those new regions, which at first blush appeared remote from mathematics, begin to disclose their deeper affinity with it and, hence, with one another (the transitivity of affinity) when seen as regions in which *mathesis* is the generic goal. It therefore makes sense—or so the early moderns are persuaded—to reenact those same procedures for genuine learning which showed themselves so fecund in the primary instance.

"Construction," then, names the "idea of modernity" by bringing to sight a coherent pattern linking together the disparate or mutually distant appearances habitually and convincingly labelled "modern" into a single constellation. Equally, it names that element within this constellation which not only emblematically reiterates the whole, but also acts as its generative source, in much the way a charged particle generates an infinite field throughout which its energy can everywhere be detected.

II   "Master of Nature, So to Speak": Consequences of Construction in Vico and Kant

The identification of mind as essentially constructive is, then, the initiating proposal through which the constellation of modernity comes into focus (and, indeed, came into focus for the radical moderns themselves.) In chapter 3 I shall be studying how this initiation is carried through in the opening phase of modernity, Descartes' elaboration of his new geometry. This chapter on its own does not tell the whole story; skepticism concerning the alleged generality and originative force of the motif of construction in its narrow sense (the construction of geometrical problems or equations), may well persist. Consequently, I have selected for brief inspection two distinguished episodes within post-Cartesian thinking that give at least a preliminary sense of the subsequent fate of this motif, a sense for the history of its permutations and amplifications, its complications with other systematic hopes (and disappointments). By examining these episodes, in the work of Vico and Kant, I can bring out more specifically how mathematics in the modern age has acted as the suggestive science for philosophy in general.

Vico's role in the history of post-Cartesian thinking is of considerable interest here, not least because of contemporary efforts to make of him a genuinely *alternative* starting-point for modern philosophy, attempts in which his rhetorical-topical style of understanding the history of human institutions is contrasted with the analytical and mathematical method of Descartes. These efforts are, in the main, misguided inasmuch as they overlook the essentially mathematical roots and orientation of Vico's conception of human knowing.[9] To put the matter as synoptically as possible: His most famous (and most controversial) proposition, *verum et factum convertuntur*, "the true and the made are convertible [that is, have identical denotation]," abbreviates this mathematical conception of knowing while at the same time disclosing its source in one and the same modern understanding of mind, of mind as the source of making, shared by Vico's Cartesian interlocutors. Hence, Vico's debate with the Cartesians concerns (1) the right interpretation of the mind's mathematical activity and (2) the paradigmatic implications of this activity for human cognitive achievement in other domains. It is in his early works, *De nostri temporis studiorum ratione* (1708) and, above all, *De sapientia Italorum antiquissima* (1710), that Vico spells out most fully his position in this debate. His case against the Cartesians is put in terms of the opposition between "synthetic" or "Euclidean" and "analytical" geometry, but Vico presupposes that this opposition is itself a tension between two versions of the thesis that the mind "makes" the objects of mathematics. Analytical geometry, which he understands quite narrowly as a technique for invent-ing symbolic algebraic equations, is the inadequate version of this thesis, since in this case only the result, the *opus*, is certain, while synthetic geometry "is most certain both in its result and in its operation" (*ideo tum opere, tum opera certissima est*). In a word, Vico's defends, in putative opposition to the Cartesians, a Euclid already brought under the aegis of the modern transformation of geometry from a contemplative to an operative science—that is, a Euclid centrally occupied with the formation or production of determinate figures and their relations. In his own words in the first of the two replies to his early critics "mathematics are commonly thought to be contemplative sciences and not thought to give proofs from causes; when, in fact, they alone among all the sciences are the truly operative ones [*operatrici*] and give proofs from causes since, of all the human sciences, they, uniquely, make their way in the likeness of divine science."[10] (As we shall see, "the likeness of divine science" lives on as an ambiguous or tantalizing residue in post-Vichian versions of the identity of *verum* and *factum*.)

Vico's *New Science* of "the human things" (*le cose umane*) does not imitate this operative, synthetic, or constructivist geometry by means of the rhetorical scaffolding of its definitions, axioms, and the like; Vico, as much as Spinoza and Leibniz before him, recognized that the *mos geometri-cus*, demonstration from axioms, plays into everyone's hands. Rather, Vico approaches all of "the human things," the "customs, laws, and ideas of the gentile nations," as constructions or *poiēmata* that must be understood by tracing them back to their originative elements. In geome-try we are in command of these originative elements from the start—whence the equal transparency of the work (*opus*) and the operation (*opera*). In forging his new science of human works Vico must work retrospectively, since prior to the articulation of this science "man by not understanding makes everything" (*homo non intelligendo fit omnia*). (*The New Science*, par. 405). His science retrospectively "makes" the truth by arranging and sending forth the elements and the guises they assume from their birth to their dissolution and recurrence. "Indeed, we make bold to affirm that he who meditates this science narrates to himself this ideal eternal history so far as he himself makes it for himself by that proof it had, has and will have to be."

"Ingenuity is given to man for the sake of knowing, that is, of making [*ad sciendum seu faciendum*]," as Vico says in *De antiquissima*. While his version of modernity remained almost wholly eclipsed until fairly recently, at least one earlier reader, F. H. Jacobi, was alive to the affinities between Vico and Kant. According to Jacobi, the Kantian revolution, comparable to the Copernician in astronomy, makes fully evident what Vico had adumbrated, namely, "that we can grasp an object only insofar as we can let it come into being before us in thoughts, can make or create it in the understanding."

Kant's "critical enterprise" is intended to set its seal on the victory of modern thought over pre-modern thought. Subsequent events in philosophy, if not elsewhere, have ratified Heine's mordantly humorous description of Kant as "the arch-destroyer in the realm of thought" who "far surpassed Maximilian Robespierre in terrorism," a description which Kant, with his proclamation of the need for a "Revolution der Denkart," a "revolution in [our] style of thinking," would not have eschewed.[11]

Kant's "revolution" has more than one meaning in his texts. While it clearly does carry the sense of an irreversible breakthrough or transformation (in the style of thinking) and thus stands to philosophy in much the same relation as political revolution stands to the empirical commonwealth, it also suggests, at least implicitly, a reversion to the original astronomical image of a completed turn. Kant turns back full-circle to the starting points, not of theoretical philosophy or metaphysics, but of modern physical science and of ancient mathematics (interpreted in a modern light) in order to retrieve for metaphysics the conditions behind their self-evident or *de facto* success and progressiveness. In this sense, his "revolution" is a matter of bringing modernity back to its senses, of forcing it to recollect, now in full philosophical awareness, its own propitious inception and hence to combat the twin perils of skepticism and dogmatism. Finally, the Kantian revolution is meant to be pacific, to put an end to the sterile contentiousness of warring schools or traditions with a "peace treaty for philosophy" in which each signatory can recognize both his own autonomy and the justice due to his rivals. In this third sense the revolution ought to be a way of making good on the promise of eventual unanimity contained in the early modern appeal to an impersonal, universal method.[12]

What is common to these connotations of "revolution" is brought out most clearly in an appendix to one of his last writings, *Der Streit der Fakultäten* (1798). Kant let a young disciple, C. A. Wilmans, speak in his behalf: ·

the old philosophy assigned to man an entirely incorrect standpoint in the world by making him into a machine within the world, a machine which as such was meant to be wholly dependent on the world or on external things and circumstances; in this way it made man into an almost passive part of the world.—Now the *Critique of Pure Reason* appeared and allotted man a thoroughly *active* existence in the world. Man himself is the primordial creator [*Schöpfer*] of all his representations and concepts and ought to be the unique author [*Urheber*] of all his deeds."[13]

In this selection from Wilman's resumé of the intentions and the impact of the first *Critique* what stands to the fore is the emphatic contrast between the mind's passivity in precritical thought and its activity, to which Kant for the first time allegedly does justice. "A thoroughly active existence" encompasses both the mind's role as the "creator of all its representations and concepts" and its obligation to be "the sole author of all its deeds," or, more simply, its theoretical and its practical autonomy. These two modes of autonomy were suppressed or obscured in "the old philosophy," by which phrase we must understand not simply *all* of pre-Kantian philosophy but, rather, premodern philosophy, together with all those manifestations of (historically) "modern" thought which are (philosophically) deficient in self-understanding. (Although this deficiency, for Kant, was endemic prior to his text, it was not congenital, as the example of mathematized science shows.) This retrospective Kantian manifesto brings home as forcefully as possible how the modern invention of "mind" is tied to the primacy of human activity or, indeed, making. Passivity, and with it any hint of the mind as *mirroring* the domain of external things, is the signature of precritical thinking.

However, the enthusiasm displayed by Wilman should not mislead us. It is equally central to Kant's "critical enterprise" that he recognizes the inevitably mitigated or compromised nature of our "active existence in the world." Indeed, construction, far from being simply an ingredient in Kant's technical account of mathematics, comes into relief as the mark of that mitigation, the sign that *human* liberation also remains in partial bondage to congitive demands or ideals we can never adequately fulfill.

Thus, in his letter to Marcus Herz in 1772, Kant claims to discover "the key to the whole secret of heretofore obscure metaphysics" by examining "the ground of the relation between that in us which we call representing [*Vorstellung*] and the object." He discards *two* possible accounts of this ground: the first, that this ground is to be found in the object, whose effect in us is our representation; the second, "that representing is itself active in regard to the object, that is that the object itself is produced [*hervorgebracht*] by the representation (as when divine cognitions are conceived as the archetypes of all things)."

Human cognition *via* representation lies between these two extremes of passive receptivity, on one side, and creative activity, on the other. Mathematics gives us our best approximations to the latter extreme, that is, to the thoroughgoing freedom to *produce* objects answering to intellectual representations. "The concepts of quanta are active in their own right [*selbstthätig*]" and therefore independent of the circumstantial deliverances of sensation or sensuous intuition. Approximation to the ideal of a thoroughly free divine or archetypal intellect yields at one and at the same time the basic sense of our "active existence" and the limits or mitigations to which this active existence is inevitably subject. Moreover, these limits become ever more constraining the further we move from mathematics into regions in which no recourse to mathematics is available. The following, necessarily brief, survey should make these claims clearer.

Liberation and its benefits are the theme of the well-known passages from the preface to the second edition of the *Critique of Pure Reason* in which Kant explores the reasons why mathematics and physics, but not, up till now, metaphysics, have been able to proceed surefootedly along "the royal road of science" (Bxii), the first since the days of Thales, the second only since Copernicus and Galileo. In both cases, the decision was taken to undo the habitual subordination of mind to the (pregiven) "object" of inquiry by making the latter's intelligibility depend on what the inquirer has inserted in the object in advance, in accordance with the relevant concept he has of it. What we learn (a priori) from "nature" is only what we have already inserted into "nature." Or, in other terms, the cognitive success of science depends on freedom or detachment from any sensory or noetic authority credited to "nature" by the premoderns.

In mathematics (and in the *rational* physics Kant only alludes to here) this a priori insertion is achieved by means of construction, the rule-governed procedure for supplying, in advance of sensory experience, a sensible (but nonempirical) intuition "corresponding to" the relevant concept (such as the concept of a triangle).

I merely note here that Kant takes his understanding of the technique of construction from *algebra* (and not, therefore, from "traditional" geometry or arithmetic). Kant's phrase "construction of a concept" is derived from the expression "construction of an [algebraic] equation," which he also employs on occasion. This latter expression, taken from Christian Wolff, refers not to putting together the equation but to the interpretation of the terms of the equation in ways that lead to the actual exhibition of a particular geometrical formation satisfying the general equation.[14]

In the present setting the role of construction is of surpassing importance since it points us towards the source of what may be called the essential *aporia* of the Kantian revolution: (1) Construction provides philosophy with a pattern to be followed if it is finally to enter on the secure path of science; (2) but, construction, in the same sense, marks the unbridgeable gap between mathematics and philosophy ("knowledge through [uncon-structable] concepts"); philosophic imitation of mathematics shows us that "examples are contagious" and that mimicry is a form of self-flattery;[15] (3) and yet, "construction" (and its successor terms) remains the name for the *telos* that philosophy, in the theoretical, practical, and aesthetic domains, continues to propose to itself without any hope of requital (save, perhaps, through the "historical" realization of the "highest good"—except that this fulfillment can only be appreciated by rational agents as an "infinite task"). There seems to be no (Kantian) passage out of the impasse at which these three routes of his thought meet.

How does Kant work himself into such a perplexity? The initial charac-terization of construction drawn from the preface to the first *Critique* did not sufficiently emphasize the obverse side of this procedure and thus of the paradigm it is meant to furnish to philosophy. What authenticates the mind's constructive or productive operations in mathematics and rational physics is in every case the successful outcome, namely, the concrete exhibition, to or in the faculty of intuition, of an individual "object" answering to the specifications of a universal concept, filling the conceptual bill. The mathematician, Kant argues in the "Doctrine of Method," can meet this standard of authenticity, since he can derive synthetic a priori knowledge from his concepts by constructing them independently of anything given to him via experience, but the transcen-dental philosopher cannot, since *his* concepts (concepts of a reality, a substance, a force), although a priori, "include nothing besides the syn-thesis of possible intuitions which are not given a priori."[16] The philoso-pher's transcendental concepts, unlike those of the geometer, can never thoroughly determine a singular object independently of some a posteri-ori perception.

Furthermore, recourse to the standard of authenticity and to this alone means that the spontaneity credited to our power of knowing (and, although in a different measure, to our power of acting) is always checked (not eliminated) by the twin necessities of having to produce (or discover) an intuitive correlate and of having to do so in a medium not of the mind's own making. In other words, the liberation from (pregiven) "nature" apparently granted by the revolutionary stance of mathematical sciences is always incomplete, and this incompleteness is itself rooted in a twofold infirmity of the conceptual mentality. First, all of our *concepts* need to be "realized," invested with a status and reference (*Bedeutung*) they cannot straightaway confer on themselves *qua* "mere" concepts; only their intu-itive or sensible correlates (whether a priori or empirical in nature) have the authority to do this. Hence, any items of mental attention or intention for which there cannot, in principle, be any corresponding intuitive representation (that is, the ideas and ideals of reason) are disqualified

from playing a full-fledged *cognitive* role. Secondly, and even more significantly for Kant's reassessment of the claims of modernity, our
minds—now in the guise of conceptual understanding and reason alike—
are not directly intuitive; they are incapable of generating at will or *ad
libitum* the sensible manifold in reference to which our concepts can (and
our ideas cannot) be "realized." Kant traces this second infirmity to the
ultimate distinction between an *intellectus archetypus,* vested, or so we can
think, with "total spontaneity of intuition," and our own *intellectus ectypus,*
bound to a givenness for which it cannot take responsibility. In a word,
we are not conceptual or rational gods, however successful we are at
legislating to nature.[17]

Nonetheless, Kant's diagnosis of this double infirmity does not cure
the aspiration to imitate the constructive aptitude of the mathematician;
it does make the conscientious philosopher aware of the risks he is running by espousing this paradigm. In all three *Critiques,* we can witness
this risky mimicry in play.

It is in fact hard to distinguish the original from its imitator in the first
Critique when Kant discusses the schemata of the pure and empirical
concepts of the understanding and their role in mediating between these
concepts and the heterogeneous intuitions which alone can give the latter
referential purchase on experience. Although Kant does not use the
language of construction in that chapter, he does speak expressly of
schemata when he is contrasting mathematical construction with philosophical knowledge in the "Doctrine of Method." In any event, Kant's
two discussions seem to be headed in the same direction: Schematization
does for concepts generally (dog, triangle, and substance) something very
much akin to what construction does for mathematical concepts (triangle,
algebraic magnitude, and so on); both procedures are aimed at giving a
concept room, or time, in which to exhibit itself, to "realize" itself in an
otherwise alien medium.

The distance between image and original grows more perceptible when
we move to the *Critique of Practical Reason* and to Kant's explicit comparison of the schematism of theoretical concepts with the "typic of pure
practical judgment." To be as brief as possible: There can never be a
schema of the moral law if by this we intend an a priori intuition or rule
of the imagination which would exhibit the *applicability* of the law *in
concreto* to a case (an action) which it subsumes. We can have, instead,
something like a schema of the *lawfulness* of the law, and this quasi-
schema, which Kant calls a "type" (*Typus*), can be concretely exhibited in
sensible objects when we expropriate the mere form of a law of nature
to serve as its symbolic surrogate. As Kant puts it: "If a maxim of action
is not so constituted that it can stand the test [imposed by] the form of a
law of nature generally, then it is morally impossible."[18]

While Kant quite clearly does not want to suggest that this "typical" test can replace, without any loss of significance, the a priori legislative force of the *suprasensible* moral law, he is concerned to locate within the domain of sensibility a platform on which the morally good can be made (or allowed) to exhibit itself.

This topic of possible exhibition is, in turn, just one expression of the even broader preoccupation in Kant's works with the relation of sensible events (*Begebenheiten*) to actions (*Handlungen*)—that is, of our empirical performances to intentions dictated by the trans-empirical moral law. What is at stake here is ultimately the manner in which the self-legislative autonomy of the practical will (the counterpart to the theoretically sponta-neous faculty of understanding) is or can become the legislator of the sensible realm in which our performances are by nature the effects of "pathological" or material motives (such as self-interest). Practical as well as theoretical spontaneity, therefore, run up against barriers not of their own making (the "pathology" of self-interest or, in more recognizable terms, the body, in the first case, the material of sensation, in the second). Mathematical construction negotiates with enviable ease a leap of intimi-dating proportions in the sight of theoretical and practical philosophy.

In the two parts of *The Critique of Judgment,* the aesthetic and the teleological, Kant attempts to reduce these proportions to a more man-ageable or at least more hope-inspiring scale; these two attempts furnish diverse blueprints for the building or rebuilding programs of post-Kan-tian philosophy. I want simply to recall, first, the doctrine of the beautiful as the symbolic hypotyposis of the moral good (par. 59) and, second, the unexpected inclusion of the rational idea of freedom in the class of *scibilia* (also called "facts" [*Tatsachen/res facti*]).

One necessary remark about the first teaching: Kant borrows *hypotyposis* from the vocabulary of the New Testament as the generic rubric for the operation of "sensualization" (*Versinnlichung*), with schematization and symbolization as its two species. The latter, based as it is on analogy, cannot produce an a priori intuition directly corresponding to a concept (or, idea—in this case, the idea of morality); in place of such an intuition it furnishes a sensible instance *and* rule or formal procedure followed by reflective judgment in constituting or appreciating that instance (here, something judged beautiful). This rule is then transferred to the original concept as though it, too, had (*per impossible*) an intuitive object corre-sponding to it. We immediately notice that symbolization, while remain-ing in orbit around the method of mathematical construction, is farther from that center of attraction than schematization proved to be.

The second teaching, concerning the class of "facts" or *scibilia,* occurs in the penultimate section of the "Critique of Teleological Judgment" and is one of the last threads woven into the fabric of Kant's painstaking attempt to legitimate thinking about nature as though its origin, career, and constitution had some final purpose.[19]

An object (*Gegenstand*) is called a "fact" (and not an item of opinion or mere belief) only with respect to our own subjective power of knowing; it is entitled to this honorific designation only if the objective reality of its concept (its having reference to something real in experience) can be proved by means of a corresponding intuition. Kant goes on to give us three cases of "facts": (1) the mathematical properties of quanta such as are exhibited a priori in geometry, (2) things (*Dinge*) and their properties which can be shown (*dartun*) by means of experience, and (3) the idea of freedom, whose reality "can be set forth or shown [*dartun*] through the practical laws of pure reason and, in accord with these, in actual deeds and hence in experience."[20] Once again the promise held out by mathematical construction seems to be guiding Kant in this passage even as he implicitly acknowledges the discrepancy between the geometer's a priori exhibition (*Darstellung*) of a concept in intuition (in a way which satisfies the demands of theoretical reason) and the idea of freedom which can give proof of its practical *reality* in experience while its *possibility* remains forever closed to theoretical reason.

Kant made a penultimate attempt to assess the prospects for the realizability of the highest good, the ultimate unity or harmony of human *nature* and human moral *freedom* (of happiness and virtue); this attempt is recorded in a series of essays and reviews written mainly in the 1790s and dealing with the philosophy of history. Since these shorter texts do not belong to any one of the three *Critiques* or to a nameless "fourth" critique with a peculiar domain of its own, Kant tends to be less occupied with questions of methodology and epistemic credentials than he is in the *Critiques* and their systematic continuations. Nevertheless, the same dualities hold his attention: autonomy and heteronomy, the rational and the empirical, moral spontaneity or self-legislation (within the community of rational personalities) and the obstacles erected by sensuous self-interest. Yirmiahu Yovel, in his excellent work *Kant and the Philosophy of History*, speaks in regard to these writings of "the problem of historical schematism"—namely, the finding of a *tertium quid* assimilating rational moral demands and the contingent empirical details of history.[21] Kant himself does not appear to have addressed the issue of schematism directly in this context; he does however, imply that its place is taken by some still "weaker" variant of the governing notion of construction.

This weaker variant is prophecy, or more exactly, that ironic mode of prophecy that makes an a priori history possible inasmuch as "the prophet [*Wahrsager*] himself makes and contrives the events he announces in advance." Kant's discussion of prophecy and its link to *a priori* history occurs in the second section of *The Conflict of the Faculties;* his principal concern is "whether humanity is continually progressing towards the better?" In the event, prophecy can answer this question in the affirmative only by becoming the art of reading certain signs in the present which "hint at" (*hindeuten*) the course of the future. Prognostication based on these signs of the times actively encourages peoples bent on improving their civil constitutions and conditions.

In the *Anthropology* Kant's title for this art of making and reading signs is a subspecies of the *facultas signatrix,* namely, "symbolic or figurative [*speciosa*] cognition." His description of the role played by "symbols" alerts us to the gap between the indirectness of historical prophecy and the directness of mathematical schematism or construction: "[symbols] are simply a means, although only an indirect one, used by the understanding *via* an analogy with certain intuitions, intuitions to which the concept of the understanding can be applied in order to provide that concept with a reference by means of the exhibition [*Darstellung*] of an object."[22]

Kant's version of modernity, which began under the auspices of mathematics, becomes in the domain of history a hermeneutic of manmade signs.

It is time to draw this sketch of Kant to a close.[23] I have tried to suggest that, and how, the mathematical operation of construction is at work in all three *Critiques,* bringing in its train and holding under its power a series of rational procedures (schematization, typification, symbolization, and what we might call "factualization"). All of these procedures have as their goal the exhibition of the mind's powers in the sensible domain, a domain not of its own making and yet uniquely qualified to give those same powers the support and sense of "reality." It is especially in the *Critique of Judgment* that this enterprise of rational self-exhibition displays its ultimate importance, for it is there that the project of effecting or producing good in the natural world through rational freedom receives its most hopeful expression. This hope is kept alive by two influences: the subjectively warranted idea of the *intellectus archetypus,* who created a world in which the actualization of human moral personality is the highest end, and the example of mathematical method, which "becomes, so to speak, the master of nature" (*KdrV*. B753). However, since the human mind, for Kant, is essentially discursive, and thus neither immediately intuitive nor capable on its own of giving reality to its highest ideas, philosophy, like humanity, lives in the grip of two unfulfillable dreams.

III   Two Zoroasters: A Sense of Endings

Kant's constructivist legacy was fought over by many would-be heirs. Fichte, for example, provides an especially clear instance of the ways the limitations imposed on constructibility by Kant might be removed, thereby giving wider range to the "active existence" promised by the

Kantian revolution. Thus Fichte, for whom the essence of man is to be a "self-creator," argues that philosophy, exactly like mathematics, can and must provide adequate intuitions of its ideas, its concepts, and, most importantly, its own activity. Equally striking is Fichte's insistence that the solution to the contest between moral freedom and constraining laws can also be found in the operation of construction. Freedom is to be understood as the autonomous source of its own limits and binding laws. Mathematics, moreover, furnishes clear evidence of how this is possible:

> We must freely construct, produce in the imagination, as we did above in the case of the triangle. In this case an evidence will take hold of us, namely, that it is only possible *in this way:* a power and thus a law will give shape to our free construction.[24]

It should also be added that for Fichte, even more explicitly than for Kant, mathematical construction is above all a lesson in mastery and free possession. Rather than taking a rule of construction from another, on good faith, each person must make himself "master of what was previously present for him in another's, alien, soul." Indeed, Fichte's extended narrative of construction as self-creation fleshes out Kant's fragmentary evocation of the self in the *Opus Postumum:* "I, the proprietor of the world."[25]

It is, however, in Nietzsche that Kant's revolution comes to its most exorbitant climax. Despite his frequent parodies of the transcendental deduction and his acerbic references to the "Chinaman of Königsberg," Nietzsche's thinking takes its bearings from Kant, as this passage from his early sketch "Der Philosoph" should show:

> It can be proved that all constructions of the world [*Weltconstructionen*] are anthropomorphisms: indeed, all science, if Kant is right. To be sure, there is a circular argument here: if the sciences are right then we are not standing on Kant's foundation; if Kant is right, then the sciences are not right. . . . Our salvation lies not in knowing, but in creating![26]

Ten years later we find Nietzsche echoing Kant's thesis concerning our a priori legislation to nature, with revealing additions:

> In the last analysis man rediscovers in things nothing beyond what he has himself inserted [*hineingesteckt*] in them: Rediscovery is called science, the practice of insertion [*das Hineinstecken*]—Art, Religion, Love, Pride.[27]

What·Kant foresaw as the definitive maturity of the mind becomes, in Nietzsche, "child's play."

We can understand this reversal in the following way. Nietzsche interprets Kant's "revolution in [our] style of thinking" as a decision in behalf of creating or *poiēsis* and against knowing or science. Or, more accurately, Nietzsche is persuaded that science and *poiēsis*, as well as all other human activities and institutions, stem from a common root: the desire for mastery. This desire, already apparent in Descartes' description of the utility of the new sciences, in Kant's assessment of the attraction of mathematical method and in Fichte's radical version of free construction, becomes in Nietzsche's analysis the single, homogeneous passion through which philosophy itself can finally be explained, "For every drive wants to be master—and it attempts to philosophize in *that* spirit."[28]

Mastery, we might say, was all along the *telos* of the project of construction; in Nietzsche this *telos* is not so much achieved as rendered explicit and unmistakable. Moreover, the desire for self-exhibition—that is, the desire to have inward thoughts and concepts acquire a worldly objectivity—is supplanted in Nietzsche by the pure energy or event of exhibition itself. Accordingly, the concepts or ideas to be constructed are now understood as interpretive fictions to be projected upon, or injected into, a "world" that is nothing other than the sum at any time of these interpretive or perspectival fictions themselves. It is in this spirit that Nietzsche reverses Parmenides' third fragment: "Parmenides said 'One cannot think of what is not';—we are at the other extreme, and say 'what can be thought of must certainly be a fiction.' "[29]

Construction, in Vico, Kant, and Fichte, did not serve only to give proof of the mind's "creativity," its aptitude for contriving fictions. Of equal, if not greater, importance is the power of construction to objectify or realize that creativity in a "worldly" way, after the fashion of sensible items. In its primal mathematical use, as well as in the metaphoric or metonymic extensions of that use, construction confirms that the internal ego or *res cogitans* can have a grip on the external or the worldly. Nietzsche ends the entry on Parmenides in this way: "Thinking has no grip on the real, but only on . . . ."

These "thought-strokes" (*Gedankenstriche*) show us how far Nietzsche has gone to deprive construction, or making generally, of the evidentiary force ascribed to it in the Cartesian-Kantian tradition. No longer does the exhibition of the will to mastery provide outward confirmation of the propriety or felicity of inward concepts, laws, and ideals. Instead, the poles of "subject" and "object," between which the modern tradition fixes the operation of construction, must themselves be dismantled.

Salomon Maimon, one of Kant's most perceptive commentators, provides an illuminating gloss on the theme of *mathematical* construction: "All of the concepts of mathematics are thought by us and at the same time exhibited as real objects [*als reelle Objekte*] through construction *a priori*. We are, in this respect, similar to God."[30] Nietzsche retains our constructive similarity to god, while exposing the entire field of mathesis as the play of willful fictions.

The practice of philosophy in the present age continues to be governed by the etiquette of construction, now stripped of the epistemic and onto-logical authority Kant meant it to enjoy. In other terms, contemporary philosophy, in both its analytical and its postmodern or deconstructive versions, proceeds under Nietzschean auspices. The superficial estrange-ment between these two versions turns out to be a family quarrel between two branches of the Nietzschean family. Differences in style and tempo—formal decorum in the one, intoxicated frolic in the other—do not wholly mask their patrilineal affinity and family resemblance.

It would be belaboring the obvious to rehearse at any length the genea-logical script contemporary, postmodern, Nietzscheans have already writ-ten for themselves.[31] Here I note only the revaluation that construction undergoes in their hands. For Kant (and his modern predecessors) con-structibility was thought to give us gripping proof that the mind was working productively and objectively, that the ontological gap between the conceptual and the sensible could be bridged, even made to vanish, thanks to the rule-governed installation of conceptual generality within the sensible or phenomenal domain. Construction, then, is what keeps the mind from being unpresentable, from looking "unseemly." Contem-porary Nietzscheans cancel or dismantle the twin notions of (subjective) intentionality and (objective) reference or representation to which mod-ern construction is anchored. (Cancellations registered in such tropes as "the disappearance of man," "il n'y a pas de 'hors-texte,' " or "every *representation* is always already a *representamen*.") Loosed from these moor-ings, construction becomes unruly, anarchic, even self-destructive. The infinite project of modernity, seen from a post-modern perspective, yields a world in which, as Maurice Blanchot writes, "the image ceases to be secondary with respect to the model, where imposture lays claim to truth, where, finally, there is no longer any original, but an eternal scintillation where, in the vivid flash of detour and return, the absence of origin disperses itself."[32]

To treat analytical philosophy as a direct descendant of Nietzsche (or of Nietzsche's potentiated version of Kant) would seem a much more delicate task. After all, Frege's attempt to define "cardinal number" ap-pears worlds apart from the Nietzsche of, say, *Beyond Good and Evil*. Nevertheless, by examining two significant analytical works conceived after the demise of Frege's logicism we can begin to see the marks of family resemblance to which I have alluded.

Carnap, in *The Logical Structure of the World*, the *Aufbau*, occasionally acknowledges Nietzsche by name. More importantly, he links his enter-prise directly with "transcendental idealism," as in the following passage:

Constructional theory agrees with *transcendental idealism* in the conception that all objects of knowledge are constructed (in idealistic language:
are produced [*erzeugt*]) in thought; indeed, the constructed objects are
the objects of conceptual knowledge only as logical forms which are
constructed in a definite way. This holds ultimately of the basic elements
of the constructional system.[33]

Carnap's understanding of a "constructional system" differs significantly from the Kantian exemplar. On the one hand, "construction" (or
"constitution") is now understood as the construction or production of
concepts (especially the systematic generation of more complex from
elementary concepts—see *Aufbau* par. 1), *not*, as in Kant, the production
of a priori or sensible *intuitions* corresponding to concepts. On the other
hand, for the Carnap of 1928 it is no longer the case that the concepts
figuring in the constructional *system* of (objective) science need to acquire
their "objective reality" from their reference to anything sensibly experienced, that is, to anything lying "outside" the system itself. The "basic
elements" on which the constructional system rests are themselves constructions (see par. 109) and, indeed, selected for this (epistemically)
privileged role *ad libitum* (pars. 59–61). The constructional enterprise
appears to have been completely liberated from any constraints exercised
by the "nature" of any pre-systematic "given" ("given" either as the event
of experiencing certain non-constructed "objects" or as the object or
datum of some such experience). Accordingly, while in Kant (mathematical) construction is at bottom a compromise through which the outward
or sensible is submitted to the jurisdiction of the inward or conceptual,
but only to the extent that the spontaneity of the latter is recognized as
incomplete (insufficient for the generation of the material of intuition),
in Carnap the activity of constructing concepts and conceptual systems
seems to have free play within a field bounded, at one end, by an
unarticulable "given," at the other, by the economical conventions or
stipulations of formal logistics. What is in large measure obscured by this
free play is the status of the "world" which the preferred constructional
system is designed to fit—that is, the *being* or autonomy of that nature
over which, for Kant, construction renders man, "so to speak, the master."

Carnap's constructions have been called "brilliant, but ill-starred."[34]
However "ill-starred" from certain technical points of view, the spirit of
Carnap's *Aufbau* nonetheless survives in the work of Nelson Goodman,
from *The Structure of Appearance* to *The Ways of Worldmaking*, the latter
a Nietzschean dénouement to the analytical drama begun in Frege's
Begriffsschrift.[35]

We could gauge the distance between Goodman's earlier and later work in terms of the criterion of "extensional isomorphism" for the legitimacy of constructional definitions. In *The Structure of Appearance* this criterion is required as a "necessary and sufficient condition for the accuracy of a constructional definition" (par. 3). In *The Ways of Worldmaking* Goodman retrenches significantly:

> Such considerations [as, for example, the existence of equally legitimate *definientia* with disjoint extensions] point to a criterion framed in terms of an extensional isomorphism that requires preservation of *structure* rather than of *extension*. Since a structure may be common to many different extensions, this allows for legitimate . . . alternative *definientia*. The isomorphism in question is global, required to obtain between the whole set of *definientia* of a system, and the whole set of their *definienda*, but is not symmetric.[36]

This change in the *requirements* associated with a constructional definition (in Carnap's sense) is the rigorously technical expression of Goodman's expansive tolerance for alternative, even competing, world-versions. In his words: "With the reconception of the nature and significance of reduction or construction or derivation or systemization we give up our futile search for the aboriginal world, and come to recognize that systems and other versions are as productive as reproductive."[37] Let me restrict myself to three brief comments:

(1) The decision to abandon the search for "the aboriginal world" is Goodman's way of rendering Nietzsche's account of "How the 'True World' became a Fable" (in *Twilight of the Idols*). In both, what is quite consciously foresworn is any essential distinction between "the world" *tout court* and "the world for/of someone" (compare *The Will of Power*, par. 556 [Kaufmann]). In a moment of what may be captious irony, Goodman remarks that the question of "a universal or necessary beginning" of world-making "is best left to theology," thereby divulging that modern celebrations of the human power of making world-versions are shadowed by something like onto-theological nostalgia.[38]

(2) According to Goodman, "recognition of multiple alternative world-versions betokens no policy of laissez-faire. Standards distinguishing right from wrong versions become, if anything, more rather than less important."[39] "Truth," however, is no longer the exclusive standard; instead, it becomes one among many modes of "fitness," enjoying a somewhat ambivalent primacy over other modes specific to other "renderings" (such as designs, drawings, organizations of diction and rhythm). It would not, I think, be an unwarranted inference to say that "fitness" is generically an *aesthetic* standard; if so, both Nietzsche and Goodman transfigure the epistemic legislation imposed on nature (Kant's

imitation of the Copernican turn) into aesthetic autonomy. For both, it is the emphatic, indeed, impassioned endorsement of the "factor of fabrication," as Paul Ricoeur calls it in his review of Goodman, which underscores the coincidence of knowing and making, of *theoria* and *poiēsis*. (Compare Goodman: "The differences between fitting a version to a world, a world to a version, and a version together to other versions fade when the role of versions in *making the worlds they fit is recognized*.")[40] That Goodman retains his analytical sensibility even in *The Ways of Worldmaking*, where Nietzsche's name is never mentioned, merely reminds us that Dionysus can beget Apollonian offspring.

(3) All differences in characteristic style and local provenance notwithstanding, the post-Heideggerian deconstructionists and the post-Fregean connoisseurs of the arts of world-construction share this key trait of family resemblance: If, for Derrida, "There is nothing outside the text" (save for other texts), Goodman's world-versions are explicitly not versions of The World—there is, we could say, nothing outside any world-version except other world-versions. Self-involution seems to be the common mark of fabricated texts and world-versions alike. (This might explain why "representation" for one branch of contemporary Nietzscheans, "reference," for the other, remain like a succubus unsettling otherwise self-referential dreams.) Is self-involution the inexorable outcome of the modern project of construction? The project which took its bearings, from the first, by the desire to master and possess nature, the seemingly ineliminable or intractable locus of Otherness? Does the desire to set the seal of inward conceptual freedom on matter, space, sensibility, in short, the "external," end by fulfilling itself as aesthetic self-satisfaction rather than technological conquest?

Husserl in his *Krisis* could still quite plausibly associate the founding project of the conquest of material nature with modern man's desire to gain "an even greater power over his own fate and thus an even greater . . . happiness."[41] What we might call the "hedonistic eudaimonism" of a Descartes (compare his account of *chatouillement* and *générosité* in *Passions of the Soul)*[42] is replaced in Kant by the infinite task of moral self-perfection (promoted in part by the ruses of social history, in part by the eschatological needs anticipated in his tract *Das Ende aller Dinge*). Marx and Nietzsche leave to their would-be heirs the choices between "the absolute movement of becoming" and "the external recurrence of the same." The modern "end of history" shows up as endlessness, as when Foucault writes of "freeing history from the thought of its transcendental subjection" by "opening it to a temporality which would not promise the return of any dawn," or when Blanchot speaks, as though impersonating Nietzsche's ultimately speechless "last man":

We assume man to be essentially satisfied; as universal man, he has nothing left to do, is without needs, and even though as an individual he still dies, he has neither beginning nor end, but is at rest in the becoming of his static totality. What awaits this ultimate man . . . is the desire of man without desire.[43]

We cannot help asking: Is there anything inherent in the genesis of modernity which would have allowed one to prophesy these (and other, analogous) consummations? It surely would not be extravagant to see in the project of construction, from its seemingly limited origins in the new mathematics to its world-embracing aspirations, evidence of the demiurgic, world-fashioning mood of modernity. But construction is also the paradigmatic agency of self-divinization, the means to satisfy and objectify what Luther, just prior to the advent of radically modern mathematical science, had detected as man's inevitable desire (in the absence of grace): "Man is unable naturally [naturaliter] to will that God be God; he wishes himself to be god and God not to be God."[44] The apparent exhaustion of this demiurgic, self-divinizing conatus leaves in its wake either incessant restiveness or inexpressible satiety or a motley amalgam of both. Feuerbach's forecast of the dissolution of theology into anthropology seems not to have reckoned with these apparently baleful results.

Let me summarize the implications of this rapid introductory survey. The constructivist project, rooted in Descartes' geometry and exfoliated in Kant's critical enterprise, took its bearings from the desire to master and possess nature, where nature was understood as the locus of apparently ineliminable or intractable otherness. Mind could aspire to master its other (in the form of extension, or the manifold of given sensations) by externalizing itself in a construction carrying the clear marks of its inward and deliberate origin. (Compare Marx's claim that in unalienated labor "Our products would be like so many mirrors out of which our essence shone back to us.")

What unifies the extensions of Kant's program undertaken by the Nietzscheans is the elimination of nature as residually other to the mind. Simultaneously, Nietzsche and his progeny demolish any unitary, inward self or subject whose energies are invested in the activity of construction and whose identity is inscribed in the outcome of that activity. Gilles Deleuze writes in his book on Kant: "The first thing we learn from the Copernican Revolution is that we are giving the orders."[45] Nietzsche's contemporary heirs have undercut the force and the sense of this revolution by erasing the "we" and obscuring, if not removing, a world to which "our" orders can be addressed.

Earlier I wrote of Kant's penultimate attempt to find a surrogate for mathematical construction in his essays on the philosophy of history, a surrogate to be identified with the covert union of prophecy and making. Kant's *ultimate* attempt may be found in his *Opus Postumum,* where he selects a poetic or mythic prototype for the highest end of human practice—"Zoroaster: the ideal of physical, and, at the same time, moral-practical reason unified in a single sense-object." Some lines later on the same page of the manuscript he writes simply, "Zoroaster, Zoroaster."[46] Modernity from Kant to Nietzsche (and his later champions) is a contest among spokesmen for Zoroaster. It seems to leave us with a choice between the infinitely deferred union of *physis* with moral will and the eternally recurrent fulgurations of artistic will.

2

The Euclidean Context:
Geometria More Ethico Demonstrata

On Attic stelae, did not the circumspection of human gesture amaze you?
                                                                    Rilke

I Preamble

Even though self-origination, the claim to have made a radically new beginning, is an insistent theme among the revolutionary moderns (see chapter 3, II, i), it is equally true that without the preservation and the revivification of certain premodern traditions the rhetoric of this claim would be either trite or vacuous. As I suggested in the preceding chapter, the logical rhythm of philosophical modernity is antiphonal, a *new* voice raised in opposition to old voices, which must have been sufficiently audible or clearly recollected for the new to define itself. The issue, then, is not one of historical continuity versus incommensurability, so long as these contrasts are meant to signify an all-or-nothing state of affairs. To anticipate themes I shall be treating in the next two chapters, for a Descartes to proclaim his invention of a *scientia penitus nova* (a thoroughly new service) or for a Hobbes to insist that no political philosophy existed prior to the writing of his *De cive* is an act of will *instituting* discontinuities and ruptures in the face of what we would be most likely to consider clear counter-evidence (such as Hobbes' translation of Thucydides and his epitome of Aristotle's *Rhetoric*). Descartes and his revolutionary companions of course had their teachers and their predecessors, but it is critically important to understand how and why they deliberately erased the traces of these past influences and anticipations. In the *First Discourse* Rousseau is still very much aware of this operation: "Verulam, Descartes, Newton, these perceptors of the human race, had none [no teachers] themselves; indeed, what guides would have led them as far as their vast genius carried them?"[1]

The questions raised by the relation the early moderns instituted with the ancients are nowhere more subtle and complex than in the domain of mathematics. Late sixteenth- and early seventeeth-century Scholasticism might have been judged irreparably decrepit by the moderns; the case of Greek mathematics was, in their own eyes, different in both degree and substance.

These differences are, in the first instance, a result of the conscious, largely successful efforts made in the Renaissance to restore and rejuvenate the spirit of ancient letters through the discovery, editing, publishing, translating, and exegesis of surviving Greek mathematical manuscripts. To a large extent this philological activity was undertaken by the Humanists and, consequently, was relatively independent of *both* ongoing Scholastic discussions of, for example, the place of mathematics vis-à-vis physics and metaphysics within the classification of theoretical sciences *and* the subtheoretical tradition of practical or commercial arithmetic that had long flourished in many of the same sites in Italy and Germany where the Humanists set to work.

While Descartes and others took over the *matter* of their new mathematics from the Humanists' editorial labors, they also transformed and indeed subverted its pedagogical and philosophical spirit.[2] With this transformation came an even more consequential metamorphosis of the understanding of the "mathematical" character of mathematics, a change in the conception of how foundational concepts and procedures figure in the elaboration and validation of mathematics. It is on this fundamental metamorphosis in the conceptual and procedural understanding of mathematics that the equation of learnability (or intelligibility) in general with learnability via mathematics ultimately rests.

To put this as forcefully and as schematically as I can: The radical moderns did not merely take over and elaborate in new directions the materials of Greek mathematics; "reception" in this case was a transfiguration of a theoretical into a productive or *poiëtic science*. At the core of this new understanding of mathematics as *poiësis* is the technique of construction. This has two essential corollaries: (1) The "invention of the 'mind,' " to which I alluded above, uses as its template the adeptness of the intelligence in solving problems by means of this technique; and (2) in its most uncompromising version, this constructivism is neither simply a methodology nor only a source of epistemological criteria, but is immediately of ontological relevance. Kepler, more usually classed with the Platonists or Neoplatonists in respect to mathematics, in fact provides an unequivocal statement of this ontological implication. In demonstrating that a heptagon cannot be inscribed in a circle—that the sides of a "regular" seven-sided figure cannot be *constructed*—he states the implication thus: "We may rightly pronounce that the side of the septangle is of the non-beings [*non entium*], since of the non-knowables."[3] To forestall misunderstanding, let me stress that *poiëtic science*, not *art*, is the salient theme here, as I argue in chapter 3, I.

All of these forceful statements of the case I am concerned to defend are open to the most serious challenges at their common root—namely, the discrimination of ancients from moderns within the mathematical domain on the basis of the alleged supremacy or paramount role of constructive techniques for the moderns. Accordingly, prior to examining the moderns at first hand, I want to inspect some broadly significant details of Greek mathematical procedure itself, for it is there that the questions at issue may be given exemplary clarity.

The main question—what status and role do constructive techniques play in ancient and modern mathematics—can be recast as a series of four closely allied inquiries.

(1) To what extent do techniques (constructive or otherwise) stay within or stray beyond the boundaries prescribed by the implicit ontology of Greek mathematics? In other words, do the results licensed by the deployment of explicitly available techniques always make sense in terms of a pre-understanding (*Vorverständnis*) at work in the mathematical tradition and, sometimes at least, documented in philosophical reflection? This very general question of technique—that is, of mathematics as in certain visible respects a *technē*—belongs to the essence of my investigation into the advent of modernity, an advent equivalent to a release of the potentialities of *technē* from the constraints imposed on the mathematician's activity by his understanding of the difference between *technē* and *epistēmē* (or *theōria*) and thus by the self-understanding rooted in that difference. Let me formulate this somewhat differently with the help of a famous passage in the work of the historian Paul Tannery referring to the Greeks' supposed ability to handle algebraic techniques for solving second-degree equations: "What the Greek mathematicians lacked were less the methods . . . than the formulas suitable for exhibiting [*exposer*] the methods."[4] If "techniques" is substituted for "methods," then the question facing us becomes: With what degree of success did the ancient geometers manage to keep various techniques, each with obvious advantages in matters of economy and generalizability, under (suitably) tight rein? On what presuppositions does this policy rest, supposing that it was in fact followed?"

This first division of the main question can be more particularly pursued in connection with certain themes and technical opportunities on view by the theory of proportions in Euclid's *Elements* (especially Book 5). I shall turn back to this text after completing the enumeration of inquiries with which I began.

(2) From *technique* in general we can pass quite easily to the topic of *construction* as a particular technique or operation. The *use* of construction (or of its closest analogues in Greek mathematical practice and vocabulary) is not in any doubt; nor is it necessary for my thesis to consign this use to some distant margin of the field of geometry. On the evidence of

Euclid's *Elements* alone, episodes of "construction" are indispensable. Thus everything turns on establishing as exactly as possible (a) what "construction" involves for the Greek geometers and (b) how its contribution to the articulation of geometry *as a science* is evaluated, both by the geometers themselves and by their philosophical colleagues.

Not unexpectedly, these two questions are only modally distinguishable from one another, since the permitted or necessary range of the *operation* of construction seems likely to be bound to a particular *evaluation* of its epistemic contribution and vice versa.

(3) A third line of inquiry, quite closely associated with the first two, would address the baffling and ever-controversial question of *analysis* in Greek mathematical practice and theoretical estimation. How did this technique or family of techniques work? Were there statable "rules" for putting it successfully to work? What epistemic credentials did it carry vis-à-vis synthesis or demonstration? What connection, if any, did it maintain with constructional procedures?

The entire scene of the battle of ancients and moderns comes centrally into focus here as soon as we recall Descartes' programmatic boast that he had worked together into a functional unity the "analysis of the ancients" with the "algebra of the moderns." To ancitipate: Geometrical analysis, as it comes to be understood and, so to speak, reinvented by the radical moderns, will be counted upon to give decisive testimony that an *ars inveniendi* is far more fruitful, more copious in works than any *ars demonstrandi*.

(4) Fourth, and finally, is the topic of the locale or medium of geometry in its entirety, but especially with regard to its constructions. It comes quite naturally to us to ask: Where do these constructions take place? Or, Where are the line-lengths, plane figures, and so forth, referred to in demonstrations actually found or installed? The dominant modern answer speaks of "Euclidean space," and, as the revolutionary impact of that answer diminishes, we tend to find it self-evident that some conception of (three-dimensional, isotropic) "space" *must* lie in the background of Greek geometry, offering its adepts an accommodatingly inclusive field in which they could exhibit and manipulate the extended items required by their definitions, postulates, and axioms.

However, as soon as we note how closely the formulation of that answer hews to the modern understanding of the question (including, crucially, the "space" required by the project of a mathematized physics of extended, corporeal entities and their motions), we ought to be put on our guard against prematurely injecting question and answer alike into an inhospitable setting. The locale of Greek geometry may be foreign to the modern conceptions of extension and space.

I am going to explore only the first two issues in the sections which follow. The result of this exploration will, however, have implications for the issues of analysis and space, and some of these will be noted here and in chapter 3.

I should also forewarn the reader that my case will often be sustained by the interpretation of seemingly trivial words or phrases in the Greek texts of Euclid and others. The approach I shall be following in these instances is captured by the French aphorism "Le bon Dieu est dans les détails."

II   Episodes of Prudence and Imprudence in the *Elements*

i   *Ratios, Multitudes, and Magnitudes*

The theory of *logos*, ratio (a telling account which tallies with how two items stand to one another in their togetherness), and of *analogia*, proportion (somehow endlessly reiterating this "same" account for new pairs of related items), provides the *cantus firmus* sustaining the ever more exotic variations played on the Greek theme of reason (*logos*), the questioning speech of souls and the answering speech of beings.

Analogical predication in Neoplatonic and medieval thought (hence, the possibility of a discursive theology) *and* the mathematics of proportions, equations, and functions in modern science (hence, the possibility of a mathematical physics) would be equally unimaginable without the Greek notion of *logos*. When Kepler prefaces his description of new techniques for forming conic sections with the words "It is helpful for us to make use of the geometrical voices/terms of proportion/analogy: for I love analogies most of all, my most faithful teachers, 'in the know' about all the secrets of nature"; when Galileo, towards the end of his life, appends a "Fifth Day" to his *Discorsi* concerned with the exact criterion for equal ratios; when Descartes couches rules for moving the mind deductively towards a solution of a problem *in the language of continuous proportions;* and when Hobbes makes *analogismus* the armature on which his new computational logic and mechanics are to turn—then we can scarcely fail to hear a resonant counterpoint in modernity's response to the ancients.[5] This alone would give special urgency to a renewed investigation of the Greek "origins" of the theory of ratios.

However, our primary and most complete "source," the Eudoxean theory reproduced and most probably reworked by Euclid in his Book 5, does not take us back to those origins, but sets us at some distance from them, since the very notion of "origin" has become quite dubious in this context. More exactly, we begin by confronting a theory "fashioned" in the aftermath of the grand dismay reportedly brought on by the collapse of the "Pythagorean" identification of number and geometrical magnitude owing to proofs of incommensurability. This puts Eudoxus and Euclid in a delicate situation. Any successful theory of ratio and proportion will have to do justice to the incontrovertible fact of incommensurability, while at the same time vindicating the pretheoretical or pretechnical experience of ratio and proportion. In other words, what will be required is a sensitive negotiation between the "natural" (what is taken to be implicit in the language of *logos* and *analogia* prior to the Pythagorean crisis) and the "conventional" or "technical" (what emerge as explicit constraints on this language in the wake of that crisis). What is needed is a way of retrieving the "natural" understanding which is neither wholly prejudiced by, nor unfairly prejudicial to, the "conventional." (One index of the delicacy required is provided by the first scholiast to Euclid, Book 10, who tells us that for the Pythagoreans "the commensurable and the incommensurable are by nature, the rational [*rhēton*, expressible] and the irrational [*alogon*] are by convention." Another is Proclus' paradoxical phrase *arrhētoi logoi*, "ineffable ratios," those in face of which the telling power of *logos* seems to fall dumb.)[6]

Euclid's unique "definition" of a logos/ratio as "a sort of relation [*poia schesis*] in respect of size [*kata pēlikotēta*] between two magnitudes of the same kind [*dya megethōn homogenōn*] "has long been the target of a certain mockery; Heath (ad loc.) quotes Barrow (1666) as saying that the definition is "metaphysical . . . and not, properly speaking, mathematical, since nothing depends on it," and Granger communicates the same scorn: "une hommage rendu par la vertu mathématique au vice du métaphysicien."[7] What are we to make of this?

Euclid (or, Eudoxus himself) appears to be appealing to the readers' familiarity with a "well-known term of common life" and, at the very same moment, to be using in the *definiens* two terms, one of which is uncommon and made less determinate in meaning by its qualifying adjective (*poia schesis*), while the other (*pēlikotēs*) is most probably a neologism. What, if anything (no matter whether metaphysical or mathematical, a distinction at best posterior to the issue at hand), is being taught here? C. S. Peirce found more than one occasion to reflect on the authorial "style" of the *Elements,* especially Book 1, and his reflections perhaps buttress one's initial sense that much is happening beneath the surface or between the lines. "They [Greek writers] took it for granted that the reader would actively think; and the writer's sentences were to serve merely as so many blazes to enable him to follow the track of that writer's thought."[8]

While in the first two definitions of Book 5 *metrical* language is emphatic (*katametrēi; katametrētai*), in Definition 3 this language is deliberately suppressed, although it leaves behind a very faint echo in the neologism *kata pēlikotēta*. Significantly, in the slave-boy episode in *Meno* the root of this neologism (*pēlikos*) appears three times, always in such a way as to suggest a feature common to commensurable and incommensurable "magnitudes."[9] Accordingly, as Heath convincingly argues against de Morgan, the abstract sugstantive in Euclid does *not* signify "quantuplicity,"—that is, "the *number* of times one magnitude is contained in the other"—but something non-metrical or non-mathematical such as "size." Similarly, *poia schesis* is an "abstract substantive" capturing in a loose but quite general manner the verbal clause *hōs α echei pros β* (as α stands in relation to β), without further indication of the quantitative character of that relation. (It is not until Definition 4 that the relevant indication is supplied, when we learn that two magnitudes have a *logos* when (in case $a \neq b$) there is some positive natural number such that $ma > b$ or $a < mb$; but, the natural numbers introduced here and in the following definition of "being in the same *logos*/ratio" are to be understood as auxiliary operators or "test numbers."[10])

The point of all this seems to be that Euclid (or Eudoxus) is intent on blocking at the source two streams of thought which otherwise might flow readily from the generalized notions of magnitude and ratio occasioned by the Pythagorean crisis. One would be the "geometrization of the numerical," namely, the treatment of ratios among numbers as simply subcases of Definitions 3, 4, and 5; the other would be a renewed attempt to "arithmetize the geometrical," that is, to institute a natural, indivisible *unit* for the measurement of line segments. No anachronism is involved in this second case, except, perhaps, in the phraseology. The pseudo-Aristotelian treatise *On Indivisible Lines* is a sustained critique of unnamed philosopher-mathematicians who assimilated numerical units to geometrical units of measurement (*holōs en hapasi tois posois esti ti ameres hōsper enioi phasi*), while in *Metaphysics Delta* 15, Aristotle himself brings all magnitudes, including incommensurables, under the sway of number: even though the relation of the excessive to the exceeded in the case of two incommensurables is "wholly indeterminate with respect to number [holōs aoriston kat' arithmon] "nevertheless, this case and all others (for example, the ratio 2:1) can be considered "relatives said in respect to number and affections of number [*arithmou pathē*]" (1021a6–9).[11]

The attractions of the first alternative—"the geometrization of the mathematical"—are also documented in a Greek source, namely, Aristotle once more! In *Metaphysics* Iota Aristotle treats number solely with respect to its measuring-function or as a particular case of the relation between magnitude (*megethos*)—(compare Euclid, Bk. 5, Def. 4)—and its measure (1053a1–5, 25–26). Pressed to an extreme, this position would lead to the identification of the line-segments pictured in the diagrams of Euclid, Books 7–9 (the "arithmetical books") with the numbers they somehow depict or signify.

Leaving aside all questions of philosophical influence or fealty, we can say that at this decisive juncture Euclid (Bks. 1–4 do not use the theory of ratio; Bks. 5–13 do) is negotiating the tension between the pretechnical or "everyday" understanding and the technically motivated understanding of *logos* so as to preserve distinctions which the latter understanding threatens to reduce beyond subsequent recognition. The crucial distinction is between discrete multitude (number) and continuous magnitude, and so Euclid, by virtue of refusing to collapse this distinction—that is, by refraining from treating the discrete as one subcase of geometrical magnitude among others—remains faithful to the pretechnical conception of *arithmos* as *Anzahl*, a definite number of discrete things.[12] The two chief consequences of this faithfulness are: (1) Euclid keeps the Pythagorean treatment of ratio and proportion (Bk. VII) separate from the post-Pythagorean treatment (Bk. 5) until they necessarily intersect in the theory of incommensurables (Bk. 10, especially Props. 5–9); and (2) this separation, together with the definition of *logos* in Book 5, insures against the temptation to look upon a ratio as a particular kind of number or, conversely, a number as a particular kind of ratio. In short, Euclid does *not* countenance the notion of (positive) rational numbers (of the form m/n, where m and n are positive integers), any more than he does that of (positive) real numbers, no matter how tempting the prospect of doing so in light of the generalization of the concept of *ratio* necessary to accommodate the Pythagorean crisis.

Let me dwell on this last point just a bit longer. The very same texts led Stenzel to applaud the widening of the Greek concept of number, while Granger, aware that all the tools are at hand for this widening in the definitions of Book 5, explains Euclid's refusal to take this step on the basis of his "style," a style that prevents him from "identifying mathematical objects possessing the same [algebraic] structure, but of different origin and construction."[13]

A quite different view of the matter emerges from the interpretation I have been sketching. Moreover, nowhere does Euclid hint at an ordering relation on the sequence of *ratios* as defined by Definitions 3 and 4, and without this relation the modern concept of rational number (or fraction with integral numerator and denominator) has no support. Consequently, a *ratio*, so far as it is made known to us by *these* definitions (see below, II, ii on Bk. 5, Defs. 9–10; Bk. 6, Def. 5), is not susceptible to being treated as though it were a more general sort of number (that is, no longer an *Anzahl* or counting number). In Ian Mueller's words "whereas [our] fractions are objects, [Euclid's] ratios are not."[14]

Instead of intimations of a Bourbakian future Euclid offers us a lesson in mathematical *phronesis*, the fitting of appropriate means to ends worthy of choosing, rather than the determination of ends by the accessibility of means. In the present instance, the end in view is the retention of a pretechnical "sense" of the distinction between discrete numerical multitudes and continuous geometrical magnitudes, a distinction showing up in the linguistic habit of discriminating "few" and "many" from "smaller" and "larger." A prevalence of means over this end will later result in the suppression of this distinction in Descartes' *Géométrie* or, to take a fascinating case, in the English algebraist, Thomas Harriot, who dubbed imaginary roots (of the form $m + \sqrt{-1}$) "noetic numbers," to spare them from reproach.[15]

ii Homogeneity

Euclidean *phronesis*, however, continues to be challenged by the potential or actual virtuosity of the mathematical *art*. (This modern geometer, as I shall try to show in later chapters, accepts the victory of virtuosity over *phronesis*.) The same definition (Bk. 5, Def. 3) provides a hint of one such challenge, with repercussions within the Euclidean text and far beyond it. This is the requirement that the magnitudes having a ratio be "of the same genus."

One could easily imagine writing the history of ancient, medieval and modern mathematics (at least through the early eighteenth century) wholly in terms of the criterion of homogeneity and the attempts either to satisfy or to transform it. In the period most critical to the aims of this study, Viète, Descartes, Hobbes, and Leibniz take a stand towards this criterion, each in his own fashion. While Viète stays within the limits prescribed by the Euclidean criterion, Descartes transgresses those limits by a *tour de force* I shall be studying in the next chapter (see chapter 3, III, vi); Hobbes calls on the criterion as a weapon in his campaign against the algebraists Wallis and Ward, while in Leibniz it undergoes a vast transformation which generates his "transcendental law of homogeneity" and the operational freedoms this law permits.[16] The modern understanding of algebraic structures and the conception of logical formalization which rests on that understanding are unthinkable without the relaxations and transformations to which the law of homogeneity was subject or subjected. What, then, does "homogeneity" mean in Definition 3, and to what extent does it act as a constraint on subsequent procedures in the *Elements?*

The text itself does not specify which genera are in question, but a scholium furnishes what would seem to be at least part of the "natural" understanding:

We have to speak of homogeneous [magnitudes] because magnitudes which are not homogeneous have no ratio to one another. For a line no more stands in a ratio to a plane-surface than a plane-surface does to a body; however, a line is comparable to a line and two plane-surfaces also have a ratio to one another.[17]

That is, the necessary condition for having a ratio at all is either that the items belong to the same genus or, as the scholiast implies, that they be of the same "dimension." The second interpretation, however, seems too liberal, since a square and a circle are of the same dimension (2) but do not belong to the same genus and, indeed, do not have a *ratio to one another,* for Euclid. Furthermore, although Euclid, following the Pythagorean tradition, is prepared to talk of "linear" and "plane" and "solid" numbers, the suggestion of dimensionality *sensu stricto* is out of place here, since no number *qua* number has a ratio to a line, plane-figure or solid. Numbers and geometrical magnitudes are, then, heterogeneous. Hence, it looks as though a pretechnical understanding of what items fall into which *genera* is a prerequisite for our grasping the force and scope of Definition 3!

This is precipitous, someone might object, inasmuch as Definition 3 requires completion by Definition 4, what has come to be called "the Archimedean Axiom": "Two magnitudes are said to have a ratio to one another which can, when multiplied, exceed one another." In modern treatments Definition 4 carries more weight than the "metaphysical" Definition 3, understandably enough, since it appears to offer a strictly operational criterion for having a ratio. Pascal, for example, takes Definitions 3 and 4 as straightaway synonymous with one another (thus, for him, one is a number in good standing, since when it is multiplied by itself [*sic!*], that is, concatenated with itself, n-times, it can exceed an integer $n - 1$! All numbers are homogeneous!)[18]

Yet, this does not remove all the difficulties, for Euclid may well mean, as Leibniz and Heath think he does, that *within* the "class" of homogeneous magnitudes those and only those have a ratio to one another which are multiplicatively comparable in the sense of the Axiom of Archimedes. Thus, a finite line-segment and an infinite line would belong to the same genus—straight-line—but would not be multiplicatively comparable. The Archimedean Axiom, then, would discriminate between two subclasses of a single genus, rather than marking off the boundary between same and different genera.

This reading, taken together with what I have already said about Definition 3, has two primary implications. First, the operation of "multiplication" (repeated concatenation) does not determine membership in a class; rather, the "nature" of the class makes room, as it were, for the application of the operation in permissible cases (as in two finite linesegments). Second, *megethos,* in spite of the *generality* of the remarks and allusions made about it so far, turns out *not* to be a natural kind with respect to ratio. Let me clarify. Any two magnitudes are not first of all magnitudes and *then* members of a particular *genus* of magnitude; if this were so, then any two magnitudes would have some *schesis* to one another with respect to their relative size and thus stand in a ratio. Instead, Euclid (Eudoxus) suggests that we must treat *megethos*, as far as *ratio* is concerned, in much the way Aristotle in *Metaphysics Gamma* insists "being" (*to on*) and "one" (*to hen*) must be understood—namely, that they have genera *without further ado* (*euthus*) (2.1004a5). (In the case of *to on*, Aristotle argues, a differentia marking off the difference of any two genera of "being" would itself have to be an *on* [see *Metaph.* 3.998b23–24]; would we have to say, in the Euclidean case, that any differentia distinguishing a line, say, from a solid would have to be a "magnitude" of some genus?)

At least one medieval commentator was sensitive to the distinctions in play in the background of Definitions 3 and 4:

> Be sure to take diligent note that a ratio is not a relation [*habitudo*] of two quantities of one genus if 'genus' is taken in a broad sense or predicamentally; rather [the definition] means that all straight lines in respect to one another are of one genus and that curves are of another genus and numbers of another; nevertheless, both straight lines as well as curves and numbers are all in the genus of quantity, when this is a predicament [the category of *megethos* or *to poson*, quantity].[19]

The behavior of homogeneity within the Euclidean corpus brings to the surface some of the deepest questions posed by the interplay or interference between the pretechnical understanding to which Euclid appeals and the technical resources put at the mathematician's disposal by the definitions and so forth. Ian Mueller has shown that, in his presumed redaction of Eudoxus' results, Euclid (if not Eudoxus himself) was concerned to reduce homogeneity assumptions to a minimum in the actual proofs in Book 5—that is, he "wished to avoid proofs depending upon propositions in which homogeneity restrictions are stronger than those of the proposition being proved.[20] Although usually successful in satisfying this wish, Euclid nonetheless opens the door at least a crack to transgressions of homogeneity when he introduces the technical operation called *alternando* (Bk. 5, Def. 12) and when he proves Proposition 16 of Book 5 without calling attention to the required homogeneity of all four terms.

Although the definition of *alternando* (if $a:b::c:d$, then $a:c::b:d$) comes after the pivotal definition of proportion (*analogia*) (Def. 5), Euclid takes pains to word it in terms of ratio (*logos*), not only to avoid assuming what needs to be demonstrated (compare Heath ad loc.), but also to imply that *ratio* considered in its own right—as the two-term relation of the variables '*a*' to '*b*'—offers accommodation to a range of terms playing different roles in different particular ratios (antecedent in one, consequent in another). In light of Definition 3, this makes sense *only* when the terms are homogeneous in *their* own right, that is, prior to the functional place they come to occupy in a particular ratio (or in a series of the "same ratio"). It is *precisely* this understanding which is undermined by Descartes, for whom the alternation $a:b::a^n:b^n$ into $a:a^n::b:b^n$ makes perfect sense (see below, chapter 3, III, vi) and which also disappears from the scene when, in the wake of Galileo's use of classical proportion theory in his mechanics, velocity (or, quantity of speed) comes to be interpreted as the function of distance and time (for example, $v \propto d/t$, which is reached by the "permutation" of $d_1:d_2::t_2:t_1$ into $d_1:t_2::d_2:t_1$).[21]

In any event, it is clear that *if* the operation of *alternando* or permutation is given universal scope *without* the requirement of homogeneity of antecedents and consequents, respectively, then this requirement can no longer draw its force from an understanding of difference in kind which, so to say, antedates the introduction of this operation. Euclidean prudence would accordingly have to exercise a twofold constraint to the extent it remains guided by that prior understanding.

It would first have to prohibit the "mixing" of heterogeneous items in a single ratio, even when that entails working out more complicated proofs, and then it would have to place limits on the *technical* formation of kinds or genera. The need for this second constraint arises from an especially thorny nest of questions which will occupy me at some length in the next chapter. For the moment, let me just point to one outstanding difficulty. When the requirement of homogeneity is relaxed, dimensional differences, too, look to be less binding. If *a* and *b* are *heterogeneous* in the spirit of Definition 3, but do obey the Archimedean Axiom implied by Definition 4, then we have to make some sense of their multiplicative comparability.

Euclid seems clearly to have had in mind the repeated concatenation of one magnitude with itself a certain *number* of times until it exceeds the second magnitude (compare Book 7, Definition 15, for arithmetical multiplication as repeated addition); but this ceases to convey any meaning when homogeneity has been dropped as a constraint. However, Euclid also uses the geometrical counterpart to arithmetic multiplication when he forms, for example, a rectangle from its two unequal sides or a cube from three equal line-segments. Consequently, if I am to be in a position to compare two heterogeneous magnitudes (a line-segment *a* and a cube *b*, for example), then it is this geometrical operation which lies ready at hand; that is, instead of adding the line-segment to itself *m*-times, I "multiply" it by itself three times and then see if the *volume* of the cube (a^3) exceeds the *volume* of the cube *b*. Once this geometrical

version of multiplication is substituted for the arithmetical version intended by Definition 4 (and still "alive" in Definition 5, where the equimultiples are positive integers and always produce a longer line from a
line or a larger square from a square), there seems to be nothing intrinsic
to the procedure followed which sets limits to the "kinds" which can be
introduced. This is, of course, the notorious problem of the "naturalness"
of the three-dimensionality of classical geometry; the operation or technique of dimensional multiplication yields a^4 or a^6 as easily as it does the
square or the cube on a. It is important to notice that this technical ease
gave both qualms and opportunities to ancient geometers, sometimes
giving a measure of both to one and the same expert. Hero of Alexandria,
for example, endorses the homogeneity-requirement *and* proceeds in his
Metrica to work out a formula for finding the area of a scalene triangle
which involves the (geometrical) product of two squares.[22]

How are we, following Euclidean teaching, to know *where* the limits on
the technical formation of kinds are to be drawn? Is *phronēsis* in this
domain tied to perceptual experience so that n-dimensional kinds, with
$n > 3$, are straightaway excluded? Does it, instead, take its cue from some
noetic insight into what kinds there are, by nature? Or, finally, does
phronesis here, as in the realm of human praxis, guide itself by reputable
opinions and settled habits of discourse, honoring those kinds and only
those already in good repute? What emerges from this investigation of a
"small" detail in Definitions 3 and 4 of Book 5 is nothing less than the
question of the manifold relations and tensions between *theōria, praxis,*
and *technē*. To the extent that this is always in its initial manifestation a
question of the public or traditional sources by which human speech is
or is not oriented, we could also speak here of the politics of mathematics.

It is now time to turn to a second, even more problematical, challenge
to mathematical *phronēsis*. Euclid so vividly fails to meet this challenge
that the crack in the doorway left open by a certain ambivalence about
homogeneity now becomes a breach wide enough to admit a sequence of
mathematical conceptions and techniques increasingly antagonistic to the
pre-understanding Euclidean teaching continued to address. As in the
first challenge, here, too, the principal difficulty stems from the interpretation of multiplicative comparison, now as it is applied to ratios themselves, rather than to the magnitudes standing in a ratio.

The pertinent textual phenomena can be readily set out. In Book 5,
Definitions 9–10, Euclid defines "duplicate" and "triplicate" ratios. In
Book 6 there is an interpolated definition (Def. 5) of what it is for a ratio
to be compounded or composed (*synkeisthai*) out of ratios. Euclid makes
no mention of this definition anywhere in the *Elements;* on the other
hand, compound ratios are expressly used in the proofs of Book 6,
Proposition 23, and Book 8, Proposition 5, while the particular cases

mentioned in Book 5, Definitions 9–10, reappear in, for example, Book 8, Propositions 11–12 and 18–19, and something quite close in looks to compound ratio is employed in the proofs of Book 5, Propositions 20–23. The device of compounding ratios, with a very marked use of Book 6, Proposition 23, in particular, becomes instrumentally vital to post-Euclidean Greek geometry; Apollonius, Archimedes, and Ptolemy would be stranded without this resource. Yet, despite, or perhaps because of, its pervasiveness, only *two* tantalizingly brief discussions of what it involves are extant in later Greek materials, one in Eutocius' Commentary on Archimedes, the other in Theon of Alexandria's Commentary on Ptolemy's *Almagest*.[23] It is as though the principal mathematicians and their commentators preserved an awed silence in the face of the feats performable with its aid. This silence had stupendous repercussions in the subsequent history of mathematical thinking, beginning with medieval Latin and Arabic attempts to make sense of the operation called "compounding ratios" through an inventive hybrid of arithmetic and geometry and including Leibniz's embrace of the strange new world of ratios of ratios and logarithms (the *numbers* assigned to *ratios*).[24] Descartes advised Desargues, the only contemporary mathematician for whom he seems to have had undiluted respect, that he could make the demonstrations in his new projective geometry "more trivial . . . by using the terms and the calculus of arithmetic, just as I have done in my *Géométrie*, for there are many more people who know what multiplication is than know what the compounding [*composition*] of ratios, etc. [sic] is."[25] The axial role played by the compounding of ratios in the technical revolution through which the moderns became "modern" should already be recognizable, without my needing to multiply instances.

What, then, *is* a compound ratio (*logos synēmmenos/synkeimenos*) in Euclid? As much as one would like to meet this question "head-on," the barriers standing in the way are formidable, if not insurmountable. The highest barrier is this: A ratio for Euclid, as I have already stressed, is a *relation* between magnitudes; it is not a magnitude or a quantity in its own right. (Hence, it is most emphatically *not* a "rational number"; see below on Euclid and Dedekind). Therefore, operations to which magnitudes are "naturally" subject (such as addition of line-segments, multiplication of numbers and its geometrical counterpart; see below, II, iii) would appear to be alien intruders once transplanted to the domain of ratios (or, indeed, the domain of proportions, as will happen in algebra when equations are added to, or multiplied by, one another, and so on). Nonetheless, on the most plausible reading of "compound ratio" in Euclid, we *are* being asked to allow some such alien operation to be applied to a ratio or, more precisely, to a pair of ratios. For example, when $a:b::b:c$, a is said to have to *c* the *"duplicate* ratio of that [ratio] which it has to *b*" (Bk. 5, Def. 9). What operation would give sense to this? Is it supposed to work in the case of *all* magnitudes falling under Book 5, Definitions 3–5?

The spurious definition in Book 6—"A ratio is said to be compounded of ratios when the sizes [*pēlikotētes*] multiplied upon themselves make something [*tina*] "—takes a stab at clarification, only to pinpoint the original problem again, *viz. "multiplicity"* can only operate between quantities, not between ratios of quantities. We should also note how the 'mysterious' term *pēlikotēs* ("size") has migrated from Book 5, Definition 3, into this non-Euclidean context. Once more, however, we need to keep in mind that magnitudes are *in a relation* (ratio) *with respect to* "size"; their relation does not itself *have* a "size," nor would the product of two relations, if we could make sense of that, *be* a "size."

Eutocius tried to resolve the issue by interpreting *pēlikotēs* as "the *number* from which the given ratio takes its name" (*parōnumōs legetai;* "denominator" in the Latin translation of 1566); thereby, he reintroduced an essentially arithmetical understanding of various different ratios which goes back to the Pythagoreans and was kept alive by other ancient sources (principally Boethius) throughout the Latin middle ages.[26]

As Heath and others have pointed out, if Eutocius meant, as he probably did, the name of the number which when multiplied by the consequent produces the antecedent of the ratio (thus "3" is the name of the ratio 6:2 since $2 \times 3 = 6$), then the nomenclature works *only* for commensurable magnitudes, that is, those having to one another the ratio of a number to a number. Thus, Eutocius' solution sidesteps fruitlessly the very rationale for the Euclidean-Eudoxean theory of proportions.

The consequences of Eutocius' construal are of the greatest moment for the genealogy of modern thought I am studying. In fact, we owe the first known use of the phrase *mathesis universalis,* so central to Descartes' plans and to post-Cartesian mathematics (see note 120 to chapter 3), to the sixteenth-century Belgian mathematician Adrian van Roomen. He explains and defends the phrase at some length in chapter 7 of his *Apologia pro Archimede* (1597), where the words *authoritate Eutocii* refer back to chapter 6 in which van Roomen has quoted from the Greek's *Commentary on Apollonius,* Book 1, Proposition 2: "Non perturbentur autem qui in haec inciderint, quod illud [Apollonius' theorem] ex arithmeticis demonstretur. Antiqui enim huiusmodi demonstrationibus saepe uti consueverunt, *quae tamen mathematicae potius sunt quam arithmeticae propter analogias.*" ("Those who come upon this will not be perturbed by the fact Apollonius' theorem is proved by arithmetic. For the ancients were in the habit of using proofs of this kind often, even though these proofs belong to mathematics rather than to arithmetic, on account of the analogies.") In brief, van Roomen finds in Eutocius' comment on compounding ratios the justification he is seeking for the idea of a mathematical science embracing both geometry and arithmetic with equal completeness. The ground was being prepared on which Descartes and others could erect their own "enchanted palaces."[27]

Let me backtrack from this prospectus to summarize what has appeared so problematic in Euclid in regard to his self-understanding as a teacher. He seems to have led his students into a quandary precisely because an operation which is required for later developments in the *Elements* does *not* have any explicit justification on the basis of operations admitted up until now or available to him in retrospect (for example, numerical multiplication in Book 7). This lack of justification means that an operation is put to work in significant contexts without being anchored either to the learner's pre-understanding of the domain of "beings" which the *Elements* is addressing or to any express amplification of that domain in the light of already achieved "results." To put it even more simply: We witness Euclid relying on "composition" as a workable technique without ever knowing *why* it works; successful performance holds sway over the desire for ontological or "intuitive" elucidation. Compounding ratios is, indeed, as de Morgan had it, an "engine of operations."[28]

Answers to the question "Why did Euclid fall into this quandary?" are bound to be conjectural. So that this episode not seem altogether fortuitous, I shall set out two sorts of possible reasons for his decision to proceed as he does. One sort is "historical" or, better, "traditional"; the other, immanent to some of the aims of the *Elements* as a whole. As for the first, keeping in mind the likely origins of proportion theory in Pythagorean music theory helps us to make sense of compounding ratios in a way that must still have held a kind of commemorative fascination for Eudoxus and Euclid. In music theory, or, more exactly, the discipline known as the "Cutting of the Canon," ratios (or integers) are assimilated to the integral values of string-lengths or intervals (*diastēmata*). To compound (the ratios of) intervals is in fact to *add* their integral lengths (hence, the use of the adjectives *synkeimenos* or *synēmmenos*, "added," in referring to compounded ratios). As the "boundaries" (*horoi;* compare Euclid Bk. 5, Def. 8) defining an interval are moved to new positions on a monochord, a series of ratios is produced, the last member of which *is* "compounded" of all the previous ones; so, for example, the ratios 4:3, 3:2, "compounded," yield the interval of the octave 2:1. The result of the operation is in many senses phenomenal.

Our primary record of the relevant technique and associated theory is indeed a text ascribed to Euclid, the *Sectio canonis*, which deals only with commensurable intervals and so is most probably earlier than Eudoxus.

The discovery of incommensurable magnitudes is just what disturbs the attunement of Pythagorean thinking to a world of pure numbers, and Euclid, in consequence, was entitled only to a much-tempered version of that thinking.[29]

Nonetheless, commemoration of a tradition, a tradition not so much discarded as repositioned in a larger whole, is certainly not the sole motive for the retention of compound ratios in Euclid. For instance, he *needs* the technique, or, at least, particular specimen results (duplicate and triplicate ratios) to insure some measure of continuity as he passes from the proofs of Books 1–4, where equality of figures is the dominant concern, to Books 6 and 11–12, in which similarity of figures is a principal concern. The generalized notion of "ratio" in Book 5, the hinge on which the distinction between the two portions turns, will permit Euclid to give greater determinateness to the relation of similarity *if* ratios can be displayed which link together line-segments and "their" figures in a "numerically" precise way. So, for example, Book 11, Proposition 33, proves that "similar parallelepipedals are in the *triplicate* ratio of their corresponding [*homologōn*] sides," while Book 12, Proposition 18, demonstrates that "spheres are to one another in the *triplicate* ratio of their proper [*idiōn*] diameters." We are led to infer that the original "Pythagorean" arithmetic of line-segments and plane and solid figures, discredited by incommensurability, can nonetheless be partially preserved as a numerically determinate relationship (duplicate, triplicate) between figures and "their" sides or characteristic straight-lines (for example, diameters of spheres). Theon of Smyrna seems to be drawing the same lesson when he remarks that compounding is responsible for a "growth" (*auxēsis*) from one dimension to the next.[30] Determinate compounds of ratios mediate between a mistaken homogeneity and a disconcerting heterogeneity of dimensions. It is consequently more than an irony of editorial compilation that Book 10, Proposition 9, the theorem establishing that squares on commensurable sides have to one another the ratio of a square number to a square number, while squares on incommensurable sides do not, depends on the Porism to Book 6, Proposition 20, that "similar figures are in the duplicate ratio of their corresponding sides." The device of compounding ratios makes the specter of incommensurability look less inimical.

We have discovered in this tangled tale of composition of ratios how difficult it is for the man of mathematical prudence to behave in a completely scrupulous way. His *Aufhebung* of a still-honored tradition also commits him to the insertion of novelties which are *technically* germane to his new stance without being, or being made, transparent to the understanding of his students.

iii   Sameness of Ratios

The final instance of Euclidean *phronēsis* I want to examine is the most familiar and, in its own way, the most perplexing; namely, the definition of "being in the same ratio" or "proportionality" (Book 5, Def. 5). In view of the many sustained discussions of this definition and its aftermath already current in the literature, I can be content with a few summary reflections concerning its nature and its connection with the two topics (homogeneity and compounding) treated more extensively above.

(1) Although never under challenge in later Greek mathematics, as far as I know, Euclid's Definition 5 was the focus of burning controversies throughout the Islamic tradition. This is a clue to the difficulties intrinsic to that definition. Much thought was given to whether the definition succeeded in giving an account of the *essence* of proportionality, whether it was intended to convey only the "common understanding" of the matter in place of the "true understanding" or whether, unlike most other Euclidean definitions, it stood in need of, or was open to, proof. What seems to stand behind these controversies is the shared conviction that we draw our understanding of ratio and hence of "being in the same ratio" from the case of numbers, so that any extension of this understanding has to be made plausible or rigorous on the basis of arithmetic, that is, in the terms of the possibility (or impossibility) of measurement by integers. In other, more technical terms: The pre-Eudoxean procedure of *anthyphairesis*—successive, reciprocal subtraction of a lesser number or magnitude from a greater—employed in Book 10, Propositions 2 and 3, is made primary and then the task, accepted by some, but not all, Islamic mathematicians, is to show that "proportionality" as established by this arithmetical procedure is extensionally equivalent to "proportionality" according to Book 5, Definition 5. Consequently, while most of these scholars are quite conscious of the interplay among the accounts of proportionality in Books 5, 7, and 10, they are also agreed, in the main, that the account in Book 5 cannot stand on its own, either as a full-dress definition (meeting Aristotelian standards) or as clear rendition of our "intuitive" grasp of "same ratio." However, what they are generally prepared to count as our intuitive grasp of this matter is in fact dependent on the prior acknowledgment that *calculation* (exact or approximative *measurements* of the "real" value of a ratio) is the principal, not to say exclusive, source of our understanding of ratio as such. Thus, for example, Naṣir al-Dīn al-Ṭūsi (1201–1274) can declare that every ratio "can be called [a] number, measured by unity [*sic!*], so that the first term of a ratio is measured by the second"—that is, every magnitude can be measured by a series of rational *fractions*, a finite series in the case of commensurables, an infinite and therefore approximative series

in the case of irrationals. Shorn of all other significant details, the figure presented by Arabic work on these questions is strikingly dissimilar to what we encounter in Euclid himself. Where the latter seems almost always to be striving to keep alive a sense of the tension between the arithmetical (hence, strictly metrical) and the geometrical versions of proportionality, so much so that he omits to state crucial links between them when the opportunity arises or presses, his Islamic commentators and critics take calculation or computations as the *telos* of mathematics and redistribute the weight of its ingredient portions accordingly. Another way of saying the same thing: Greek theoretical geometry and arithmetic, as consolidated in the *Elements,* are made into means in the service of practical (or, technical) *logistics.*[31]

(2) In more recent times integrity has been restored to Definition 5, but only by virtue of a massive transformation in the understanding of number which is a continuation of the Islamic beginnings mentioned just above. I have in mind, of course, the assimilation of the criterion of "same ratio" to Dedekind's theory of real numbers as described by "cuts" in the set of the rationals. Dedekind himself inspired this assimilation when he expressly gave as the ancestor of his theory "the most ancient conviction . . . set forth in the clearest possible way in the celebrated definition which Euclid gives of the equality [*Gleichheit*] of two ratios. . . ," although it is very important to note that he also qualified this reference later on in a letter to Lipschutz: "the Euclidean principles by themselves [*allein*] without the assistance of the principle of continuity, which is not contained in them, are incapable of grounding a complete theory of real numbers as ratios of magnitudes."[32]

At first blush, it should suffice to point out that, since Euclid does not countenance "rational *numbers,*" a fortiori there is no place in his theory for "real numbers." On reflection, however, the disanalogy between Dedekind and Euclid proves even more instructive, especially in light of the former's clue in the letter cited above. For Dedekind (a) the rationals are strictly ordered by the trichotomous relation ($>$, $=$, $<$); (b) the "real" line is everywhere dense, and any point on the line either is or is not defined by a member of the class of rationals. Neither requirement can be satisfied by anything within the ambit of the Euclidean understanding of ratios, points, and lines. As regards the first requirement, we must observe that Book 5, Definition 7 ("greater ratio"), only establishes that, for some pair of integers (m, n) serving as testing numbers, *one* ratio is greater than (or, less than or equal to) a *second* ratio ($ma <\!/\!> nb$, $mc <\!/\!> nd$); it does not provide a way of ordering *all* ratios with a single stroke.[33] As far as the second requirement is concerned, Dedekind's own epistolary comment goes far towards settling the issue. The principle of continuity (of the real line) is *not* contained in or entailed by Euclid's theory of ratios.

Indeed, Euclid's definitions of point and line (Bk. 1, Defs. 1 and 2) seem to be purposely directed against any temptation to interpret a line as a "set" of indivisible points somehow already in or on the line. The lesson learned from Zeno's paradoxes by Eudoxus, by Aristotle, and, it is fair to suppose, by Euclid as well, was that the infinitely divisible continuum cannot be reconstituted out of a (denumerably) infinite number of discrete and indivisible points. Dedekind, guided by previous attempts by Weierstrass and others to "arithmetize" the continuum, presupposes that the geometrical line is essentially analogous to an ordered arrangement of "numbers," each of which is uniquely definable on the basis of other "numbers."

Far more important for the most general issues of this and the subsequent chapter is the way Dedekind's commitment to an extremely radical version of mathematical constructivism shapes his theory of all numbers, including the reals. His well-known thesis in *Stetigkeit und irrationale Zahlen,* that counting is "nothing else than the successive *creation* of the infinite series of positive integers," is amplified and elucidated by his later injunction to a correspondent: "Please understand by 'number' not the *class* itself, but something *new* (corresponding to this class) which the mind creates [*was der Geist erschafft*]. We are of a divine race and undoubtedly possess creative power not merely in material things (railroads, telegraphs) but quite especially in intellectual things."[34] The positions of Dedekind and Euclid on this point are as clearly antipodal to one another as one could imagine (note that in Book 10 Euclid consistently speaks of "finding" [*heurein*] the various irrationals.)

The disproportion between Euclidean and Dedekindian "styles" also bears on another troubling feature of Book 5, Definition 5: Does Euclid furnish an effective or decidable "algorithm" for determining *when* two pairs of magnitudes are "in the same ratio"? His phrase, in Heath's translation, "any equimultiples whatever" (*kath' hopoianoun pollaplasiasmon*) is at the root of the difficulty. Does Euclid require that we check *every* pair of integers (m, n) to see whether the relation $ma >/< nb \Rightarrow mc >/< nd$ holds? If so, how, in "practical" terms, could we ever test *all* pairs of integers? Is there an unbridgeable gap here between the *theory* of what constitutes proportionality and the practice or technique for establishing in concrete cases when proportionality obtains (since merely hitting upon the right pair of equimultiples by chance is presumably a leap, not a bridge)?

Appeals to a tacitly assumed "principle of complete induction" or to its scholastic ancestor, the *dictum de omni et nullo,* do not dissolve this problem, since chancing upon the *first* pair of integers (m, n) with the requisite effect on two ratios $(a:b, c:d)$ is itself a complex affair, especially if the ratios happen to be ratios of incommensurables.[35] When the ratios are of commensurables, then a *single* case of two integers which produce equality ($ma = nb; mc = nd$) suffices to guarantee sameness of ratio. So, too, when one ratio is greater than the other in the sense of Definition 7: a single pair of integers, for example, those whose ratio is (decidably) the same as $a:b$, suffices to show that $a:b > c:d$ or not, for all cases. When, however, we are dealing with ratios of incommensurables, the case of equality can never occur, since if it did, then, *per impossibile*, the incommensurables would have to one another the ratio of a number to a number. Thus, Definition 7 is decidable for both commensurables and incommensurables once a single case has been found, while Definition 5 is only decidable in that way for commensurables. Hence, a decision procedure for "same ratio" in the case of incommensurables cannot be derived from Definition 5. As Samuel Kutler puts the result, with one eye deliberately on Dedekind, "two ratios of incommensurable magnitudes are in the same ratio if there is no ratio of commensurables between them."[36] Euclid seems to be capturing sameness of ratio for incommensurables, *not* by an effective technique of induction but *viā negationis*. In other words, Definition 5 reiterates in its own manner the original perplexity in which the generalized theory of ratios has its source.

(3) This perplexity and its repercussions are palpable in one last aspect of Euclid's presentation. The early modern "heir" to the seat of power occupied by proportions in Greek and medieval mathematics is the equation, and an equation, as the early moderns were always aware, is nothing other than a series of ratios which can be combined and, for practical purposes, set equal to zero. Viète speaks of an *analogism* as the transformation of an equation into the original form of a proportion [*analogia*]; similarly, Leibniz talks of the *conversio aequationis in analogiam vel contra* ("the conversion of an equation into a proportion or the reverse").[37] The entire format of algebraic analysis and geometrical construction which is based on the use of equations presupposes a single meaning for the Euclidean criterion of "being in the same ratio," namely, that the two (or more) ratios are *quantitatively* equivalent. In no other way could one make sense of the transition from an equation back to a proportion, with the product of the extreme terms understood to be equal to the product of the mean terms ($a:b::c:d$, that is $ad = bc$; thus, $ad/bc = 1$). This "reading" is arguably the one which leaps most readily from the page to modern eyes; it also disguises a far from superficial plurality of meanings and echoes of meanings suggested in Euclid's own work.

To make the ensemble of these meanings audible once again, we need first to keep in mind that Euclid himself seems to have avoided defining *analogia* on its own, although he does tell us to call "magnitudes having the *same* ratio" "*analogon*" (Def. 6). Perhaps to compensate for the resulting lacuna, a later hand interpolated between Definitions 3 and 4 of Book 5 the words: "*analogia de hē tōn logōn tautotēs*" (Proportion is the sameness/ identity of ratios). Another manuscript insertion occurs after Definition 7: "Proportion is the similarity [*homolotēs*] of ratio." This second interpolation also shows up in the main medieval Latin translations in the same position in the text occupied by the first addition in the Greek original. "Proportionalitas est similitudo proportionum." Aristotle defines *analogia* as the *isotēs* (equality) of ratios, while much later Pythagorean sources try to strike a compromise, at least at the lexical level: "*Analogia de esti pleionōn logōn homoiotēs ē tautotēs*" (Analogy is the similarity or the sameness of several ratios).[38]

These variations, it seems to me, are not just fortuitous or unmotivated, nor do they merely enshrine unconnected historical traditions (such as the Pythagorean and the Euclidean). Euclid's own usage, or even, if one prefers, the usage he chose to retain in his compilation of earlier texts, bears witness to the manifold of concerns to which the core notion of *analogia* remains addressed: the "similarity" of rectilineal figures (for example, two rectangles, the corresponding sides of which are "in proportion," according to Bk. 6, Def. 1); the simultaneous "equality and similarity" of solid figures (which are contained by the just-mentioned similar planes; Bk. 11, Def. 9–10); and the "sameness" of metrical relations secured by the definition of proportional numbers (Bk. 7, Def. 20), to name only a few.

Our question comes to this: What is the sameness or identity which is made evident in Book 5, Definition 5, when magnitudes are said to be "in the *same* ratio"? (*Tauto* is undefined in the *Elements*.) Strict *equality* seems to be excluded on the following ground: The first axiom of Book 1 asserts the unrestricted transitivity of equality, whereas Euclid has to *prove* the transitivity of sameness of ratio (Bk. 5, Prop. 11). *Similarity* seems to be too weak, since two ratios which are the same do not merely resemble one another; each is somehow a replica of the other while remaining nonetheless distinct. (2:4 is the same ratio as, say, 3:6, and both, in music theory, yield the same octave, yet their makeup is distinct). The fairest response to the question appears to be another question: How must we understand the dialectical interrelations of sameness, equality, and similarity in order to become adept at the recognition of informative proportions (*analogies*) in different contexts and dimensions? "Informative" is the apt word here, since the question as now posed is nothing less than the mathematical expression of the all-absorbing issue of forms, of *eidē*, and their seeming identity within difference. I shall be returning to aspects of this issue in later sections, especially in connection with the Cartesian transformation of the "natural" *eidē* into mathematical equations. At this juncture I want only to replace the Euclidean formulation above within its native philosophical settings, so that the idea of a "dialectic" in the *Elements* might look a bit less outlandish.

Aristotle reckons with this same trio of terms in natural language—namely, *tauton, ison, homoion*—on two significant occasions in the *Metaphysics*. In Book Delta he is surveying the ways in which "relatives" (*ta pros ti*) are articulated in speech; notably, his first set of examples is drawn exclusively from mathematics and includes, as I mentioned earlier, the case of incommensurable magnitudes. At the end of this part of his survey he turns to "equal," "similar," and "same" and tells us that they are articulated with respect to "the one" rather than to "number." And he concludes: "Those of which the being/essence (*ousia*) is one are the same; those of which the quality (*poiotēs*) is one are similar; those of which the quantity is one are equal."

Where this passage from Book *Delta* reflects the accepted or reputable usages to which primary philosophy is initially indebted (cf. *Topics* A2), the second passage, in Book Iota, blurs the relatively clear lines of distinction to which accepted usage may have habituated us. Once more discussing "the one" or unity, Aristotle now sets out the relevant connections of the same, the similar, and the equal in this manner: "The Same is said in many ways. In one mode, we sometimes speak of the same in reference to number; the same is also said if something is one both in definition and number, for example, you are one with yourself both in form and in matter; again, if the definition of the primary being is one, for instance, equal straight lines are the same, and equal quadrilaterals with equal angles . . . in these examples equality [*isotēs*] is oneness [*henotēs*]. Similars are beings not the same simply, if not being indistinguishable in their composite being, they are the same in form [*eidos*], for example, the larger square is similar to the smaller and unequal straight lines are similar; they are similar, but not simply the same" [1054a32–66].[39]

Among the many nuances which should occupy us in this complicated passage the most pronounced are (1) the attempt to "derive" equality and similarity from the multivocity of 'the same'; (2) the coalescence of sameness and equality (hence, in light of *Delta* 15, of essence and quantity) in the example of equal quadrilaterals with equal angles; (3) the introduction of the very rare word *henotēs* (oneness) to capture this coalescence both in that case *and* in the case of your "self-identity"; (4) the distinction drawn between "composite being" and *eidos* in such a way that allows Aristotle to treat similars as formally the same (and, thus, without limiting similarity to the category of quality as in *Delta* 15). To say the least, distinctions which earlier seemed fixed along familiar lines (substance, quantity, quality) now exhibit a wondrous mobility in keeping, or so it appears, with differences inseparable from the multivalent "logic" of identity and "being the same."

Aristotle, in this second passage especially, is offering us a deliberately more pedestrian version of the dizzying play among the same, the similar, and the equal enacted through the hypotheses of the *Parmenides*. (See, for example, *Parm.* 161c–e). I won't try to delve into those deep waters here. Nor is it necessary to work out the possible historical or chronological affiliations between the *Parmenides* and the Eudoxean theory of same ratio, on the one hand, and Aristotelian dialectic and Euclid, on the other. What is needed is the recognition that Euclid, from the evidence assembled so far, is disinclined to substitute technical univocity for the kaleidoscopic "analogies" among his fundamental terms since it is the latter, not the former, that best communicates to the learner the multifarious character of the "same" *mathesis*.

iv   Conclusions

I have spent a good deal of time on the tangled issues raised by a handful of definitions, theorems, and manuscript interpolations mainly taken from Book 5 of Euclid because these tangles seem to me to fall centrally within the scope of classical Greek mathematics and thus begin pointing to the basic differences separating it from radically modern mathematics. This separation is not mediated or bridged simply by the apparent continuity of subject matters (such as the theory of proportions), for everything depends on how this appearance is assayed. Does the Cartesian theory of equations, for instance, merely keep the *Schein* of Eudoxean proportion theory, or is it, to continue in Hegelian terms, the *Erscheinung* of the latter's essence? Or, conversely, is the Eudoxean theory the *Erscheinung* of which modern theory has all along been the *Wesen*?

Many contemporary interpreters opt for this last view, often using the language of "anticipation" and depicting early Greek theory in a reconstituted form which is tailored to fit the proportions of technically more advanced achievements. Common to these interpretations—so common that their differences in detail and orientation become quite secondary—is the mostly unspoken conviction that a Eudoxus or a Euclid would have resorted to improved techniques to repair deficiencies if they could have been brought to his attention. This implies that uniformity and univocity as brought about by refinements, innovations, or extensions of *technique* are among the essential goals shared by mathematicians throughout the ages.

Posed in these terms, the thesis of continuity usefully makes the status of technique the gravamen of discussion. Has success in achieving incontrovertible results always weighed more heavily in the balance than the delicate enterprise of matching technically achieved results to a pretechnical "understanding" of the domains in which *mathēsis* seems so much at home? Clearly, I have been arguing that habits of discourse detectible in Euclid let us see how a case can plausibly be made for a negative answer to that question. It should be equally clear that this argument is not intended to imply that technical refinements and so forth are excluded or denigrated by the Greek mathematicians who have learned those habits.

This would obviously be wrong-headed, as even a cursory look at, say, Book 10 of the *Elements,* Apollonius' *On Conic Sections,* or, perhaps most challengingly for my position, any of Archimedes' treatises, would reveal. As I said earlier and have been trying for some pages to illustrate by reference to Euclid, the issue at hand is not whether technique (or technical method) is absent or present in Greek mathematics, but whether manifestly resourceful techniques inevitably have the upper hand over what I have been calling a pre-understanding to which the mathematician *qua* teacher of *mathēsis* remains duly sensitive, just as the learner is obliged to acknowledge those points at which his "pre-understanding" is strained and, possibly, must be either expanded or abandoned. The incomplete portrait of Euclid I have been sketching discloses features more in keeping with a teacher who retains the upper hand over the technician as far as he can (the case of compounding ratios in II, ii above, shows one limit to this kind of didactic mastery). For such a teacher ambiguity and equivocity, when they belong to elements figuring in the student's pre-understanding, must be woven into the web of learning, not discarded by *fiat.* This strand in the exercise of mathematical *phronēsis,* so resistant to precise delineation exactly because it is not governed by *rules* of discovery and invention, is central to the design of a science in the Greek sense of *epistēmē* and thus must come first in the order of teaching, prior to matters of completeness and consistency, prior to special arrangements concerning permissible or necessary techniques.

Aristotle uses a rare poetic word to describe the "occupational hazard" to which the mathematician (or philosopher *cum* mathematician) is exposed; alluding to participants in the Academy he says: "the things said [in mathematics] are true and fawningly seduce the soul [*sainei tēn psychēn*][11] (*Metaph.* N3.1090b1; compare *Odyssey* 10.217). *Phronēsis,* as I am speaking of it here, consists in knowing when and why the blandishments of mathematical *technē* are to be resisted.

III   The Powers and Idioms of Construction in Greek Geometry

Of all the techniques mobilized in the *Elements* none is more conspicuous and none has, in the last analysis, become more perplexing than the family of procedures usually known nowadays by the collective name "construction." As I have already indicated in the opening chapter, it is from this mathematical technique that a certain self-understanding of modernity draws its impetus. But this way of speaking immediately directs us to the basic dilemma: Is it right to refer to "*this* mathematical technique" as though it remained essentially one and the same from Greek geometry through, say, Kant? If we are entitled to speak this way, then it is no longer clear why radically modern philosophy found so patent an inspiration or paradigm in a procedure indifferent to the polemical distinction between ancient and modern thinking. If we are *not* entitled to speak this way, then the reference of "construction" when used of Greek mathematics becomes opaque. It will take us some way towards dissipating that opacity if we begin by seeing how the issue of "construction" in Euclid arises in the most central way for the moderns themselves.

IV   The Establishment of the Kantian Orthodoxy

Attention to the philosophical implications of the *detailed* design of Euclid's texts first became critically acute once the stage had been set by theoretical debates in the eighteenth century concerning the ontological commitments and the epistemic credentials of mathematical propositions in general. The chief parties to these debates were members of the Leibnizian school (along with one importantly idiosyncratic sympathizer, J. H. Lambert) and Kant.

Stripped to the barest essentials, these debates were addressed to two interlinked questions: (1) Are all mathematical assertions derivable from self-evident definitions (which, in turn, can be reduced to the principle of identity or non-contradiction)? and (2) On what grounds can it be claimed that the *defienienda* of a mathematical theory *exist*? The first, of course, simply is a restatement of the Kantian question "Are all mathematical truths analytical a priori judgments?" The second presses close to the heart of the cardinal problem of modern philosophy, namely, "What, if anything, connects the conceptual order to the phenomenally or sensibly 'given'?" Together these two questions determine the parameters within which Euclidean 'construction' could be given legitimate and interesting meanings.

Let me confine myself to a few of the 'high points' of this debate, which is, I should add, so far without an obvious victor or dénouement.

Wolff, the leader of the Leibnizian school and the "scholastic" authority on whom Kant's early lectures in mathematics largely depended, followed Leibniz (a restricted version of Leibniz) and, so he was convinced, Euclid, in making *all* the propositions of geometry and arithmetic either immediate or deductive consequences of the initial definitions, while the latter are "elucidations" (*Erklärungen*) of the meanings of the terms ingredient in them and can ultimately (or, in principle) be reduced to statements of the analytical identity or inclusion of one term with, or in, another. "*All* propositions"—that is, the axioms and postulates, as well as the theorems and problems. This is the crux of the debate; according to Wolff, the postulates in Euclid, Book 1 must be interpreted as "indemonstrable *practical* proposition(s)" the truth of which can be seen as "flowing" straightaway from a single elucidation (for example, the drawing of a straight line between two points flows from the definition of a line). These postulates are more idiomatically named *Heische-Sätze* (imperative-propositions) since "they show that and how something can be made or done," while the axioms and theoretical propositions or *Ergwägungs-Sätze* (ponderables) assert that "something inheres or does not inhere in a thing." "Problems" and, in turn, "theorems" are derivable from more than one definition. We get, therefore, the following schema:

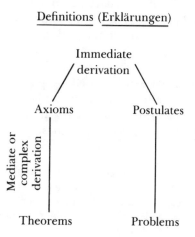

This schema's most striking features are the grouping of axioms *and* postulates as "indemonstrable propositions" and the obscurity surrounding the logical derivation of a postulate (an instruction to make or do something) from a definition. The second point can also be made by saying that a postulate, although officially called a "practical proposition" (*propositio practica*), shows little sign of having an origin in practice or mathematical "activity"; its embeddedness in a systematically deducible *theory* makes its epistemic status uniform with that of axioms.[40]

Lambert takes note of this second feature when he criticizes Wolff's term *Heischsatz*, as "inept" (*unschicklich*) inasmuch as it suggests that a postulate has an exclusively propositional character. In his essay on the

Criterium veritatis Lambert records his own surprise at reading Euclid long after having read Wolff: "instead of a theorem he [Euclid] begins with a problem [*Aufgabe*]. What is this, I thought, shouldn't the theory come first before one moves on to putting it into practice [*Ausübung*]? Euclid must have thought much further about the matter!"[41]

Lambert's own response to this Euclidean surprise was twofold: (1) to incorporate the *practical* or active force of a postulate into the very meaning of all "basic concepts" of a mathematical theory and (2) to make the "positive possibility" of a basic concept (such as extension in the case of geometry) dependent on our ability to produce it [*sic!*] by means of constructive action (satisfaction of the principle of non-contradiction being only a negative criterion for a basic concept). All complex concepts are certified by tracing their logical genesis back to simple or basic concepts; likewise, all more complicated figures are to be authenticated by construction out of simple figures. In further criticism of Wolff, he remarks in a letter to Kant (February 1766) that *definitions* must come at the end, not at the beginning, of a systematic presentation; elsewhere, he says they ought to encapsulate the *mode* by which the defined object can be constructed.[42]

The main outcome of this engagement with Wolff is Lambert's insistence on the existential weight carried by postulates and transmitted through them alone to definitions, axioms and theorems; "What is thinkable is nothing (a dream), if it cannot reach existence [*Existenz*]." The passage from the *Criterium veritatis* continues:

> The main and most dexterous artifice [*Hauptkunstgriff*] consists in this, that the possibility of an equilateral triangle proves itself [*sich . . . erweist*] on its own, so to speak. We have no better way of refuting someone who believes it to be impossible than to show him how he himself can effect it, bring it about [*ins Werk setzen könne*].[43]

This certification of existence through a construction trades on our power of sensible intuition (since we must be able to perceive what we are doing), and, at the same time, eschews any help from intuition as ordinarily understood or experienced. He asserts both that "logic and geometry require nothing more than a thinking being" *and* that signs, which should replace the *figural* diagrams of geometry, "do us . . . the further service that with them all our thinking is transformed into an unbroken series of sensations [*Empfindungen*] and representations [*Vorstellungen*]." "Symbolic cognition is an intermediate [*ein Mittelding*] between sensation and genuinely pure thought." Somehow, the postulates of geometry, instructing us in the way a figure is to be constructed so that its existence might become manifest, are postulates of a science "independent of all existing things [*von allen existentibus*]."[44]

Kant enters this picture on a determinate occasion. The then-renowned mathematician A. G. Kästner published in 1790 a series of essays in the second volume of the *Philosophische Magazin* edited by J. A. Eberhard, one of Kant's principal adversaries and a staunch defender of the "Leibnizian" thesis that mathematics consists solely of analytical truths. One of these essays "Was heisst im Euclids Geometrie möglich?" provoked Kant into composing a reply to Kästner; this was published above the name of his student Johann Schulze as part of the latter's long review of the whole volume.[45]

Kästner is generally careful to stay above the philosophical fray. He is chiefly concerned to clarify how "possibility" functions in the *Elements*. It means, according to him, first, "what one assumes as possible" and, second, "what arises from this assumed possibility, in ways whose possibility one assumes or demonstrates [*darthut*]."[46] The constructional problems in Euclid fall under the second meaning of "possible" since in their case Euclid "shows how one can make the thing of which he is speaking, if the assumed possibilities are conceded."[47] The postulates make up the first class; to show that *they* do not contain any contradictions the concepts involved in them must be evolved [entwickelt] further than Euclid has done. Kästner, chastened by his own dismaying failures to prove the Parallel-Postulate, leaves this task to others. Thus, he leaves undecided the question whether "one merely assumes the possibility" of combining the relevant concepts in a contradiction-free manner or "has experienced [*empfinden*] this possibility."[48]

Kant's reply predictably centers on the question of the grounds of acceptance of a geometrical postulate. Let me simply quote the most salient passage:

> The description [*Beschreibung*] which occurs a priori through the imagination in accord with a rule and is called "construction" is itself the proof of the possibility of the object [*Objektes*]. Mechanical drawing . . . which presupposes the latter as its model is not relevant here. However, that the possibility of a straight line and a circle can be proved, not *mediately* through proofs, but only immediately, through the construction of these concepts (which is not, to be sure, empirical), stems from the fact that among all constructions . . . some must be the first.[49]

My excursion into the three-sided debate among Wolff, Lambert, and Kant should have brought to the fore the two decisive preoccupations not merely influencing but indeed dictating the context in which the issue of "construction" in Euclid can be, and has almost always been, raised by modern interpreters. Reformulated in the light of the last few paragraphs these are: (1) What contribution, if any, does sensibility or imagination make to the *rational* enterprise of (systematic) mathematics? and (2) How is the "reality" of mathematical concepts authenticated? That is, what sort of production via reason (Wolff), via deeds (Lambert) or via the perceptual and imaginative powers (Kant) furnishes *proof* that the conceivable also "exists"?

My contention is that neither question is even formulable *apart* from the particular circumstances of philosophy (and mathematics) eventually brought about by the advent of modernity. It makes sense to raise the question about ancient mathematical construction as a question about the "existence" of "objects" satisfying definitions only when (natural) sensibility has been dissociated from reason and only when this dissociation is taken to entail that rational concepts and ideas as mental "entities" (or as the intentional correlates of mental activities) are essentially impotent to decide whether there is anything "in the world" answering to their contents.

Let me be as careful as I can in stating anew what this extravagant claim does and does not imply. Modern reason divorces itself from sensory awareness not because of any particular misdeeds on the latter's part, not even one so flagrant and upsetting as its long affair with Ptolemaic astronomy. The ground for divorce is rather the recognition that pretheoretical, premethodical perception keeps reason in thrall to a putative source of knowledge over which it cannot, in principle, exercise systematic command. This putative source is "Nature" understood as the ineluctable and never completely exhaustible correlative to natural or naive perception. Reason wins its freedom from the tutelage of nature only at a considerable cost, however. Kant gives an exact accounting of this cost when he notes, in the preface to his *Metaphysical Starting-Points of Natural Science:*

> *Essence* is the first inner principle of everything which belongs to the possibility of a thing. Thus we can ascribe to geometrical figures (since nothing is thought in their concept which expresses an existence [*ein Dasein*]) only an essence, but not a *Nature.*[50]

Essences and the concepts in which they are resident as principles of *possibility* are not (in any immediate way) anchored to the actuality or natural "existence" of the "cases" they either formally subsume or materially (*inhaltlich*) determine. Hence, the freedom of reason with respect to the formation and the imposition of concepts is simultaneously a new and more exacting bondage to preconceptual "Nature" since the latter alone holds seemingly unchallengeable authority over what can or cannot be "real," even when "real" means only the givenness (or materiality) of phenomena to which reason systematically addresses itself. In Hegelian terms, the struggle of master and slave is played out as the drama of reason's desire to "objectify" or "realize" itself (as freedom), although the measure of what it is to be "real" is, so to speak, borrowed or usurped from pre- or arational nature. "Construction," as I have already suggested in chapter 1, comes to hand as a mediator in this struggle; but to select construction as the mediator is already to have foresworn any relation between the "rational" and the sensible, the essential and the phenomenal, other than that between mind and mindless "nature." In other words, the choice of a mediator is predetermined by the characterization of the struggle furnished by *one* of the two parties at odds with one another.

This claim is not intended to dismiss the question of "construction" in Greek mathematics out of hand as though it were merely a mirage produced by the modern self-understanding. On the contrary, a great many aspects of ancient self-understanding can only be brought into full view in connection with that question, as I shall soon be trying to show. At a minimum, we cannot simply assume that these aspects coincide with or overlap those features with which the modern way of posing the question of mathematical construction has become fatefully entangled. Perhaps only by freeing ourselves (in speech) from that entanglement can we come to see a quite different "struggle" being enacted on the terrain of classical Greek mathematics. Euclid's *Elements* might then be read either as a battleground in its own right or, as I shall suggest, as a recollective scenario, a *mimēsis* of significant actions in that struggle, carefully giving to each side its "proper shape," its *idia morphē*, as Aristotle says the successful poet and portraitist will do (*Poetics* 1454b12).

That such a struggle did occur is well documented in Proclus and Pappus and strongly suggested by passages in Plato and Aristotle. All along its key terms of reference are "problems" and "theorems"; throughout what is at stake is the understanding of the respective role (or "non-role") of each, both in the acquisition and presentation of mathematical knowledge and in the ontological constitution of the *mathēmata*. The hallowed locutions of the end of Euclidean propositions, "Q.E.D."/ "Q.E.F." (*hoper edei deixal/hoper edei poiēsai*) epitomize the issue in two very provisory ways: (1) What share should or must fall to making, *poiēsis*, within the discursive progressions of mathematics? That is, *what* is it that is "made" or "done" in the course of the mathematician's speech and might seem to remain inaccessible to his speech were it not for his *poiētic* aptitude? (2) How does the motion and, hence, the temporality of making (an event in human speech or practice) bear on the "being" of the *mathēmata* themselves? Or, in other words, does the temporality endemic to mathematical "making" (and, indeed, to mathematical "showing" or "proving" since this, too, moves in speech from step to step) mean that the *mathēmata* are temporal as well—that is, are most faithfully addressed in the language of coming-into-being, of genesis?

These provisory, yet already encompassing ways of establishing what is at stake in the question of *poiēsis* and "construction" should serve to orient us as we seek to align ourselves with the particular directions taken by Euclid (and by the traditions he commemorates). As always, these particularities are not simply cases to be subsumed without further ado under ready-made generalities; rather, the details of their texture elicit reflections potentially capable of letting us speak pertinently about a "whole" not so much illustrated as enlivened by "its" particulars. This circle of explication cannot, as we seem to have learned, be rectified.

Let me begin by keeping the vexed and indeed focal question of the postulates at arm's length so that I might return to it (in II, ix) with resources gained by examining some of the traits of Euclidean style *in situ*. The most promising resources are potentially at hand in the lexicon, semantics, and syntax of "construction," but to tap them, one must proceed with care.

Three lexical "slots" ought to be distinguished: (1) terms designating *problems* as distinct from *theorems;* (2) terms in the initial enunciations designating procedures or goals associated with problems; and (3) terms in the body of propositions designating procedures and operations. Once these distinctions have been made and, so to speak, mapped onto the language used by Euclid, some unexpected and illuminating results start to appear forthwith. First, there is no generic "term" for (1) unless it is the verb *poiēsai* in the tag *hoper edei poiēsai,* "what it was required to do," at the conclusion of some of the propositions, but not at the conclusion of all of those with which modern interpreters almost automatically join the idea of construction (and, inevitably, the idea of "existence-proof"). For example, Book 10, in which Euclid reproduces Theaetetus' theory of irrationals, does not contain a single proposition (out of 115) ending with this tag. Second, there is no *single* term in the enunciations corresponding to procedures and so forth, but rather a family of interestingly different, only occasionally overlapping terms. Third, within the enunciations we find, predictably, that the members of (2) recur, while occasionally an operational-term appears which has not already been used in the enunciation.

The reader of Heath's magisterial translation will not have been put in a position to detect these discriminations since he prefers to render a number of distinct terms from classes (2) and (3) by forms of the same English word "construct." Similarly, a reader steered by expert accounts such as van der Waerden's will have been led to "see" constructional activity taking place in areas which Euclid marks out with a distinctively different vocabulary.[51] At the very least, then, more study is needed before we can confront the overarching question of the relation between Euclidean and modern "construction." Some observations based on such a study follow.

(1) Proclus does furnish a generic term—*kataskeuē*—customarily translated as "construction" in his enumeration of the six constituents of any problem *or* any theorem aiming at completeness: the enunciation (*protasis*), the exposition (*ekthesis*), the specification (*diorismos*), the *kataskeuē* (Heath: "construction or machinery"), the demonstration (*apodeixis*), and the statement of conclusion (*symperasma*). The *kataskeuē* is something occurring *within* the working out either of a problem or a theorem; its pretechnical sense, "preparation or making ready," may be more helpful here, since its function is to get things ready for the climactic demonstration (namely, that such-and-such holds true of the specified kind of figure[s] or number[s] or that such-and-such a figure or number is of the required kind). These preparations do involve certain operations over and above what has already been done in the exposition or setting out; for instance, extra lines must be drawn, joined, cut in a specified ratio, and so forth in order to get a "diagram" ready for the ensuing proof. When the frequently repeated phrase *tōn autōn kataskeuasthōn* occurs within a proposition we should understand it in light of this preparatory function common to theorems and problems. It retards our insight into nuances to translate this and other locutions uniformly by "construct" or "having been constructed."[52]

(2) This last point is importantly connected with the variety of operational terms in use in enunciations and propositions alike: *synistanal, anagraphein, paraballein, katagraphein, ekballein* (see below (II, vi) on the preferred tense and voice of these verbs). *Kataskeuazein*, as far as I have been able to determine, *never* appears in the enunciation of a problem (or theorem), nor does the highly suggestive term *syntithenai*, "to put together" or "to compose" (a figure), frequent in post-Euclidean mathematical texts but restricted in Euclid to the transformation of a proposition by *Componendo* (Bk. 5, Def. 14). What, then, are we to make of this variegation in wording? Has Euclid merely compiled materials of different ages and authorship so that variations in vocabulary are coincidental? Or, under the same hypothesis of "mere" compilation, did Euclid nonetheless strive to preserve informative variations? The second conclusion begins to seem by far the more plausible as soon as we observe how the different terms in the (partial) list above are respectively associated with different *kinds* of figure or other geometrical item. Thus, *paraballein* (to apply, literally, to throw upon, so that the base of the figure coincides with a given line-segment) is used characteristically of parallelograms, even though *we* might feel free to say that any rectilineal figure can be

"applied" to a given segment as its base. We can "extend" or "produce" straight lines only (*ekballein*—literally, to throw out from a given terminal-point), and both inscription (*engraphein*) and circumscription (*perigraphein*) are possible only when we are dealing with circles (or spheres) and polygons (or polyhedra).

Matters are less simple when we come to the two most frequently employed operational-terms, *synistanai* and *anagraphein;* the first is invariably translated by Heath with "construct," the second, with "describe." Heath remarks that *synistanai* was the customary word for constructing *triangles* on a given segment, and Proclus, who often notes the significance of this distinction, fills out the other side of the picture by linking *anagraphein* to the "construction" of *squares*.[53] A syntactic detail now becomes relevant as well: *synistanai* takes the preposition *epi* (upon) with the dative, while *anagraphein* is followed by *apo* (from, *not* out of, which would of course be *ek*) with the genitive. The primary root of these distinctions is, in all likelihood, the way the given line-segment is or is not constitutive of the very nature or look of the figure for which it furnishes the starting-point. Given a line-segment *as* the side of a square, I can immediately derive the square figure (and its area) from the length of its base; the initial segment is potentially (*dynamei*) that square and, *in this setting*, no other figure or species. The triangle, on the other hand, does not issue as just that figure immediately from a line-segment of a given length; the two remaining sides must be brought to stand together, one at each of the terminal-points of the line and at angles guaranteeing that they will intersect at a third point.[54] That these lexical and grammatical differences reflect differences in the nature of certain figures, *prior* to their being "constructed" in a particular setting, is in some measure corroborated even when the pertinent locutions are allowed a wider extension. Thus, in Book 6, Proposition 25, when Euclid shows how to organize (*systēsasthai*) a rectilineal figure similar (in look or shape) to a given figure (and equal in magnitude to a third figure of whatever shape), the *anagraphein* + *apo* (a line segment) construction is legitimate since we already know how the remaining sides of the required figure have to stand to one another. Similarly, we find the same syntax in several of the theorems concerning parallelograms and parallelepipedal solids (as in Bk. 6, Props. 27–28; Bk. 11, Prop. 37); Proclus reminds us that of all isoperimetric parallelograms the square has the greatest area since "the rightness of the angles and the equality of the sides have all the power (*to pan dynatai*) as far as the increment of areas is concerned."[55] The square, we could say, is the paradigm of parallelograms.

What is coming into view along with these nuances of grammar is nothing less than the implied presence of forms, of *eidē*, in the background to the operations permitted to the geometer. The word *eidos* is quite rare in the *Elements* (see Bk. 6, Prop. 19, Porism), but pervasive in its companion-piece, the *Data,* where it is first used in a definition: "Rectilineal figures are said to be given in form or look (*tōi eidei*) when their respective angles are given and the ratios of their sides to one another are given" (Def. 3).[56] These looks or kinds appear to be irreducible and, for that reason, capable of entering into manifold determinate relations with one another. The geometer tutored by Euclid is being schooled in respect for those kinds, a respect activated by his various operations and not compromised by their forcefulness. (A circle is not, for Euclid, an infinite-sided polygon even though "circles are to one another as the squares on their diameters" [Bk. 12, Prop. 20].) The impact of this respect for kinds will be felt much later, almost nostalgically, when Leibniz tries to craft his *analysis situs* in order to retrieve in an immediate way (which still means, for him, via characteristic symbols) the shapely figures obscured in geometry by the equations in Cartesian specious algebra.[57] At the moment, however, we are only witnessing the influence of that respect on Euclid's variegated language of operation. We have not yet encountered the question of the source and status of the respective kinds or looks. (See below, on the postulates.)

(3) I have already mentioned that Book 10, for many moderns the high-point of the *Elements,* does not contain a single "problem" among its 115 propositions; moreover, the operative word in those propositions aimed at exhibiting various relations of (in)commensurability/(ir)rationality among line-segments and figures is *always heurein,* "to find or discover," exactly as in the arithmetical Books 7–9, and never "to draw, extend, produce, etc." These propositions of discovery end with *hoper edei deixai* (Q.E.D.), not "Q.E.F." When *poiein is* used (as in Props. 95–102) it signifies the *action of a line* (or a figure) in "producing" an area (or a breadth) falling into a designated class; these base-lines or "sides" are vested with a power, a *dynamis* to make an area of a certain sort (see Def. 4 and Heath's note ad loc. on *hai dynamenai auta*).

All of these facts must prove disconcerting to anyone wanting to see in Book 10 a set-piece of sophisticated constructional technique (yielding a subset of the "algebraic irrationals")! On the contrary Theaetetus (and Euclid) seem to have avoided relying on what may look to us like an altogether "natural" or naive procedure—namely, using a straight-edge alone to generate an infinity of new irrational lines expressible as the square root of the sum of squares *via* the Pythagorean theorem. Moreover, Wilbur Knorr has made a quite plausible argument that Theaetetus may have had on hand purely computational (logistical) means for approximating the "values" of at least some of the different classes of irrationals treated in Book 10; if this was the case, then we should be all the more struck by the omission of these heuristic tools from the final redaction of the theory.[58]

If, then, the theme of Book 10 is not the "construction" of irrationals, what might it be? We can draw on the ancient sources for clues. The first is the passage in the Scholium to Book 10, which I have already cited (II, i), where we are told that incommensurable magnitudes are incommensurable by nature, while irrational magnitudes are irrational by convention. The second source is a fragment of Apollonius' (lost) treatise "On Unordered Irrationals" (*Peri ataktōn alogōn*) preserved or paraphrased in the Arabic translation of Pappus' Commentary on Book 10. Euclid, so we are informed, gives us the theory of "ordered [presumably, *taktoi*] irrationals . . . which are not very far from the rationals; then Apollonius took up the unordered irrationals, where the distance between them and the rationals is quite great."[59] These two pairs of contrasting terms—"nature"/ "convention" and "order"/"absence of order"—should part some of the clouds darkening the surface of Book 10, even though they cannot be found explicitly in Euclid, who, as we have been seeing, is habitually content to portray traditions in the best light without *obiter dicta* of his own.

To reap the benefit of those clues we must first remind ourselves of the crucial distinction Theaetetus/Euclid draws between commensurability and rationality (Bk. 10, Defs. 1 and 3). The first is an intrinsic condition of magnitudes such that they have to one another the ratio of an integer to an integer and thus share in the unity naturally at the base of the numbers. The second is a condition of magnitudes (here, straight lines) in respect to a line set forward to represent a unit-measure. Lines commensurable with the latter either in length or "in square" (*dynamei*) only are to be called "rational," that is "sayable" (*rhēton*); all other lines incommensurable with the designated unit-length are called *alogoi* ("deprived of an exact reckoning"). Euclid (or his source) appears to have avoided quite deliberately the earlier term *arrhētos* ("unsayable") for this second class; indeed, the principal aim of Book 10 is to teach us we can indeed "say" something quite determinate about the "irrational." We can name its fundamental types and put them in an intelligible order, even though we can never eliminate the last residue of irrationality.

If the irrationals are so only thanks to human *thesis*, the "choice" of the unit-length, nonetheless, they will display surprising integrity when their kinds and their order have been set forth in full. It is almost as though the human convention allows them the privilege of having a nature.

This interplay of convention and nature comes out most conspicuously in the threefold classification Euclid achieves over the course of Book 10 (Props. 21–114). There is only and only one type of *medial* irrational; twelve and only twelve types of *binomial* irrationals, twelve and only twelve types of *apotomes*. (The same scholiast cited above calls these the simplest kinds, *eidē*.)[60] Moreover, there is a discernible order, a *taxis*, in which we find ourselves moving from the *medial*, through the *binomials*, to the *apotomes* and thereby removing ourselves farther and farther from the rational and, ultimately, the commensurable.

It is not at all difficult to recognize in this pattern a response, indeed, a successful response, to Socrates' exhortation in the *Philebus* that we also proceed from the One to the Unlimited by way of all the determinate or limited types in between. Euclid's theory is an object lesson in how we can keep the limitless (and the irrational) within steady and "sayable" bounds. It is not altogether surprising, then, that one ancient editor placed the following theorem at the very end of Book 10: "From the medial there come to be an infinite number of irrational straight lines and none of these is the same as any of the preceding."[61]

(4) Euclid's lexicon of terms connoting operations either dictated by the enunciation (as in a problem) or required for a proof showed us what role might be allotted to *nature* in the constitution of geometry; the structure of Book 10 has suggested that the distinction between nature and *convention* is not in every case hard and fast: The "conventional" irrationals show up in an intelligible order reminiscent of nature. And yet, in both of these instances I have referred to *operations*, for example of applying a parallelogram to a given straight-line, of bisecting a line, and so forth. It still seems that motions, of hand or mind, are indispensable to the presence of those figures and those relations in which *mathēsis* is meant to instruct us. A line-segment might be said by Euclid to have a *dynamis* to produce a certain figure (as in Bk. 10, Def. 4), but it does not make sense to us to claim straightaway that the line does the job all on its own. We have, minimally, to draw the lines, squares, rectangles, and so on whose properties might nonetheless be independent of our operations. Furthermore, the temporality or discursivity immanent in these operations—they take time—may or may not be transitively inherited by the lines and figures in which they issue; in any event, the question of time and movement is inexorable.

V   Genesis and Knowing

Proculus, in a much-discussed passage in his *Commentary on Euclid*, records for us a debate among students and associates of Plato over just this question. Speusippus (and the otherwise unknown Amphinomus) is represented as insisting that all the propositions in geometry be called "theorems," while Menaechmus, the pupil of Eudoxus, counters that all those propositions should be called "problems." (In this stark form, a prophetic anticipation of the battle between ancients and moderns in the seventeenth century and after!) Speusippus took the strong "Academic" line that, since all matters of knowledge are timeless, any implication that a construction brings something into being or makes something which previously was not (*genesin . . . kai poiēsin tou mēpō proteron ontos*) ought to be "scotched."

"Therefore," he says, "it is better to assert that all these things [isoceles triangles] *are* and that we observe the coming-into-being of these not in the manner of making, but of recognizing [*ou poiētikōs alla gnōstikōs*], treating the timeless beings as though [*hōsanei*] they were presently coming into being; consequently, we shall say that we treat them theoretically, not problematically."[62]

Menaechmus, a pupil of Eudoxus *and* Plato, introduces an extremely relevant distinction into the otherwise bald assertion that all geometrical propositions should be called "problems":

> "Problem" is two-fold [in meaning]. Sometimes a problem is to furnish what is being searched for [*porisasthai to zētoumenon*], sometimes, taking something [or: this] as already bounded off [from others], to see [*idein*] what it is, or what sort it is, or what its affection is, or what relations it has to something else.[63]

No doubt Proclus has to some extent "stylized" the accounts he, or his likely source, the Stoic Geminus, reports, in order to sharpen the contrast between the one-sided positions which he will go on to conciliate. (Nonetheless, the wording of Menaechmus' qualification, especially the phrase *porisasthai to zētoumenon* is probably drawn from accurate sources; these details will become extremely significant once I turn (in VIII through IX) to the "existential" relevance of constructions and postulates.) For my present purposes, it is the Speusippean position that needs amplification, especially in regard to the way two manners of conceiving *genesis* in mathematics are discriminated.

For Speusippus the language of *genesis* has an "as-if" character; in order to avoid running afoul of Eleatic strictures against the generation of "what is" from what previously was not yet in being *and* in order to preserve the permanence or exemption from change supposedly intrinsic to the being of what alone can be incorrigibly known (*ta noēta*), we must not treat the constructions and motions on display in a geometrical proof as "makings" in the course of actual performance, as time-consuming just for the reason that first this is done, then afterwards that is done, and so forth. (Compare Aristotle, EN X4 on the time-consuming and thus unconsummated nature of a *kinēsis*, as distinct from an *energeia*.) To treat these *gnōstikōs*, in the manner of recognizing or knowing, must involve, then, a suspension or supercession of their time-consuming character. What, taken literally, seems now to be coming into being for the first time (note the present participle *gignomena* in the "Speusippus"-passage translated above) must be regarded figuratively as having already been accomplished all along.

This interpretation of the *poiētikōs/gnōstikōs* constrast finds some welcome corroboration in a passage in Aristotle's *De caelo* dealing with contemporary attempts to construe the cosmopoietic tale of the *Timaeus* metaphorically, that is, such that the references to "before" and "after" (as in the case of the disorderly versus the ordered motions in the Receptacle) can be understood as a way of talking "for the purpose of teaching" (*didaskalias charin*), a way which makes matters more distinctly knowable (*hōs mallon gnōrizontōn*). Geometrical diagrams (or diagrammatic proofs) are brought in to illustrate the procedure ascribed to the *Timaeus,* but, for Aristotle, the analogy will not hold true since:

> In the making of diagrams, all the initial components having been set out [*pantōn tethentō*], the same [*to auto*] comes about simultaneously, but in the demonstrations of these [philosophers] the same does not come about simultaneously; but this is impossible. For the things assumed earlier and subsequently are contradictory. They say that ordered things came to be out of disordered, but it is impossible for the same to be simultaneously disordered and ordered. There must be a genesis involving the separation [of things] in time as well. In the diagrams, nothing is separated in/by time. [*De caelo* I 10 280a4ff.]

Brimming with difficulties though it is, this Aristotelian text clearly stands in close connection with the Academic discussion reported by Proclus and helps us to grasp some of its nuances more perspicaciously.[64] First, the collocation *hōs mallon gnōrizontōn/poēsei* in *De caelo* is echoed by *gnōstikōs/poiētikōs* in Proclus; apprehending *genesis* in a knowing manner also involves the discrimination of distinctive elements or constituents. *Hoi gnōrimol* are the distinguished citizens, the ones I can pick out easily in a crowd; similarly, at the start of the *Charmides,* Socrates, hard upon his return from the Battle of Potideia, can pick out in the Palaestra of Taureas many men familiar (*gnōrimos*) to him, while others are unrecognizable (*agnōtas*). The denominative suffix *izō* in *gnōrizō* accentuates this sense of the activity of familiarizing oneself with something by discriminating each "part" from other "parts." We can easily see the bearing of this on teaching by way of geometrical diagrams for here, especially when we come to quite complex formations (say, the theory of regular solids in Euclid, Bk. 13), our success in grasping the truth of a theorem depends on our acquainting ourselves with smaller and distinctive constituents of the figure, *one by one,* until the whole makes sense. The first lesson drawn by the "Platonists" to whom Aristotle is adverting here is that didactic generation, as we might call it, does not entail ontological genesis. We could compare this to the justification of Platonic mythopoiesis given by Plotinus: "But myths . . . must separate in time the things of which they tell and set apart from one another many beings which are together . . . at points where rational speeches also make generations of the ungenerated and separate things which are together" (*Enneads* 3.5.24ff.) Both examples—didactic generation and mythopoietic timing—serve to remind us of the pleasure we quite naturally take in hearing how things came to be what or as they are. Cosmogonies, theogonies, genealogies and histories play on this natural pleasure, perhaps in acknowledgment of Aristotle's own claim that "nature spoken of as genesis is the path to nature" (*Physics* B1.193b12–13).

A second nuance of the *De caelo* passage is more obscure, but correspondingly more important to an understanding of the densities it foreshadows. In prohibiting the assimilation of cosmopoiesis to diagrammatic genesis Aristotle distinguishes between necessary simultaneity and necessary non-simultaneity. We might presume that his interlocutors would be quick to agree that, in the case of geometry, the "result" is simultaneous with *and identical to* all of the initial "components" (*if* these are what Aristotle has in mind when he uses the phrase *pantōn tethentōn,* as Simplicius thought).[65] Consequently, the result remains fundamentally the same as its components. The contrast Aristotle goes on to draw between the pre-ordered and ordered condition of the cosmos in the *Timaeus* makes it plain that "sameness" must mean something like ontological homogeneity, rather than the sum of particular features. Thus, a triangle is not itself a line, but it *is* determinate and "ordered" just as its constituent sides and angles are. Moreover, it is in the *genus* of divisible magnitude just as they are. Didactic genesis does not affect a transition from the nongeometrical to the geometrical. Hence time makes or marks no difference between a whole and its "parts" with respect to this homogeneity (not, of course, to be confused with dimensional homogeneity), unlike the necessarily temporal, albeit a priori, synthesis through which a Kantian schematization or construction is achieved in order to allow a passage between different genera (that is, the conceptual and the sensible).

This anticipation of Kant makes one final comparison almost irresistible, even if it seems to take us briefly off the main track being pursued. In this discussion Aristotle sets geometrical genesis side by side with cosmic genesis, only to deny any genuine analogy between them. Kant, in the previously mentioned review of Kästner's article on Euclid, institutes still another parallelism: construction of the concept of a suprasensible being which joins together in itself "all markedly heterogeneous realities"—namely, the idea of God as the *omnitudo realitatis*. For Aristotle, the question of diagrammatic genesis is misleadingly tied to the question of the *modus essendi* of the cosmos; in Kant, the rationale (and limits) of constructibility turns out ultimately to be an onto-theological matter. In Aristotle, the being of the cosmos does not take time, although it gives everything natural time to be what it is; for Kant, *we* don't have enough time in which to "construct" an intuition answering to the idea of a perfect being. Geometry encapsulates the theoretical problem of theology but can never mete out its solution in sufficiently powerful doses.

VI   The Perfect Imperative

Where have these analyses and excursions brought us in respect to Euclid, who, as I have been indicating from the start, never stands back from his presentation in order to furnish philosophical self-commentary, but, instead, puts on display what is requisite and appropriate to the learner's reenactment of *mathēsis*? The controversy between Speusippus and Menaechmus, as stylized by Proclus, was not without its influence on Euclidean discursive practice. This confluence comes through most clearly in the tense and mood of his verbs for "operation." With only a few notable exceptions, Euclid chooses to put these verbs in the perfect passive imperative. Bisecting a line-segment at a point is expressed as "let it have been cut in two" (*tetmēsthō . . . dicha*); describing a square on a line is "let it have been described on *AB*"; "contriving" (Heath's idiom) that *A* is to *B* as *C* is to *D* is "let it have come about that" (*gegonetō*). The importance of this stylistic trait is twofold: First, Euclid does not give instructions or permission to a reader to carry out a specified operation but casts the operation into impersonal, passive form; second, the perfect tense tells us that the relevant operation has already been executed prior to the reader's encounter with the unfolding proof (of a theorem or of a problem; the use of the perfect is uniform in these two classes of propositions).[66]

Now we are indeed in unfamiliar territory. As though to strengthen the hand of the Speusippeans, Euclid invites us, not to perform an operation on our own, nor to observe him performing the operation before our eyes, but rather to consider the operation as already anonymously performed before the "present moment" in which we are following and taking stock of the movement from enunciation to conclusion. This verbal operator does not so much suppress time as shift it backwards, into an unnoticed past during which neither teacher nor student was necessarily on hand. Thus, where Aristotle in the *De caelo* passage appeared to concede to his interlocutors the intelligibility of diagrammatic genesis, with the understanding that in this instance the teacher is simply showing the student how the result can be evolved step by step via certain motions, Euclidean practice for the most part takes an even subtler line. In this case the *genesis* required to prepare a figure or set of figures for the eventual establishment of a conclusion has already taken place before the proof gets under way.

Let me set out the same line of thought much more simply. In a Euclidean proposition nothing moves or is moved save for our eyes and, perhaps, minds as we follow the transitions from step to step. Nor are we called upon to do anything to move the sequence of transitions on its way. The diagram we see exhibits the antecedently executed operations the outcome of which is now confronting us. In this respect, a Euclidean proof and the accompanying diagram are immune to the grave and unsettling perplexities induced by their counterparts in Galileo's *Discorsi*. There the phenomenon under mathematical scrutiny is of course motion, with its varieties of laws; but, *nothing* is moving (Galileo always says "il mobile," avoiding "il corpo") and in the diagram we encounter only the completion of the path traced by a mobile, never its being-on-the-move.

Compare what has happened in the very first proposition of the *Elements*, the "construction" of an equilateral triangle: "with the center *A* and the interval *AB let the circle BCD have been described* [*gegraphthō*] and again with center *B* and interval *BA* let the circle *ACE* have been described and from the point *C*, in which the circles cut one another, to the points *A*, *B*, *let the straight lines CA, CB have been joined* [*epezeuchthōsan*]." The underscored phrases invite us, perhaps, to recollect or imagine someone's describing the circles; they do not oblige us to imitate these recollected actions *in statu nascendi*. The time it took actually to describe the circles is not in any way recorded in the diagram or the ensuing demonstration.

These are not idle details or unmotivated conclusions, as the textual history of the *Elements* reveals at certain junctures. As Malcolm Brown has shown, some of the variant versions of proofs in Book 12 printed by Heiberg in his appendix differ from the canonical text insofar as they "expurgate *poiēsis* language from the arguments and . . . inflect many of the terms suggestive of motion or activity into perfect or aorist tenses."[67] Some go so far as to substitute *hoper edei deixai* for the received reading *hoper edei poiēsai*.

Brown sees the two versions as products of a Eudoxean and a non-Eudoxean response, respectively, to the controversy over *poiēsis* in geometry. One should note, however, that the supposed Eudoxean version (printed as the authentic text by Heiberg) is by no means free of the perfect passive inflections. On the contrary, the "Eudoxean" text sometimes shows evidence of a desire to neutralize the implication of present or future activity demanded within the body of the proof by starting out not only with a perfect passive imperative, but with the imperative of the verb *noein* (Bk. 12, Props 17 and 18: *nenoēsthōsan* "let two spheres have been entertained in thought").

If this last detail can be taken to betray traces of Euclid's own hand, we can add it to the stylistic evidence accumulated earlier as support for a now-familiar inference. Euclid is a moderate, a *phronimos*. He does not follow the radical Speusippeans, themselves perhaps prone to taking at face value certain fulminations against insinuating the language of doing and making in the Platonic dialogues, to the point of eliminating *any* reference to the *poiētic* dimension of mathematics; he retains, for example, the distinction between "Q.E.D." and "Q.E.F." On the other hand, he shows no inclination to emulate the "Menaechmeans," as described in Proclus, in letting that *poiētic* dimension overshadow the theorematic. Of the total of 465 propositions in the *Elements* (not counting lemmas and porisms), only 51 are *problēmata* in the accepted sense. Numbers apart, it is more significant that Euclid's style, especially the pervasive inflection of verbs of operation into perfect passive imperatives, undercuts any temptation to confound the already enacted *genesis* appropriated for his proofs with procedures or motions taking place "now." The temporality figuring in the student's coming to know the truth of a proposition by moving through its parts is not, or so it seems, inherited from a temporality intrinsic to the "beings" on which Euclidean *mathēsis* is focused.

Is this, though, the whole story? Even if someone grants that "constructions" generally occur under the sign of this strangely reticent imperative addressed to no one on the scene ("let the lines *AB, CD* already have been drawn") it would not follow that *no* construction occurs in the present. If there is nothing on which the already executed imperatives at least once were able to operate, then Euclidean *mathēsis* would appear to be eternally deprived of reference to any entities of whatever sort. Must there not be some, perhaps only a few, *primordial* constructions to be performed at the outset and whenever needed thereafter, constructions no longer ancillary to proofs (as in the case of *kataskeuē* in the narrow sense) and no longer bounded by the presupposed forms or natures of figures, but active performances without which there would be nothing at all for geometry to speak about? Isn't the necessity of primordial constructions recognizable in the structure of Euclidean *mathēsis* as genuine mathematics no matter what philosophical or quasi-philosophical considerations might have affected his self-understanding?

VII   The Evidentiary Force of Constructions in Greek Mathematics

With these questions we are in sight of the end of a long and sometimes circuitous path which began with a summary account of the *modern* conception of the role played by constructions in Euclid. This conception received its historically most influential and philosophically most elaborate formulation in Kant, for whom a construction carried out in pure or in empirical intuition gives proof of the "objective reality" of a mathematical concept. The general challenge I posed to myself at the start of this section remains in force: If construction can be plausibly shown to occupy the same systematic or structural place in Euclid (selected as a paradigm of ancient mathematics) as it does in Kant and, for that matter, in Descartes, Hobbes, and Leibniz, then any talk of a *radical* reconceiving of mathematics in the modern age becomes idle and any attempt to discern the constellated array of modernity from the ruling idea of "construction" will prove abortive.

A. G. Kästner's interpretation of "possibility" in Euclid as the possibility of "making" the lines and figures required in theorems and Kant's endorsement of this interpretation as conformable to his account of constructibility in the first *Critique* worked together to establish a consensus among almost all post-Kantian historians of Greek mathematics. This "Kantian Consensus" received its most influential expression in the works of the famous Danish scholar H. G. Zeuthen. In his 1896 paper "Die geometrische Konstruktion als 'Existenzbeweis' in der antiken Geometrie" Zeuthen concluded that "construction, together with the requisite proof of its correctness, served to insure the existence of what had to be constructed."[68] Others have reached this conclusion by different paths and with additional evidence, but rarely has it been challenged in any fundamental way. Heath, to take one highly distinguished example, supports Zeuthen fully in his account of the transition from Euclidean nominal definitions ("the subjective definition of names") to "real" definitions ("the objective definition of things") via constructions. "Exists" and "can be constructed" are synonymous in this sphere.[69]

Agreement that they are synonyms may nowadays have become an almost automatic response; for Kant and the Kantians, the synonymity rested on a complicated argument, each step of which embodies fundamental decisions concerning the evidentiary bearing of an intuitable construction, the existential relevance of a mathematical concept, and, above all, the necessary tie between these two. Without this argument and the decisions its executes this "Kantian consensus" would be lifeless and arbitrary.

At the risk of preempting a thorough analysis, let me set out the basic steps of this argument in very summary fashion:

(1) A construction is a deliberately engendered instrumental or mental *operation* or movement (Zeuthen distinguishes these as "actual" and "formal" constructions, respectively);

(2) The result of a construction is, in each case, an *individual,* accessible to inner or outer intuition;

(3) This intuitable individual bestows *objectivity* on a mathematical concept, shows it to be more than merely possible (free of any contradiction among its constituent marks) or, in other words, confirms the relevance of that concept to existence (its existential *Sinn* and *Haltung* as Kant says) and does so in a way unattainable by the conceptual understanding when left to its own devices. (This Kantian argument reiterates in a highly condensed and stylized form the essential agreement among Descartes, Hobbes, and Leibniz, all their intramural dissensions notwithstanding. Descartes on the transition from an equation to the line-segments which are its roots [that is, the shift from essence to existence, in the idiom of Meds. 5 and 6], Hobbes on the "exposition to sense" of the result of a geometrical synthesis, and Leibniz on construction as the proof of the "real possibility" of a mathematical concept are ringing three idiosyncratic variations on a common theme.)[70]

It is not the merits, but the physiognomy of this argument that must concern us when we face the Euclidean texts. Could it be that this Kantian consensus, however dim its once sharp profile has become in the course of the last two centuries, continues to act upon our reading of Euclid with results somewhat like those of the familiar optical illusion of the involuntary transfer of an image on which we have been made to concentrate to a second sheet of paper? If we learn to break the hold of this reflex, will that second sheet remain a total blank?

A delicate etiquette has to be followed if we are to reach insight into this domain. At first we have to proceed apophatically, *viā negationis,* in order to judge the fit or lack of fit between the models generated by the Kantian consensus and Euclidean (and other Greek) practices. In case no such model does fit it will become necessary to investigate, with the greatest caution, the grounds on which a disparate self-understanding of the "working" Greek mathematician may be based. The road from the initial *apophasis* to a possible *cataphasis* is obviously longer and fraught with many more perils than the following "bird's-eye view" suggests. Moreover, to travel along this road at all we have to divide it into more passable segments, each with its own signposts, even when these divisions are ·finally artificial and thus place the wholeness of the route in some jeopardy.

I begin by prying apart the two inseparable coordinates by which the Kantian position has been located and made to serve as a point of reference for attempts to figure out (or to transfigure) Greek constructional practice and its presuppositions. These two coordinates are the "evidentiary" and the "existential." I shall treat the latter in the section on "Constructibility and the 'Existence' of Geometrical Beings."

With regard to the evidentiary coordinate an initial distinction will be useful. On the one hand, we can consider the actual (or imagined, "formal" in Zeuthen's phrase) construction as it is taking place, in its mode of ongoing operancy, as we might say; on the other, we can attend to the outcome, the *trace* left by the operation of tracing. The question of evidentiary weight applies to construction both *in statu fiendi* and as a *factum*. What kind of evidence does a (product of) construction yield? Why do we turn to it for evidence, if indeed we do? How do we gain the confidence to make this turn when we make it?

i   Instrumental Constructions

An initial family of answers to questions such as these brings us face to face with still another distinctive phenomenon which has to be reckonded with if we are to understand how some of the radical moderns conceived their new attitude towards mathematics and *mathēsis*. I have in mind the phenomenon of "mechanical" or instrumental constructions and the function they might be given both in generating and in legitimating mathematical explanations. Other possible modes of construction—"free-hand" (but, then, the hand is the instrument of the soul, as Aristotle says) or merely "imagined"—share much of the same "logic" with the mechanical mode, especially when the *trace* and its existential relevance return to the picture. Nonetheless, throughout the tradition of thinking about these modes the mechanical has drawn attention to itself in exceptionally persistent ways. As I shall show in the next chapter, Descartes begins the pivotal second book of his *Geometry* by mocking the ancients for their habit of discarding many intelligible curves from the body of science solely because they can only be produced with the aid of instruments. He goes on to introduce us to the famous Cartesian compasses designed to produce an infinity of curves of different degrees in a purely mechanical way. In the preface of the *Principia* Newton reaffirms this Cartesian disposition and, indeed, carries it to the point of founding geometry on "mechanical *praxis*"; "Geometry is nothing but that part of universal mechanics which exactly proposes and demonstrates the act of measuring."[71] The intimate collaboration and then fusion of disciplines we now know as "mathematical physics" seems to be nothing other than the working out of this Newtonian redistribution of disciplinary powers, with the attendant critique of the ancients for having peremptorily consigned mechanics to the status of a manual art (not a science) and for charging the art, and not the artisan, with inexactness and error. Part of the proof that universal mechanics grounds geometry is precisely that geometry does not teach but demands "the description of those lines [straight lines and circle]" and this demand can only be met prior to geometry by the "art" of exact mechanical construction.[72]

Having plunged so far ahead into a much later story, I want to make one last comment on the state of affairs programmatically intended by Newton's *Principia*. The "mechanization of the worldview" featured so prominently in accounts of the seventeenth century no doubt depends in essential ways on the conception of the universe as a machine and on the idea of God as a divine artificer or mechanic (at least in Newton's *théologie grise*). As my discussion of Descartes and others will intimate, an equiprimordial factor is the institution of the "mind" as mechanically the most adept faculty. For Descartes and his corevolutionaries it is not so much that the "mind" proceeds, or can be habituated to proceed, mechanically in the fashion of an incorrigible computing device (although this view is also held in prospect by Bacon in the preface to the *Novum Organum* as well as by Hobbes and Leibniz when they advise us to make *calculemus* replace *cogitemus*); rather, the primary import of this mental mechanization comes to the surface in the promise of endless inventiveness. Man, Vico will write, is "the God of artifices." The mechanization of nature advances *pari passu* with the machinations of the mind.

Let us return to the issue more directly at hand, although now with a recognition that we are *not* dealing with historical or terminological marginalia. I said that mechanical constructions have been especially conspicuous in modern accounts of the Greek tradition, whether they are given pride of place or demoted to second rank behind the purely "formal" or, as in Kant's case, a priori constructions, undertaken in pure imagination.

Thus it comes as quite a surprise to learn that among the extant ancient sources Euclid is totally silent (in the *Elements*) about the instrumental means employed or permissible in his constructions while others, including Archimedes, downplay the epistemic significance of their own or others' inventions, their heuristic value notwithstanding. All the constructions in the *Elements* are in fact performable using only ruler and compass; and yet, not only do Euclid and his commentator Proclus say nothing about this, but on the evidence of the text further restrictions on the allowable use of these simplest instruments were part of Euclidean strategy. We have already seen one instance of this in regard to Book 10, where using a ruler alone to "transport" the length of the hypotenuse of a Pythagorean triangle would have permitted the "construction" of an infinity of new irrationals not falling into the three classes brought into view in that book. Similarly, as some scholars have observed, the Greek compass was not opened freely to allow the transfer of distances corresponding to the radius of a circle with a given point as its center, even though this would have rendered the construction required in Book 1, Proposition 2 otiose.[73]

In a related vein, we might note that even so "practical-minded" a mathematician as Heron of Alexandria (*ho mēchanikos*, as he was known) sometimes took pains to reduce the number of auxiliary "constructions" (*kataskeuai*) needed to reach Euclid's results, while Pappus faults geometers who solve a simpler problem (for example, a plane problem) by using more complicated means, most often involving mechanical constructions (such as the so-called "linear loci"; see below, chapter 3, III, iv on these distinctions and on Descartes' loyalty to Pappus' rule).[74]

None of this implies, of course, that the Greek geometers were deficient in "know-how." Even the early Pythagorean Archytas was able to "solve" the notorious problem of doubling the volume of a cube with the use of extremely technical constructions in three dimensions, which was virtually impossible without employing mechanical devices. (van der Waerden points to one facet of the paradox of geometrical "motion" when he exclaims, "In Archytas' diagram, everything is in motion; he thinks kinematically.") Archytas' solution was only one among many collected and reported in late antiquity by Eutocius, including a very mechanical solution attributed to Plato.[75]

We are therefore brought before the question of the uses and the standing of this "know-how." The question is not parochial to the Greeks, nor is it merely technical, since it involves both the delimitations of what is to count as thoroughly knowable as well as the habits and, hence, the *ēthos* of those who have "learned" (*hoi mathēmatikoi*). With respect to those delimitations I choose the word *ēthos* to indicate, once again, the decisive significance of the practical domain for the separability or collusion of the theoretical and the technical-productive. One could also say that the question of knowing the proprieties and possible improprieties of "know-how" is a matter of an *existentiale*, in Heidegger's sense, that is, of the understanding informing the *habitus* and comportment of the mathematician with regard to the being of *mathēmata*. (This is why a history of the placements and reclassifications of mechanics and the *artes mechanicae* in antiquity, the Middle Ages, and early modernity would throw so much light on this *existentiale*. To give one hint of these potential illuminations, a hint quite relevant to the next chapter, al-Farabi, whose interests in advanced mathematics were considerable, situated algebra in that subdivision of the mathematical arts (or sciences) which also includes the art of making pneumatic and other mechanical devices and has the collective name ʿilm al-hiyal, *scientia de ingeniis* in the medieval Latin translation, as though the translator had a premonition of Descartes).[76]

What, then, is the *habitus* to which the Greek mathematicians may characteristically have been accustomed? At this point I shall limit myself to some brief descriptive comments, reserving until the end of this section any attempt to probe the deeper significance of their attitude.

The task of duplicating the cube, the so-called "Delian Problem," played a cardinal role in Greek mathematical (and "methodological") controversies over a very long period indeed (from Hippocrates of Chios in the fifth century B.C. to Eutocius in the sixth century A.D.); it is worth noting that Descartes, in Book 3 of the *Géométrie,* is proud of his ability to show how the solution to this classical problem fits in smoothly with the entire sequence of solutions yielded by his new approach. Of immediate relevance in the present context is the way proposed solutions continually raised questions concerning the epistemic suitability of the use of instruments in geometry. Our main ancient sources (Plutarch; Ps.-Eratosthenes *apud* Eutocius) place the origin of these questions within the Platonic circle and thus remind us once again that the issue of the powers and limits of technique (here, specifically, mechanical or instrumental technique) is inscribed in a more inclusive sphere of reflection on the nature of *mathēsis* as such.[77]

Plutarch on two occasions has Plato reproaching Archytas, Eudoxus, and Menaechmus for devising "instrumental and mechanical constructions" in response to the Delian Problem. No doubt there is more than a suggestion of *parti pris* on Plutarch's side since he explains this reproach in the language of an orthodox or "vulgar" Platonism for which the noetic and the aesthetic are sharply divided. Mechanical solutions represent, then, an untoward descent into the somatic and banausic which spells ruin for "the good of geometry," while abetting the liaison of mechanics with the arts of war. (Ps.-Eratosthenes may be said to give a more sober report of the same episode; he attributes to all three geometers success in "proving demonstratively" [*apodeiktikōs gegraphenai*] their solutions, failure, when it came to practical implementation, "except for Menaechmus, in small part, and even then with difficulty." Nothing is said here of Plato's "reproach.") Plutarch's interpretation may well both exceed his 'factual' sources and fall short of much more subtle indications of Plato's judgment of the "productive" arts in the *Dialogues.*[78] Be this as it may, one detail of his account of Plato's criticism stands somewhat apart from his own "ethical" condemnation of Eudoxus et al., namely, the charge that "they were trying to get hold of two means in continuous proportion by irrational means, in whatever way was practicable." Might the deliberate antithesis *di'alogou* . . . , *analogon* be more than a play on words? If we take into consideration (1) the equivalence of the "Delian Problem" to the problem of finding two mean proportionals between two given straight lines $(1:x::x:y::y:2)$ and (2) the formal resemblance of the latter to the task put to the slave-boy by Socrates in the *Meno* (that is $1:x::x:2$, where 1 and 2 are the areas of the respective squares), then the mathematical core of Plato's rebuke may well have been the demand that any proposed solution do more than simply exhibit or point to the lines which have been constructed so as to have the required ratios to the initially given bases. It must envelop, so to speak, these particular relationships within a comprehensive account (*logos*) of ratios in general, including those between irrationals (*alogoi,* as in Euclid, Bk. 10, Def. 3). Only under this condition can we hope to understand *why* the instrumental construction works out the way it does and thus furnish what in the *Meno* Socrates calls the *aitias logismos,* the account that reckons with the source responsible for something's being as it is. Otherwise we would only have "true opinion" that this construction yields the desired linesegments; in Descartes' view, we would only be "groping," proceeding, as he says, *à tâtons.* Viewed in this light, Plato would not have been upbraiding Archytas and company for using "organic constructions" on the grounds that these could only give *inexact* solutions to any particular problem, for exactness (or, inexactness) would not be the salient point, any more than practicality or impracticality. Instead, Plato would be reminding an Archytas or a Eudoxus of the more exacting task of learning how to see things in their togetherness, hence as enveloped in an account which is no longer focused on problems presented piecemeal, but on the connectedness and kinship of the principles underlying particular solutions.[79] Plutarch's reference to "the good of geometry" can in this way be restored to its less conventional and more provocative Platonic homeground, where "The Good" is an ingathering of distinct forms into a unity which enables each to do its work in fitting concert with the others.

Archimedes returns us to the heart of the issue of instrumental constructions and the *ēthos* of their users. He does so for self-evident reasons. First, his achievements represent the apex of Greek mechanics as well as the direct route towards the modern goal of a mathematized physics; second, the rediscovery of his tract on method allows us our fullest glimpse into some of the actual heuristic procedures of ancient geometry, at least in its later phases. Archimedean geometrized mechanics *in praxi* deserves just as much attention to detail as I tried to give to Euclid's *Elements,* but these details must be set aside in favor of a few very general comments.

We must not assume that Archimedes' widely celebrated talent for contriving mechanical inventions is of a piece with his readiness to use arguments drawn from mathematical mechanics in a "pure" geometrical context. On the one hand, we know from the testimony of an admirer such as Carpus, himself an "engineer," that Archimedes was somewhat disdainful of these mechanical appliances and entrusted only one case— the making of armillary spheres—to writing.[80] In his geometrical theorems talk of "transporting" (*metapherein*) the center of gravity of line-segments or circles from one point to another is not to be read as the record of real manipulations via instruments, but as a hypothetical "idealization" (What would happen *if* one were to . . . ?) drawing its plausibility, to be sure, from propositions and principles of rational statics.[81]

With this more likely reading at our disposal we can better understand Archimedes' discussion of his way of discovery in the preface to his "Method Concerning Mechanical Theorems," discovered by Heiberg in 1906. Archimedes is most careful to distinguish here between demonstrating (*apodeixis*) a proposition and investigating (*theōrein*) a proposition by means of mechanics prior to the elaboration of a demonstration. As he puts it, "for we can more readily provide a demonstration if we have got hold in advance of some acquaintance with what is being sought than if we tried to find this without any advance acquaintance."[82] As the sequel shows, advance acquaintance with what is sought comes to us when we treat geometrical relationships (for example, in Proposition 1, the relation between the area of a segment of a parabola and an inscribed triangle) as though the lines "making up" the relevant figures were the arms of a balance with centers of gravity, equilibrium-points, and so forth. Once an investigation (*theōria*) in this style has been completed we have, as he says, only a sort of indication or impression (*emphasin*); but, we can be guided by this as we set out to give a full-fledged demonstration of the same proposition using only axioms and propositions for which we have nonmechanical evidence.

Consequently, Archimedes' *Method* is neither a procedure for "mechanical" theorem-proving, nor a thoroughly rule-governed heuristic (as some latter-day commentators on "analysis" in Greek mathematics clearly wish to assume), nor, finally, an intellectual justification for using mechanical instruments such as levers as a means of grasping *why* figures stand to one another in various complex relations. Taking advantage of his idiom, we could say, instead, that this use of hypothetical "idealization" has "brought to light by mechanics" (*phanen dia tōn mēchanikōn*) what we are in search of; the devices we "imagine" in the investigation provoke the figures into manifesting certain of the connections we shall then have to draw fully into the light (*apodeixis*) by strict proof. This picture of the vaunted mechanical method, and only this picture, as far as I can make out, is in harmony with another express declaration of Archimedes on the topic of geometrical knowing. In the letter to Dositheus prefacing his treatise "On the Sphere and the Cylinder," referring to the properties (*symptōmata*) he has *proved* to be true of certain figures he remarks that "they were there by nature in these figures" (*tei phusei prouperchen peri ta eirēmena schēmata*), although unknown to previous geometers.[83] The Archimedean *ēthos* seems to involve, then, the same respect for the independent *nature* of the *mathēmata* we saw to be inherent in the Euclidean *ēthos*.

So far I have only been establishing a negative point. Instrumental constructions do not appear to be especially privileged as evidentiary sources by the Greek mathematicians or by their philosophical compatriots. A *tour de force* such as the constructions devised to solve the Delian Problem has to be integrated into a connected body of learnable truths before its intelligibility, as distinct from its occasional success, can be assessed. For Archimedes, as we just saw, method as a heuristic proceeding is not a matter of physical contrivances but a circumspect *metaphor*, an intellectually mobilizing transfer of statics to geometry in order to free our thinking about complex geometrical relations from the paralysis to which Meno's paradox would have condemned us.

ii   Construction as Operation (Sensory and Imaginative)

The question of the evidentiary status of construction as process (*in statu fiendi*) thus can and must be addressed univocally to all varieties (including "imagined" constructions), not only to those worked out with instruments. What is it, we ask once more, in or about an operation which supposedly gives it the quality of evidence, possibly even self-evidence? Is its authority, with respect to *mathēsis*, original or borrowed? What legitimizes the selection of some constructive operations as elementary, others as derived? Since mathematics, at least on the view being considered here, presupposes both the prior enactment and the intelligibility of those founding *activities,* no matter how dimly they are remembered later, we can also ask: How could the passage from the latter to the former be accomplished? All of these issues circle around one central enigma: What reasons are there for us to think that *mathēsis* should (or should not) take its rudimentary bearings from premathematical human activities (of which certain "constructive" motions are a subclass)?

Let us make a tangential assault on this enigma. First, it should be clear that the questions coalescing into this ultimate formulation do *not* have to do with the historical or biological *primacy* of human practices over theorizing or *mathēsis*. Aristotle's account of the coming-to-be of the theoretical sciences in *Metaphysics* Alpha is sufficient in this place to show us that it would make perfect sense to say that the practice of measuring the overflow of the Nile antedates the articulation of geometry as a science. But, as I already mentioned at an earlier stage of my analysis, *nature* (*physis*) signifies not only the coming-to-be, but the what-is-has-become of something, and *now* the question is whether the historical, biological primacy of practice is also constitutive of the "finished" nature of geometry, either wholly or in crucial part. Put still differently, we might all easily agree that the "idea" of a demonstrative geometrical science would never have arisen, for us, in the absence of prior practices such as surveying, mensuration, and architecture, but this agreement does not *eo ipso* entail that the latter's identify *as* as a science is also due to remembrance of, or appeal to, earlier or even contemporaneous *practices*, including bodily motions.

To argue that some such entailment does hold and to do so without simply begging the question by asserting that it must hold (otherwise, there would be no sense at all in trying to *ground* geometry on constructive evidences) require that we can discern traits in some human operations which are passed along as a "genetic" legacy to the geometer who sets about instituting his *mathēsis* as a science. More simply, we must be able to detect certain pregeometric activities satisfying criteria which would qualify them to be the original source of the evidence supposedly intrinsic to the constructions taken up by geometrical science. The relations we are looking for ought to be logical, not psychological.

No potential criterion seems simultaneously satisfied by a pregeometrical movement *and* attested by Greek sources which speak even indirectly to this issue. The criterion of *achievability*—that an operation *can* be performed—will not get us far, since it is satisfied by a multitudinous class of bodily and instrumental motions only a very few of which come to figure in *mathēsis*. If we restrict ourselves to "free-hand" drawings we are still at the same impasse, since an extraordinarily winding curve is "achievable" just as much as a free-hand circle is. This suggests that we look, instead, to the criterion of *simplicity;* free-hand circles are more simply achievable than complex curves. Appeal to simplicity alone, however, is already a complex affair. In Euclid the *definitions* in Book 1 pick out simplicities (the partlessness of a point, the evenness of a straight-line or a plane-surface, the equality of the radii of a circle); are these "simple" in the same way selected actions are now said to be? Even if we say that the Euclidean *postulates* (read as imperatives to action) constructively fulfill some of the definitions, it is the latter, not the former, that specify what it is to be simple in each relevant case. (See chapter 2, IX, for more on the interplay between definitions and postulates.) Hence, we would need still another reason for saying that the simplicity of definitions, too, merely commemorates simple pregeometric actions (*which* action in the case, say, of the partless point?), and, at all events, the criterion of simplicity loses its own simplicity as it rushes us into an unsettling, even if not truly infinite, regress.

A *third* criterion now suggests itself, the criterion of familiarity, whether of some of our own movements or of other "natural" movements. Almost-rectilinear and almost-circular motions might be said to be more familiar to us in our experience without our needing to authenticate them by an additional appeal to simplicity. However, if these familiar operations are to be enshrined in the initial conditions on which geometry as *mathēsis* is claimed to rest, then they must lend themselves to a "normative idealiza-tion," as some have argued; that is to say, they must not only be familiar to us, they must also be legitimized as initial conditions (and hence as normative) and be open to some kind of purification or "abstraction" guaranteeing their (infinite) repetition in just the same way.[84] (Otherwise, of course, we would have a potential infinity of inimitably singular actions, no one of which could in principle lay claim to normative authority.) But the disposition to satisfy these new requirements of legitimacy and enduring sameness does not seem to be inherent in the *familiar* operations *per se;* recognition of the need for normativeness and repeatability comes from a "metapractice" which cannot be reduced to inference from a set of unrepeatable "first-level" actions. Nor can we explain the institution of this metapractice in terms of reflective mimicry of natural movements already experienced as recurrent. Not only would this hold only of one kind of movement, the circular, if it held at all; still worse, the experience of these natural movements includes the experience of recurrent anoma-lies in the case of the planets and therefore already requires its own variety of idealization (perfectly regular orbital velocities, and so on) if geometrical constructions are to be educed from it as a normative model. (Compare Euclid's astronomical work, *The Phenomena*.)[85]

This survey of these three potential criteria leaves us with a negative result; namely, that only a deliberate leap could take us from the realm of human and natural movements to the "ideal" constructive operations of geometry. These, in their turn, would have to acquire their evidentiary status elsewhere than in the simple and familiar feats by which are lives are punctuated. However, we were looking not for a leap, but for a continuum of evidence.

Perhaps, then, it is in the *traces* of a constructive operation and not in the act of performing it that we can find a convincing rationale for giving constructions evidentiary primacy within Greek mathematics.

The trace of a construction, however the latter came to be produced, is in Greek a *diagramma*, "what has been marked off by lines," a "diagram." Nothing is more suited to bring home to us the exotic perplexities we confront in this entire domain than the question of the "being" of a diagram and its bearing on the "being" of the *mathēmata*, for this is, after all, a rehearsal in mathematical terms of the grand dilemma set to us by the world's imagability, the incalculable alliance of "being" and "not-being" thanks to which human teaching and learning are possible (and possible false). The most radical differences between ancient and modern "ontology" seem to be concentrated in this dilemma. This formulation, however, is much too precipitate at the moment. Let me begin again, much more simply.

Can it be the case that a diagram carries *on its own* the weight of evidence required by Euclidean *mathēsis*? How much weight is presumed to be involved here? Surely enough that geometrical teaching and learning can take place in the presence, or on the occasion of, "each" diagram accompanying a proposition; but do these presences or occasions exhaust the meaning seemingly lodged in geometrical teaching and learning? Could the latter do without diagrams (or, I should stress, any other kind of "image" including a letter or "abstract symbol")? If not, why are diagrams indispensable, not to say ineluctable? And, are we looking for evidence in diagram-types or diagram-tokens? What relation do types have to their tokens?

It is critically important to recognize that these questions and the many corollaries they spawn are "context-sensitive" in the extreme. For a Kantian—and here as elsewhere Kant consummates a line of thinking initiated by Descartes—these questions may be only in appearance identical with their Euclidean predecessors, inasmuch as the former presupposes that diagrammatic occasions bring together *concepts* and *intuitions* (a priori or sensible). Precisely this presupposition must be held in suspense when pre-modern geometry is on view. What I called earlier the Euclidean "respect for (naturally distinct) forms" serves as a reminder of how powerfully this presupposition would impinge upon Euclidean practices if it were preemptorily released from suspension.

About these matters Euclid is once again as fundamentally silent as his Greek geometrical successors. We can, however, reasonably conjecture that he was well aware of the perils to which *mathēsis* would be exposed if its teachers were to appeal, without further ado, to direct sensory perception of isolated diagram-tokens. Protagoras, for example, trained his persuasive skills on this misapprehension, as we know from Aristotle's report in *Metaphysics* Beta: "For the circle is not touched by the ruler at [only one] point, but rather [at more than one], as Protagoras used to declare in refuting the geometers."[86] This brand of rhetorical skepticism, it seems to me, has less to do with the unavoidable imperfection or inexactness of a perceptible diagram than with its intrinsic contradictoriness. The circle drawn before our eyes is *both* breadthless *and* wide enough to make contact with a line at two or more points. (A Cartesian "graph," on the other hand, is free of this contradictoriness as well as of any inexactness, not because it is absolutely exact *qua* perceptible token, but because the technique for constructing it is, in principle, error-proof; see below, chapter 3, IV, iii). Moreover, simple ostension, even if it were a part of the geometer's didactic repertory, would never be quite so simple, since its success would already presuppose the learner's ability to put out of account the sensory presence of the diagram in favor of its intended significance, a significance which is never genuinely fulfilled by or within that sensory presentation. This is the lesson brought across to us in the slave-boy episode in the *Meno*; merely pointing to the diagonal drawn in the diagram and giving it its accepted name does not convey the incommensurability of this line with the side of its square; hence, the boy cannot be said to know what it is he is seeing.[87]

If not individual diagrams, but "space" were proposed as the object of our immediate intuition, then we should have to note that there is no term corresponding to or translatable as "space" in this generalized sense anywhere to be found in Euclid's *Elements*; *to chōrion* for example, is the area enclosed within the perimeter of a specific figure, while *topos* and *thesis* in the *Data* have functions determined by the contextual aims of that work as a "dialectical" foil to the *Elements*, not by a physics of space hidden in its background.[88]

The immediate issue is still the extent to which direct perception of general "spatial" relations, and no longer of isolated diagrams, might underwrite the evidentiary credentials of diagrammatic constructions. The stumbling-block to this new move is precisely the evidence we have from Euclid himself of the "maladjustment" of key *optical* phenomena to the geometer's desires. Thus, in Theorem 8 of the *Optics* Euclid proves that "equal magnitudes at unequal distances are not seen as proportinate to [the ratio of] these intervals" (but, in fact, to the ratio of the visual angles), while in Theorem 9 he demonstrates that right-angled magnitudes from a distance appear to be rounded (*peripherē*; note the impact this has on any alleged *spatial* confirmation of Postulate 5!).[89] Since diagram-tokens must convey the "information" supposed intrinsic to them *independently* of distance or angle of vision, on pain of becoming irrelevant to *mathēsis* if this condition were not met, it appears that "space" fares no better than isolated diagrams as an *immediate* source of geometrical evidence.

Let me try to make the point at which I have arrived as clearly as I can. The staged versions of geometric practice we meet with in the Platonic dialogues (such as *Meno, Euthydemus, Republic* 6, *Theaetetus*) and, more allusively, in various Aristotelian writings (such as *Topics, Posterior Analytics, Metaphysics* Theta 9) make it plain that visual exhibitions were part and parcel of mathematical instruction; my comments have not been meant to cast doubt on this historical datum. On the contrary, the task is to make sense of this datum once immediate sensation has been discounted as the ground of the didactic persuasiveness of the exhibited diagrams. Perhaps what is needed is a *mediated* source in virtue of which the *trace* left by a construction can be made to yield evidence germane to the geometer's insistence that matters must stand as he has been showing them to stand with the help of his diagram-token. A mediated source, in other words, that would aid us in comprehending that elegant pairing of inscribing (*graphein*) and showing forth from an inscription (*apo-deiknunai; apo-phainein*) which proved so fertile for the Greek mathematical tradition.

One such mediated source has been invested with a show of great likeliness by philosophers' discussions of mathematics (if not so emphatically by practicing mathematicians), namely, *phantasia*, the "imagination." So much turns on how we understand this least straightforwardly illuminating word (and how we chart its destiny as it passes into *imaginatio*/imagination/*Einbildungskraft*) that a summary account of its possible place in mathematical learning is self-defeating. In defiance of this caveat, I want to bring into this deliberately narrowed field of inquiry—the question of the evidence mediately gleaned from constructed traces—at least one set of pertinent considerations.

Imagination, in Kant as much as in Descartes and Vico, is primarily understood as something artful, a technical flair which is *born with us*, but needs discipline and regimentation. This union of the innate with the acquired is at the root of many of the ambiguities to which modern conceptions of the imaginative faculty seem to be prone. For example, in a remarkable passage in the Schematism chapter our imaginative facility in forging a schema or image as a *tertium quid*, a point of rendezvous for concepts and intuitions, is said to be "an art hidden in the depths of the human soul, an art whose true knacks [*Handgriffe*] we shall never guess at from Nature save with difficulty or be able to set before our eyes unconcealed" (KrdV. B181). Kant thereby calls into question the Cartesian or Hobbesian "technological" conception of imagination, without in any way repudiating its artfulness. Indeed, his paradoxical notion of a "hidden art" matches in ambivalence prior attempts to grasp the imagination as a natural power disciplined by precepts to carry out its artful works. I have in mind, as one instance, Leibniz's oscillating stance vis-à-vis the status of imagination within *his* version of *mathesis universalis*. On the one hand, mathematics is defined as the "logic of the imagination," concerned with "whatever is subject to the imagination"; on the other, it is the aim of the *ars characteristica* to "disburden" or "alleviate" the imagination.[90] The relevant Leibnizian texts seem to lead us to something like the following inference: Left to itself, the imagination is naturally the fount of indistinctness and confusion, it continually adds to its burdens by compelling itself to attend to shifting manifolds of restive images; put under the dominion of a strangely higher art—the at of forming univocal, but nonetheless sensible, characters—the imagination need only attend to signs of its own making. These signs are "left, so to speak, [as] visible traces on the sheet of paper and can be examined [and re-examined] at leisure." Thanks to this higher art, or more precisely, this imaginative art raised to a higher power, we can "touch incorporeals as though with the hand."[91] It is only a slight exaggeration to say that for Leibniz, as much as for Kant, the art of imagination in its schematizing (and hence, constructive) employment stands midway between the taxing frenzies of *furor poeticus* and the willed rigidities of deductive logic.

These brief suggestions set the tone for a possible appreciation of the distinct register in which *phantasia* plays its parts within classical Greek thinking about (and in) mathematics. Simply stated, radically modern thinking about imagination takes its bearing from the phenomenon of productive arts, including especially those arts adept at fashioning internal, mental images and then embodying these elsewhere, by design. "*Concetto,*" a *terminus technicus* in later Renaissance "aesthetics," stands at the head of a path leading to Baroque extravagances in the arts *and* in philosophers' theories of concept-formation. For the ancients, *phantasia* is not borrowed from the arts and crafts to be put to "theoretical" use, but is primordially rooted in the experience of the (re)appearances of light (*phōs*; see *De an.* 3.3.429a3–4), an experience, we should be quick to add, of awesome duplicity, since "false" dreams and deceiving mirages are also luminous. Human beings, according to Aristotle, are most likely to act in accordance with imagination "when their intelligence is covered up [*epikalyptesthai*] by affection, by diseases or by sleep" and, as a later example adds, by madness.[92] "Imagination" seems to come most potently to light when *nous* is hidden behind a veil.

This is not the whole story, however, since *phantasia* does perform tasks which need not be either pathological or oneiric. One of these tasks is expressly associated by Aristotle with the drawing of mathematical diagrams, and this association returns me to my main theme. Lest it be engulfed in the deep waters of his various accounts of *phantasia*, let me try to keep it afloat by sticking as closely as I can to the surface of a key passage in *De memoria et reminiscentia*:

> It is not possible to think [*noein*] without a phantasm. For the same affection occurs in thinking as in the drawing of a diagram. In the latter case, even though we are not also using the triangle's being determinate in quantity, nonetheless we draw it determinately as to quantity. In just the same way, the person thinking, even if he is not thinking of quantity, places [a] quantity before his eyes, but does not think of it *qua* [*hēi*] quantity. Even if the nature [of what he *is* thinking] *is* among the quantities, but indeterminate, he places before him a determinate quantity, but thinks of it *qua* quantity only. (450a1–6)

This suggests, among other things, that every phantasm has some determinate magnitude and that the intellect has to make accommodations to the respective *phantasmata* in order to fix its own intended theme. Thus, the ubiquitous Aristotelian *qua* (*hēi*) is once again "reliably" in play. I think, say, quantity *as* the indivisible category (*to poson*) or *as* the genus of continuous, divisible quantity (*to syneches*) by way of entertaining a phantasm of some determinate, continuous magnitude, just as I entertain triangularity as such by drawing a particular equilateral triangle, the determinate magnitude and figure of which can be disregarded. While

Aristotle does not explicitly say that I can or do carry out the drawing *in imagination*, rather than in the sand or on a papyrus roll, the claim that the same *pathos* is involved in that case as in the entertaining of a phantasm suggests that we could treat a diagram as a transcription of what has already made its presence known to us in (or as) a phantasm. Does this amount to saying that Aristotle is a Kantian *avant la lettre*?

At least three related topics would have to be fully explored before we could in good conscience succumb to that temptation: the generality of *phantasmata,* their iconic being, and, finally, their source.

(1) If we had reason to think that Aristotelian phantasms (or, imaginative diagrams) are *general* images, we would have moved from the diagram-tokens considered earlier to diagram-types, and to these we might now be inclined to assign the evidentiary force supposedly inherent in the traces of certain constructions. All along the central issue has been the possible transition from singularity (of a construction) to generality (of a proof or solution) and vice versa; perhaps phantasms are sign-posts directing those twin transitions.

Unfortunately, Aristotle, in the passage just cited, does not treat the phantasm as general or universal, but as determinate, although allowing the intellect the freedom, as it were, to disregard its determinateness. That is, I *think* the phantasm as indeterminate or, more precisely, I *think* indeterminately via the phantasm, where "via" should be read more as "through" than as "by the agency of." The phantasm, because always determinate, is a bar to the generic or categorial indeterminacy (and hence unity) it is the business of *nous* to appreciate. *Nous* desires, so to speak, to appreciate quality as the immediate, indivisible unity of a category (see *Metaph. Eta*6.1045b1–7); nonetheless, a quantitative and quantitatively determinate *phantasm* insinuates itself between noetic desire and its desideratum, since "no thinking takes place without a phantasm." We now have the perplexing situation in which the lucidity of the phantasm occludes the "object" of noetic desire so long as we do not know how to negotiate the *qua* so as to render that phantasm diaphanous. When we do know how to negotiate the *qua* in this way, the "imagined" triangle of such-and-such determinate size and angles lets something else (the indeterminate triangle) shine through. And this means that there is no generality or universality in the phantasm as such; thinking it *as* indeterminate when it is in truth unavoidably determinate is the work of *nous,* not of *phantasia.*

(2) So, too, is the appreciation of the iconic being of a *phantasm.* Aristotle situates the discussion of diagrams within the more inclusive topic of memory. He also distinguishes two regions within that *topos*: items memorable *per se* and items memorable *per accidens* (450a24–25), the latter being all those items which "are not without phantasia," that is, the items of intellectual appreciation (*noēmata*) of which he has spoken in *De anima*

3.8. This distinction indicates that the psychic presence of a phantasm as an enduring and thus memorable impression (*typos*) left by the movement of sensation is *not* essential to its being (indispensably) present for thinking; when it is present in this second way, its first mode of presence merely accompanies the solo performances of *nous*.

In both modes, however, a phantasm appears to have the being of an icon (*eikōn*). Aristotle, it seems, is prepared to collapse the distinction made by the Eleatic stranger between *phantasmata* and *eikones* for the sake of inserting iconic being (and not-being) into the very texture of a phantasm. Thus he says:

> For just as what is painted [*gegrammenon*] on the panel is both a picture [*zōon*] and a likeness [*eikōn*] and both are one and the same, yet the being of the two is not the same, so that it is possible to contemplate it both as [*hōs*] a picture and as a likeness, so, too, we must take the phantasm in us both as an item contemplated [*theōrēma*] in its own right *and* as the phantasm of something else. Insofar as [*hēi*] we take it in its own right [*kath' hauto*], it is an item contemplated or a phantasm; insofar as we take it as *of* something else, for instance, as a likeness of, it is also a memorial record.[93]

More than one mystery is shrouded by this difficult passage. We can, however, make fairly direct sense of at least this much: The being of a phantasm is intrinsically dual, such that one and the same entity has being in its own right *and* has being only inasmuch as it is of something other than itself. The same pigments and figure on the panel are both the painting and the likeness or portrait of the sitter. The shift from one "style" of being to the second once more invokes the lability of the Aristotelian *hēi*—"insofar as." Now our question becomes: *Of* what is a mathematical phantasm a likeness? With this question our difficulties begin anew, for, although we could say that the sitter sits for his portrait, it is by no means clear what is being portrayed by a phantastic likeness. When the answer to that latter query is "Nothing," we get the case Aristotle goes on to record of Antipheron of Oreus and other lunatics who treated their phantasms as records of what never happened to them. If we prefer to think that geometers are not characteristically madmen and that their phantasms *are* likenesses of something, we are then as a crossroads. On the one hand, we could conjecture that the iconic phantasm "triangle" records previous experiences of particular triangles drawn "before our eyes"; but then, each of these would have been the diagram of a triangle with sides and angles of determinate magnitude and these proportions are *not* reproduced by or in the determinate phantasm except by accident. Since the phantasm, as we saw, is in each case already determinate, the inference from its particularity to what has to hold true of all equally determinate sensible triangles or triangular-shaped surfaces is not motivated or authorized by anything intrinsic to the phantasm *qua* likeness. Alternatively, we could speculate that the phantasm may be a likeness *of* what-it-is-to-be a triangle, the essence of triangularity; but this speculation encounters new obstacles when we try to have it jibe with an earlier passage in the *De anima*.

In Book 3, chapter 4, once Aristotle has achieved the passage from sensation (*aisthēsis*) to intellection (*noēsis*), he pauses to consider the difference between the ways *nous* discriminates and judges magnitude and what-it-is-to-be-magnitude (or, similarly, between flesh and the essence of flesh). This difference is marked off by one of Aristotle's habitually compact phrases—"either by another or by something in another relation" (*ē allōi ē allōs echonti*); for comparison, consider a standard English rendering: "either by different faculties or by the same faculty in different relations" [M. S. Hett]—and then by a comparison which has proved exceptionally impregnable to unanimous analysis: "The intellect judges the essence of flesh by a power other than and separate from the sensitive, or as a straight line which has been inflected is related to itself when it has been straightened out."[94]

Contemporary geometric and optical usage (see *An. Post.* 1.9.76b9; *Physics* 5.4.228b24 229a2; *Meteor* 377b22; Euclid, Bk. 3, Prop. 20) of "inflected line" (*hē keklasmenē eutheia*) discloses that the underlying "image" is of a straight line (or visual ray) meeting a surface or second line and being turned back in the direction whence it began so that an angle is formed:

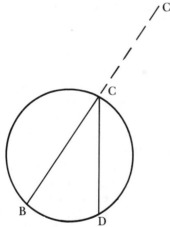

The "path" *BCD* is thus a "broken" or inflected line, while *BBC'* would perhaps be the line *BCD* "stretched towards itself " (*pros hautēn ektathēi*), that is, extended in its original direction past the inflecting surface. This usage, together with the immediately preceding illustration in *De anima*, should allow us to say that *aisthetic* surface of a being deflects the original intention of *nous*. What we identify indirectly or superficially as "flesh" by virtue of a complex of sensed-qualities is more truly a certain ratio of those qualities; it is this ratio that we retrieve when the intellect is returned to its direct intention towards *what* a composite being essentially is.

However this illustrative comparison is to be more finely understood, one salient implication emerges with enough clarity to serve my present purposes. The essence of magnitude has no magnitude, just as the essence of straightness is not a continuous magnitude (see *De an.* 429b18–20 about essences of mathematicals [*epi tōn en aphairesei ontōn*]). Since, as we learned from *De memoria*, all phantasms (even though free of sensible matter) are of a certain determinate magnitude, no phantasm can be a likeness *of* an essence. while a noematic form is not *of* an essence but actively identical with it. Thus, it follows that the iconic quality of a phantasm does not qualify it to share or usurp the responsibility of intellect when it is a question of thinking the essence of a mathematical item.

Consequently, the phantasm *qua* likeness is not like an essence, nor is it the general likeness of (all) particular sensible triangles. Whatever evidentiary weight, if any, it might still carry on its own is *not* germane either to the generality ascribed to a constructed diagram or to the essentiality of the mathematical item which is, as it were, being dia-grammed. This also means that phantasms have no business to conduct in the sphere of demonstrative science, where knowledge of essences is presupposed and attributes must hold true universally and *per se* of all items within a specified class. (*Phantasia* and its cognates do not appear anywhere in the text of *Posterior Analytics*.)

(3) Much the same conclusion can be obtained along another route, one searching for the source of phantasms. Here the difference between the ancient and the modern view signalled above acquires additional meaning. Images, for Descartes and Hobbes, characteristics (or signs) for Leibniz, arise in the main as the deliberate products of an interior art or a regimen of calculated prudence (as in Hobbes—see *Leviathan*, ch. 3). It is the secret of this art in its various versions to rule the production of images in such a way that each is invested with the generality and essential-ity that alone can give an image evidentiary prominence. For an Aristotle, in contrast, phantasms are unruly; they do not arise in the soul as a result of *art* even though they can be retained or re-evoked by a kind of art for the sake of memory and recollection. (Early modern algebra will, in fact, exhibit significant analogies with mnemonic techniques reaching back to the ancients and to Ramon Lull and revived by Ramus and Bruno.) As the accompaniment of a sensation active in the present (see *De an.* 3.3.429a1–5), the Aristotelian image illumines or throws light on the sensible form, placing it before the aperture of potential noetic focus

(*De an.* 3.8.432a3–6, especially *en tois eidesi tois aisthētois ta noēta* ("the intelligibles are in the perceptible forms"); as the remnant of a present sensation the phantasm through its lucidity makes up for the absence of the sensed form. In neither case is the phantasm artfully fashioned or produced by the intellect as a vehicle or replica of its acts of knowing.

Phantasms are at their most unruly when they occur in dreams. In this respect the treatises *De somnis* and *De divinatione per somnium* are vital sequels to the earlier texts documenting Aristotle's physiology of knowing. In dreams the *iconic* relation of a phantasm is greatly attenuated, since the phantasms are like reflections (*eidōla*) in stirred water and it take someone capable of "theorizing likenesses" to judge them and their possible, but improbable, semiotic bearing on future events (see *De divin.* 464b6ff.). It is no wonder, then, that ancient dubieties are reawakened and even fortified in medieval Jewish and Islamic "philosophical theology" for which *the* central issue is the relation of prophetic imagination to theoretical intellection.[95] It is also no cause for amazement that the motif of prophetic imagination is almost utterly erased in radically modern texts, to be replaced by the palimpsest, as it were, of "philosophies of the future." In any case, the following quotation from Leibniz should suffice to fix the terms of comparison:

> When an idea is absent in us [not apperceived], some sensible image, or definition, or aggregate of [symbolic] characters takes its place; these need not be similar to the idea. Some phantasm, the whole of which can be perceived simultaneously, always takes the place of the idea.[96]

For Leibniz the phantasm is not only a surrogate for (sometimes inacessible) mental ideas; in the most desirable instances, it is something made to come about as a "characteristic" of an idea or ideas correlative to it.

iii   Noēsis and "Mental" Construction

Having tested the qualifications of directly perceived and imagined "traces" to serve as the primarily evidentiary locus of mathematical constructions, I can now turn to the only remaining candidate with some modern support, a trace left behind or inscribed in the mind itself as the result of some (equally mental) constructive activity. It is important to note that pure versions of this final position are much more difficult to find either in historical accounts or philosophical extrapolations from these. Customarily, the alleged movements of pure mind are yoked to a second faculty, whether this is called "(transcendental) imagination" or "pure" or "sensible intuition," and it is only in tandem with any one of the latter that constructive activities can produce a durable trace. In other words, the mind requires some medium other than itself on which to inscribe or project its constructions (*in statu fiendi*).

This should not come as a surprise since the modern practice of mathematical construction as well as the reflective interpretations and extensions of that practice are all along preoccupied with the passage from the "mind" understood as an interior concentration of energy to something "other" than the mind and counting as the field in which its energies are potentially displayed. The "other," at the level of principles, is characterized by exteriority, extendedness, multiplicity, and phenomenality. Consequently, under certain conditions of reflective self-understanding, the mind can also find this "other" in itself, that is, can contrapose to its own inwardness and unity the outwardness and multiplicity of *its* imagination or pure intuition. Even with this insinuation of otherness into the self-identity of the mind, however, the activity of construction continues to traffic in the same polarity between inward and outward.

As I have been arguing since the start of this section, modern interpretations of construction in Greek mathematics are generally worked out under the auspices of this polarity and its various figurations, from the Cartesian to the Kantian and beyond. The interpretive stance in all of these is the posture of *Vorstellung*, in its Hegelian sense.

My investigations of Euclid and allied Greek sources have so far shown, I hope, the perplexities in which we are caught when, imitating that interpretive stance, we try to recover, *in the terms it dictates to us*, the evidentiary force supposed to reside in Greek mathematical constructions. These can be understood as acts or as products of construction; in both cases, the explicit or implicit logic of the modern interpretation seemed to cut across the grain of the ancient texts, distorting, rather than revealing, their constitutive commitments and responsibilities. With the final candidate for assessment—mind as constructive on its own—we have come full circle. Most simply, if advocates of this last candidate habitually assume that mind must have a psychic or physical medium wherein it exercises its constructive powers, then the sought-for evidence either belongs to the nature of such media (such as the sensory or imaginative presence of the constructs as *facta*) or reverts to the active operations performed autonomously by the mind. The first disjunct has already been examined and found wanting; the second reinstates the thesis that (purely mental) construction in *statu fiendi* is the bearer of evidence. And these are mutually exclusive disjuncts.

As in the instances I examined earlier, here, too, Euclid is fundamentally silent about the role noetic movements or constructions might play in his own practice. He breaks this silence on a very few occasions and, even then, only by verbal allusions rather than direct discourse. I have in mind the eight occurrences of the verb *noein* I have counted in the

Elements, all of them appearing in five propositions of Book 12, which has already been singled out for the vividness of its idioms of activity. Of these eight occurrences, five are in the now-familiar perfect passive imperative (*nenoēsthō; nenōesthōsan*), as if to suggest that the requisite noetic energy has already been exercised; one is a present passive imperative and two are present active subjunctives. It is more than a little curious that all eight concern either spheres or cylinders, that is, two of the three (and only three) geometrical items in the whole of *Elements* which are defined by way of a generative *motion* (see B. 11, Defs. 14, 18, and 21). It almost seems as though Euclid, ready to repeat the same phrases time and again, breaks with this pattern in these propositions in order to mitigate, if only partially, the kinetic and *poiētic* connotations of the definitions; thus, for example, in Proposition 17, two spheres are to have been thought (*nenoēsthōsan*—"let two spheres be conceived [Heath] around the same center *A*") when Euclid could presumably have used a slightly more elaborate phrasing involving semicircles and rotations about their diameters (for example, *perienechthentōn*; compare the wording of Def. 14). At a minimum we could infer from this evidence that even when *kinēsis* is most patently demanded, it still takes place under the auspices of *noēsis* in its perfective aspect and (generally) anonymous mode.

We do not know, however, how Euclid might have understood *noesis* itself. Is it *kinetic* in its own right or does it somehow outstrip movement while also being the consummation of movement? Only in philosophical writings anterior to Euclid can we find clues to possible answers to this and cognate questions, and it is not my intention to begin deciphering those endlessly various clues here. Let me substitute an asymptotic approach to the lines of investigation laid out by the issue itself.

Of all recent philosophical interpreters of Greek mathematics it is Oskar Becker who was most insistent on the kinetic and thus primordially temporal character of geometrical (and arithmetical) thinking for the ancients *as well as* for the moderns. His Heideggerian inspirations, especially in *Mathematische Existenz,* joined forces with his defense of Brouwer over Hilbert to yield the internally most consistent picture of the history of mathematics from the viewpoint of noetic or dianoetic *genesis.* To take a representative illustration:

> "Becoming" (*kinēsis* in a general sense), which we observed in the "emergence" of figures through construction . . . is for the Academics only the sensible image [*Abbild*] of an inner movement of "discursive thinking" [*dianoia*] (more precisely, of a "deeming" [*Vermeinens*] which progresses in a determinate process.)[97]

Becker calls upon both Proclus and Aristotle to testify in behalf of the primacy of inner, dianoetic movement in mathematical constructions.

Proclus presents very complicated problems with regard to this and all other associated issues in the theory of ancient mathematical practice; in particular, his notion of *kinēsis phantastikē* (*In Euclid* 94.26 [Friedlein]) demands prolonged attention, especially when studied in tandem with his thesis in the *Commentary on the Republic* that *phantasia* is a "form-giving intellection of the intelligibles" (*noēsis . . . morphōtikē noētōn*: 1.235.18 [Kroll] and compare 1.121.2–3: *tais phantastikais . . . kai morphōtikais kinēsesin*). Detailed study might show that in crucial respects Proclus is the first "modern" precisely because of his exaltation of *phantasia* as form-giving or, as Becker is even willing to say, "creative."[98]

Aristotle is another matter, as his painstaking efforts in *De anima* to associate and to dissociate movement and noesis remind us. If any movement is an imperfect performance (*energeia atelēs*), while thinking is the very perfection of performance (see *De an.* 3.7.431a1–7), then our understanding of noēsis remains moored to our experience of movement and alteration, especially in sensation or perception, *until* we come to see that in those cases, too, actualization is more truly an advance into self-sameness than a becoming-other (*De an.* 2.5.417b6–7: *eis auto . . . hē epidosis*). We can also remind ourselves that on Aristotle's test for the interchangeability of verbal aspects (tenses), "to be seeing" and "to have seen," "to be thinking" and "to have thought" mean the same. (Compare *EN* 10.4.1174b12–13, where the act of seeing, a point and a monad are allied as not admitting any coming-to-be.)

Becker does not discuss these passages any more than he deals with the equally intricate discussion in *De An.* 3.6 of the intellect's thinking of formal indivisibles in an indivisible time (430b15ff.). Instead he refers to the intriguing collocation of time and inner, psychic, motion in *Physics* 4, with particular emphasis on 219a4–6: "For even if it were dark and we were suffering nothing through the body, but a certain *kinēsis* were present in the soul, straightaway [*euthus*] it would simultaneously [*hama*] seem to us that some time had taken place [*gegonenai*]." Is this "trenchant understanding of 'inner' movement [*Bewegtheit*] in the soul," as Becker calls it, sufficient warrant for overturning the results of Aristotle's analysis of the unmoving and temporally indivisible nature of noēsis as the consummation of movement?

Becker does not try to situate the cited sentence within the movement of *discourse* initiated at the start of chapter 11, nor does he pursue the topical clues in which this portion abounds. If one were to take these steps, then it would become apparent that Aristotle contrasts time as it "seems to us to have taken place/to have passed" owing to movement in the soul, with time such as we know it to be "when we think [*noēsōmen*] the extremes as different from the middle and the soul declares that the 'nows' are two, the prior and the posterior" (219a26–29). That is,

Aristotle's topical speech passes from opinion and perception about time as it has passed (*gegonenai*) to *noēsis* of time in its being (*einai*), and the latter is not itself characterized as a passage, an interior movement or the like. Time, we could say, is brought to a halt when its essence is noetically grasped as the twoness of the discrepant "nows."

We can conclude, therefore, that while there surely are movements (or, more precisely sources of movement) "in the soul"—desires, judgments, deliberations, for example (see *De an.* 3.9)—*noēsis* is not among them; its freedom from movement is not a privation but a perfection, a being-concentratedly-at-work, matched by the invariance of essences. For the most radical of the moderns the mind principally concentrates on staying in motion, while essences survive primarily as laws or histories of motion.

VIII   Constructibility and the "Existence" of Geometrical Beings

i   Introduction

My survey of the possible meanings and potential cognitive sources of construction in Greek mathematics has now passed through the penultimate stage. So far I have been deliberately holding at bay exactly that component of the Kantian consensus without which it remains ontologically inert. For mathematical constructibility to carry the weight it supposedly does, it must be an unambiguous and indubitable proof of the *existence* (or "objective reality") of the constructible mathematical concept. Lambert's declaration, cited above, must now be allowed to body forth its entire force: "The thinkable is nothing (a dream) if it can not come to existence." The statements of a Zeuthen or of a Heath accordingly become something more than historiographic reconstructions only when they are read as a précis of the ontological commitment voiced summarily here by Lambert and ornamented with a filigree of distinctions by Kant.

What is needed, then, is an elucidation, however synoptic and cursory, of the *place* of the question of mathematical existence *per se*, something like an aerial map of the region in which mathematical existence is naturally at home as an ultimately hospitable issue for reflection. If this survey should disclose that "existence" is at home in a specific domain, *but not in others*, for whatever reasons, then the questions of mathematical existence and of its ties to mathematical constructibility can be most congenially posed *only* when its proper domain is in view. Outside of this domain we are bound to be ill at ease with these "same" questions; we come sooner or later to experience that sense of displacement which arises when we no longer know how to go on in habitual ways. Just such a displacement occurs when we try to situate the question of "existence," in the sense given to that term by Kant and his heirs, in the domain of Greek thinking about *mathēsis* and the "being" of what it studies.

Accordingly, we must now ask whether and under what guise the question of existence enters into the texture of Euclidean *mathēsis*. A later formulation by Zeuthen will give us some initial help towards bringing these matters into focus.

> In the Platonic theory [sic] the primitive geometrical notions are our own creations, we give them an existence either by our definitions or by the postulates, which completely enumerate the properties we want to attribute to them without demonstration. Their remaining properties, as well as the existence and the properties of the figures we construct by using the postulates, require demonstrations resting on the postulates and on certain general notions enunciated at the start together with the postulates.[99]

While it would be misguided to expect of a Zeuthen the same degree of philosophical perspicacity we encounter in, say, a Leibniz or a Kant, neither is his statement philosophically naive. On the contrary, it records in the language of historical scholarship a cluster of theoretical convictions of great moment for the issues now hanging in the balance; among them are the thesis that the primary *concepts* of geometry are "our own creations" (presumably *ex nihilo*, for Zeuthen), the claim that we give these creations existence via definitions and/or postulates which would seem equally to be our own work and, finally, the assertion that all else in the body of geometrical science takes life from these primordial creations given existence by our *fiat*. As to this third conviction, since Zeuthen had earlier identified constructions with existence-proofs, it seems fair to infer that our *fiat* is in fact either a directive to construct (either "actually or formally") the respective concepts or an assurance in advance that possible constructions do correspond with the primary *definienda*. Zeuthen's view is often narrowed to the single proposition that, for Euclid and for Greek mathematicians generally, to assume that the straight line, for example, as defined in Definition 4, *exists* is to postulate the possibility of constructing a straight line. If we now persist in asking what "existence" is intended to signify in this setting, we suddenly find ourselves entangled in a web of riddles. If we postulate that we can construct *this* straight line between *these* two points, have we shown or assured ourselves that *this* straight-line exists (imaginatively or in material fact) or that the concept "straight-line" exists, that is, has at least one instance? Why should this one, or for that matter, a numberless succession of single constructions, establish the existence of the concept or of its *denotation*? If a construction is assumed to be "formally" possible (as Zeuthen designates it) is this also a proof that the "same" construction is actually possible and that the actually constructed individual will sometime exist as a thing locatable in space? Can we make existential assumptions about any and every non-self-contradictory concept figuring in geometry or only about some? What allows or compels us to select this subclass (the postulates) on the hypothesis that we do make some such choice or run up against some such limit? Finally, when we do postulate the possibility of an elementary or primordial construction, what are we asking the student to accept? Is his acceptance meant to be an insightful acknowledgment of how things possibly stand, or is it rather a concession to the intellectual authority or superior will of the teacher? In either case how has the teacher gained assurance that the "objects" about which he is knowledgeable "exist"? All the filaments of this web meet at the center.

To extricate ourselves from this web we cannot appeal to Euclid himself (or to Archimedes or to Apollonius), since it is from his text and its apparent silence regarding these issues that the web is spun. We need to apply pressure upon the center holding its filaments together, and this center is nothing other than the received opinion that the concept of "existence" *is* at home in the economy of Euclidean *mathēsis*. If this opinion yields to pressure, the web of riddles it holds intact will unravel. This does not mean that no new riddles will succeed it; on the contrary, we should expect it to be replaced by a network of enigmas as daunting as the originals. Our only advantage in this altered state of affairs is that the new riddles encountered in the home-territory of *Greek* mathematics might no longer be outlandish.

ii   Einai and Existence in Aristotle

Why would anyone expect the center of the first web to give way to pressure? An impressive body of philological evidence has been accumulated by Charles Kahn (and others) which matches in many salient respects the insights of Heidegger into the sense of the verb "to be" in Greek thinking. The key outcome of Kahn's studies as far as the present issues are concerned is that in classical Greek there is no *distinctively* "existential" use of *einai/estin*, nor is there a separate verb obviously carrying the kind of force we are in the habit of ascribing to locutions of the form "*x* exists": "there are/exist Φ," "es gibt . . .," " or "il y'en a . . ." Moreover, it would be mistaken simply to attribute to Greek philosophers a naive confusion of logically independent functions (predication, identification, and assertion of existence) somehow latent in one and the same standard verb. Instead of chastizing Parmenides, Plato, and Aristotle for their inability to discriminate the separate members of this Fregean or Russellian trichotomy, we ought to regard them as remaining loyal in their various manners to "the decisive use of the verb in the creation of Greek ontology . . . [namely] the veridical use, in which the verb *esti* means 'is true' or 'is the case.' "[100] (It is Kahn's insistence on the decisively veridical use of *einai* which inevitably puts the reader in mind of the Heideggerian juncture of *on* and *alētheia*, even though *Unverborgenheit* is patently remote from Kahn's allegiance to a Tarskian notion of sentential truth.)

Controversial details aside, Kahn's linguistic analyses put us in a position to appreciate how elusive the classical Greek texts prove when questions of *existence* (and of reference) are posed to them directly, that is, as an automatic reflex conditioned by the putative rigors of a Russell and others. One instance of surpassing importance for our understanding of postulates (and the *being* of mathematicals) in Euclid is the Aristotelian discussion in *Posterior Analytics* of the difference between *ti estin* questions (What is . . . ?) and *ei estin* questions (If/whether . . . is?). Questions of the second sort have habitually been taken as questions of existence (as in "Do centaurs exist?") and, in turn, projected onto the existential assumptions believed to underlie, or to be theoretically equivalent to, Euclid's postulates.[101]

Closer and I think more persuasive analysis of Aristotle's arguments has shown us that (1) his claims cannot be fitted to the grid provided by the standard "existential-quantifier" of the first-order predicate-calculus ($[\exists x]$) and (2) that the intended focus of his *ei estin* question is, in the idiom of a contemporary logician, the *predicate*, not the *subject-variable* bound by an existential quantifier. In other terms, Aristotle is asking "Are some 'xs' truly Φs?" rather than "Are there 'xs' *simpliciter*?" His question concerns the *propriety of identifying* something as having the nature or characteristics referred to by a predicate (such as "human being," "god," "centaur") rather than the "existence" of that something or of its congeners, that is, the admissibility of its/their name to a designative position.

This analysis, developed by Gomez-Lobo and Mignucci independently of one another, almost certainly coheres with the details of these particular Aristotelian texts (*Post. An.* 11.1–2) as well as with more general features of the Aristotelian design for a science of *to on hēi on*.[102] Let me single out only a few major signs of this coherence, before attempting to assign its lessons their due place vis-à-vis the Euclidean enterprise.

First, Aristotle's compound expression for what is customarily construed as an "individual," *tode ti*, all along contains a sortal element, as J. A. Smith was able to prove in what must still be the most important one-page article on Aristotle ever published. In his formulation *tode ti* would mean 'anything which is both a this and a such-and-such,' the two characterisations being co-ordinate. "x is *tode ti*, if it is both (a) singular and so signifiable by 'this' and (b) possessed of a universal nature, the name of which is an answer to the question *ti estin* in the category of *ousia*."[103] Without extending any further the enigmatic matter of the grounds of individuation (and, hence, of demonstrative reference, the target of the *tode*), we can be content here with emphasizing, as Smith did, the indispensability of a prior recognition that an "individual" is of such-and-such a kind or sort. Accordingly the *ei estin* questions singled out in *Posterior Analytics* are characteristically asking whether a specimen item, a *tode ti*, *already* identified under some one sortal or specifying term (such as "animate being"), is or is not truly identifiable as instancing some other sortal term, usually one of narrower extension ("god," or "centaur," for example). Particulars or individuals bare of any specifying identity do not, it seems, figure in Aristotle's ontology. Consequently, there could not be any Aristotelian answer to the question "How many 'things' or 'individuals' exist or are present in a certain locale?" in the absence of differentiated eidetic counters (how many humans? how many gods?).

Second, these two examples of "eidetic counters" should serve to remind us that the *ei estin* questions are not without their bearing on predemonstrative concerns; indeed, Aristotle's own samples at 89b32 (centaur, god) give us an initial and lastingly pertinent clue to the way the "existential" question, even when reconstrued as "veridical" in the manner of Kahn, remains theo-locative, that is, directed towards placing, or displacing, the divine or semidivine in, or from, its traditionally represented locale. This link is not as outlandish as it might initially seem, inasmuch as the availability of a *genos hypokeimenon* to *apodeictic discourse* is decided, not by *apodeixis*, but in the light or under the sway of reputable opinions (*ta endoxa*). Aristotle's *ei estin* questions occur within a fourfold pattern (*ei estin, ti estin; hoti estin; dia ti estin*) meant to yield the architectonic of a demonstrative science; at the same time answers to at least some *ei estin* questions (Are these animate beings gods?), if available at all, must be sought extramurally and thus independently of the cognitive assurances or allurements of a science already instituted. In this respect Aristotle is no foreigner or metic; as can be seen from the preponderance of uses of *einai* in an apparently absolute context—that is, where "exist" would be the canonical modern translation—the status of the "gods" (or, of souls after death) is never far from the scene. To cite only one early example, from the sixth-century poet Alcman: *Aphrodita men ouk estin margos d'Erōs hoia [pais] paisdei/*"Aphrodite is not, but rampant / lustful Eros like [a child] sports." The question of existence slips from its onto-theological moorings only to become a vagrant on the limitless sea of formal logic.[104]

Finally, to descend from the onto-theological to the mathematical, Aristotle himself provides more than a minor clue to how *einai* might be understood in a demonstrative setting akin to the Euclidean. In *Posterior*

Analytics 1.12 (78a9ff.) he is discussing the possibly mutual entailment of premises and conclusions (*p,q⇌r*); he puts the case in view formally as follows: "Let *A* be the case [*estō gar to* A *on*]. If this is, then these are the case, which I know to be the case, for example, *B*. From the latter, then, I shall show that the former is [*hoti estin ekeino*]." Since he proceeds to say that mutual entailment (or, reversibility of inference) is more likely to be found in mathematical than in dialectical arguments, we can securely conclude that the phrase *to* A *on* has to do with a complete premise (as in this [type of] figure being isoceles), rather than with an individual whose existence, apart from type, is being either postulated (hypothetically) or simply asserted. Moreover, in the *Prior Analytics* (A24.41b14ff.), on one of the occasions when he uses the verb *axioun* in a specifically mathematical context, it is clear that inferences from postulates referring to singular cases (here, *this* angle *AΓ* in a semicircle = *that* angle *BΔ* in the same circle) involve a *petitio principii* unless one also makes explicit the general or universal "premiss" that all angles in *any* semicircle (formed by lines drawn from the end-points of the diameter) are equal. Aristotle places inferences to particular or individual cases under the aegis of the universal in which a necessary and natural connection is entailed, just as the affirmative response to an *ei estin* question would warrant our describing (some) members of a general type as truly being of a more specific kind (e.g., "Some mammals are centaurs" *contra veritatem*) or, no doubt more controversially, "Some ones are monads, i.e., arithmetic units" (cf. *An. Post* 1.2.72a23–24 with *Metaph. Iota* 1.1052b31ff.). If both of these examples are plausible candidates, this has signal relevance for the accessibility and status of "individuals" in any of the now-customary senses of ostensibly singular "objects." When Aristotle clears the geometers of the charge of laying down false hypotheses (*Post. An.* 1.10.76b40ff.), he is not concerned with existential assumptions (such as "a straight line exists") but with their use of single diagram-tokens to deputize for general types (as in "let this _____ be a straight line, one foot in length"). Evidentiary force stems not from the single token on its own and hence not from its constructibility as a criterion of the existence of the type, but from its intelligibility as a vicarious instance of a type already acknowledged and defined (for example, straight line). As he adds: "The geometer comes to no conclusions on the basis of its being *this* line which he himself names; he concludes only from what such diagrams make manifest [*dēloumena*]."[105] How these constructed tokens succeed in making something other than themselves manifest,—that is, how it comes about that I know to disregard everything in the token which does not jibe with its vicarious role (as in its not truly being straight or one foot long)—is not explained by Aristotle here or elsewhere; he does capture the enigma inseparable from these successes when he asserts that paralogisms (arising from an undistributed middle-term) can be easily detected in the case of mathematical arguments for there such things "can be as it were seen by intellection, while in dialectical (or sophistical) arguments they escape notice [*lanthanei*] (*Post. An.* 1.12.77b31ff.). Somehow the tokens drawn by the geometer serve as visible icons provoking thought to its work of knowing the natures for which those icons have been fashioned to deputize.

At all events, we have seen that the *einai* involved in Aristotle's *ei estin* questions is predicative, *not* existential, and that when such a question is raised, the items under scrutiny are not single individuals but types or general natures, consonantly, of course, with the mission of *apodeixis* to display necessary and *per se* connections among those natures. And, as I have already remarked, the *being* of the genus studied by any one science is not a question for that science; either perception or reputable opinion serves as evidence of the discriminable presence of that genus (compare *Post An.* 1.12.76b16ff.).

If these results are basically correct, then the Aristotelianizing interpretation of Euclid's postulates and constructions favored by Heath et al. loses one of its mainstays. That is, when the supposed strict parallelism between Aristotle's hypotheses construed as existential assumptions and Euclidean postulates breaks down, the claim that the latter are existential claims (in disguise—see below) needs to be supported by independent arguments; needless to say, the loss of support from Aristotle does not *eo ipso* mean that no such arguments can be furnished. To see whether they can be furnished, let me defer for the time being a direct encounter with Euclid and return to the question of the linguistic and conceptual matrices in which the notion of "existence" retrospectively applied to Euclid originates. Since Euclid never stands outside his text to offer a metatheory of its semantics and truth-conditions, we are compelled to proceed with the greatest caution as we go about examining (a selection of) those matrices and assessing their fit with the implicit strategies of that text. If it should turn out that the concerns to which the former are addressed are egregiously alien to Euclid's text, then there will be good reason to conclude that they are misfits and that a different route towards understanding Euclidean procedure must be sought.

Efforts to trace the origins of a distinct notion and idiom of "existence" have been focused on at least three *premodern* traditions, namely, Stoicism, the Neoplatonism of Marius Victorinus and Boethius, and medieval Islamic debates over the "accidental" status of existence vis-à-vis essence. The *modern* tradition finds its climactic expression in Kant's thesis that " 'being' is not real predicate." In order to see whether the question of existence can be domésticated in the Euclidean domain, we must first examine, however summarily, the main features of these post-Euclidean traditions.

iii   Hyparxis *and* Deixis *in Stoicism*

In the Stoic tradition, attention has inevitably been drawn to the technical use of the verb *hyparchein* to mark off a mode of being belonging to some, but not all, items admissible into true discourse (via canonically formed expressions signifying them). Unlike Aristotle, for whom *hyparchein* (and its variants, such as *enhyparchein*) primarily serves to indicate that an attribute "stands under the precedence" of the substance or subject to which it belongs and which, consequently, carries as many senses as the categories of being in which the respective "subjects" are brought to speech (see *Pr. An.* 1.36.48b2–4; 1.37.49a7ff.), the Stoics appear to have employed *hyparchein* primarily (even univocally) to pick out "the presence at the moment [*Gegenwärtigkeit*] of a happening or a process, the actuality of something's having an effect which results from a cause and, therefore, [designates] an activity.[106]" Syntactically, this present-moment, causally effective happening, is signified by a predicate; thus, the complete proposition "Dion is prudent" is true if and only if Dion possesses some active material quality (namely, *phronēsis*) in virtue of which he is rightly described, now, as exercising prudence (or, being white, taking a walk, and so forth).

Contested points of interpretation and nuance aside, Stoic usage (which passes over into Latin *existentia* via Marius Victorinus' rendering of the noun *hyparxis*)[107] is in harmony with at least three fundamental features of Stoic ontology and epistemology: (1) corporealism, (2) nominalism, and (3) impressionism.

Concerning (1): According to Plutarch, along with numerous ancient reporters, for the Stoics "bodies alone are beings [*onta*]." However, since they took *to ti* (the something), not *to on* (that which is), to name the comprehensive class, items other than material bodies can be said to be "somethings" without enjoying the status of full-fledged "beings." These *incorporeals* are not unambiguously catalogued in the ancient sources; among their number, however, are the void, time, place (*topos*), and the ubiquitous *lekta*, the inner mental, but somehow intersubjective *significata* of corporeal sounds and written expressions.[108] Less clearly, anything derivative from a material body, without being itself a body, is rightly placed under the rubric *asōmaton*; thus, the qualities proper to a body, those it shares with other bodies, and, finally, the situation in which a body finds itself vis-à-vis other bodies would all count as incorporeals. Accordingly, *hyparchein* signals a distinguished fashion of being something, but one of still derivative or secondary status in comparison with the full-bodied *onta* which are individual material things.

Concerning (2): It is fully consistent with this ontological gradation that singulars or individuals count in the world, whereas anything said in common or generally of a class of individuals amounts to a manner of speaking. In a key text from Alexander of Aphrodisias' *Commentary on the Topics* (= *CAG* 11.2.359.12ff., ed. Wallies), a mental formation intending what is common to a class, that is, an *ennoēma*, is placed outside even the otherwise all-embracing class of *tina* (somethings); it is, says Alexander, neither corporeal nor incorporeal (*SVF* vol. 2 329; compare *SVF* vol 2., 330). Stoic "nominalism" is, in essence, the claim that concept-formation (or construction) inevitably leaves behind the individuality of the entities principally in place in the world for the sake of mental phantasms or "general ideas," whose content is unreal since there is no body corresponding to the ideas of men, dogs, etc.[109] Universals are *anhyparkta*, the presumably derogatory label attached to the Platonic *ideai* or to Aristotelian *eidē* by Zeno of Citium (see *SVF* vol. 1. 65).

Concerning (3): Stoic corporealism and nominalism converge in what I have termed "impressionism"—that is, concrete, needless to say, *not* abstract impressionism. The workings of bodies leave their impressions on us in the shape of *phantasiai*, some of which put within our grasp what it is that is or was appearing (these are the *katalēptikai phantasiai*) and thus allow us to express articulately what has impressed itself upon us (these furnish us with *logikai phantasiai*). Knowing is a transaction between bodies sufficiently apparent to make an impression and bodies—in our case, the so-called *hēgemonikon*—apt to being excited by such an impression; they can in turn register this impression to themselves by internal speech (*logos endiathetos*) via words and the incorporeal *lekta* they signal (the *sēmainomena*) and which can subsequently be shared with other sensitive bodies through external utterance (*logos prophorikos*). Intersubjectively meaningful language is an exchange of impressions arising and taking their credibility from the "existing" qualities, states, and activities of individual bodies.

How do these three fundamental features of Stoic thinking bear on the questions raised by the orthodox identification of constructibility with proof of "existence" in Euclid? Since for the Stoics to *exist* (in the restricted sense of *hyparchein*) is to be an event causally occasioned by a three-dimensional body (compare *SVF* 2.341), the two-dimensional configurations brought onto the scene in the *Elements,* even when they are icons of three-dimensional shapes (such as cylinders and spheres), *fail* to qualify as *hyparchonta*; at most, one could say that the motions involved in producing these schematic configurations or the impressive effects of the completed diagrams, or both, might meet the Stoic tests for existential status. However, since the figures diagrammatically inscribed are not themselves full-scale bodies, neither are they *onta* in their own right, even if constructed to deputize for the shapes of bodies (see Posidonius, *apud* Diogenes Laertius 135).

This reasoning is, however, tangential to the main course of inference along which Stoic treatments of presentation, concept-formation, and expressive language impel us. What ought to strike us most deeply in these is the thoroughgoing sundering of sensibility from intelligibility. To be is to be corporeal or to be directly caused by a body and hence to be capable of making a sensible impression on another body; to be intelligible and thus communicable is to be an incorporeal *lekton*, a meaning or conceptual intention signified by a corporeal sound or sign. (The Stoic *lekton* is one of the direct ancestors of the Arabic *maᶜna* which in fateful fashion is called upon to translate *lekton* as well as *eidos*, in its Platonic or Aristotelian senses. It is *maᶜna* as the transcription of *lekton* which passes into medieval Latin as *intentio* (or *conceptus*) and thereby prepares the ground for the Cartesian-Kantian transvaluation of the *eidē*.)[110] It is, then, scarcely a matter for amazement that mathematics, to judge by the extant texts and testimonies, was shunted to the margin of attention by the Stoics, yielding pride of place to their versions of dialectic (topical argumentation) and rhetoric.

This dislocation becomes apparent in the Stoics' own handling of the truth-conditions for geometrical propositions and of the sundered ontology the latter presupposes. A geometrical proposition of the form we encounter in Euclid—that is, one purporting to express a truth concerning all instances of, say, equilateral triangles—is for the Stoics an *axiōma aoriston*, an indeterminate proposition, and thus neither true nor false on its own. It takes on a determinate truth-value (becomes an *axiōma hōrismenon* or *katagoreutikon*) when a singular demonstrative pronoun (*houtos*) replaces the general class-name in the subject-position; the result—as in "*This* [equilateral triangle] is such-and-such"—is confirmable by an act of pointing, *deixis*, to the referent of "this" so as to evoke in oneself or in another the impression or presentation to which the referent gives rise. As Pierre Pachet puts it, "La deixis . . . révèle à la fois son objet et, ce qui est plus important pour le logicien, sa propre existence: d'où sa valeur d'index."[111]

This Stoic emphasis on the indispensability of ostension for providing a proposition with truth-value points us in a number of relevant directions at once. First, the generality seemingly immanent to Euclidean propositions (as well as the typicality of Aristotle's answers to *ei estin* questions) gives way to the individuality of a token which alone, as a body uniquely qualified in such and such a way (by, say, its distinctive shape and magnitude), can carry evidentiary weight; the object pointed to by a gesture of *deixis* is either a *soma idiōs poion* or, in retrospect, the singular mental presentation which it occasioned. When either of these is replaced by a general or universal conception, an *ennoēma*, the mind loses its (immediate) grip on existing particulars. Chrysippus, using a geometrical example, suggests how we can retrieve this grasp by taking the general *ennoēma* (which is neither corporeal nor incorporeal, as we recall) as a rule or template for the generation of a potentially infinite number of (relevantly similar) individuals; according to Geminus (as reported by Proclus) Chrysippus likened a theorem such as Euclid 1.35, proving the equality of parallelograms on the same base and between the same parallels, to *ideas* (*ideai*) (see *SVF* 1.65 for *ennoēmata* as a synonym), since both cases embrace, hold in their grasp (*perilambanousin*), the generation of infinite instances within determinate limits. Euclid's theorem, then, is not about the nature of *the* parallelogram in general, but refers indirectly to all the individual parallelograms (or mental impressions of parallelograms) which can, in principle, be ostensively adduced or produced in light of this theorem.[112] Finally, the paradigmatic instance of *deixis* is the first-person singular pronoun (see *SVF* 2.895, from Chrysippus' *Peri psychēs*), the referent of which leads us back from the individual object to the individual source of the utterance or gesture. Just as apprehensive presentations are in each case *my* presentations, so, too, the objects of the final terminus of a deictic and hence determinate proposition are objects for *me* as the initial terminus. I can try to convey my impressions through external language (corporeal sounds or diagrams) in the hope of impressing similar images upon the attentive auditor. Zeno of Citium's definition of geometry, quite recently reconstructed by Jaap Mansfeld, has it that geometry furnishes a *deixis phantasiōn* "a pointing out of presentations," *ametaptōton hypo logou*, "impervious to change through argument." Restored to its inclusive setting in Early Stoic thought as a whole, Zeno's definition tells us that, in Mansfeld's words, "the objects of geometry are the imaginings of our mind; constructions ultimately rooted in our perceptions of the world around us, but processed by the mind, which is able to call them up from its treasury of such *phantasiai*."[113]

In every respect the Stoic construal of geometry transports us, quite impressively and emphatically, from the philosophical world in which the core of the *Elements* was first elaborated (whether we think of this world as predominantly Platonic-Academic, as evidenced by the contributions of Theaetetus and Eudoxus—to say nothing, here, of Proclus' identification of Euclid as a Platonist—or as mainly Aristotelian, with respect to its discursive format). What we are allowed to witness in the articulations of Stoic thought is a series of sunderings, the two sides of which are then (retroactively) reunited by the force of mediating elements impossible to place squarely on either side of the gap they are intended to bridge. Thus, the aesthetic-somatic "beings" in Stoicism are sundered from the incorporeal *lekta*, which are nonetheless the bearers of intelligibility; a somatic individual is not, as such and all along, either evocative of an intelligible *eidos* or somehow present for *nous* as immaterially informed and thus potentially knowable as well as actually sensed. The causal agency of bodies links them to the presentations which the *lekta* make sense of, even though this agency cannot be understood as a full-fledged body in its own right. Thus, too, individual bodies are sundered from universal conceptions only to be re-wedded to the latter either through acts of ostension or of production of particular impressions (for oneself or for another). Finally, the publicity of outward speech is sundered from the privacy of silent meanings or intentions, even though, for the Stoics, the *lekta* are meant to be intersubjective. It is scarcely an accident that Greek Stoicism had its greatest impact on Roman forensic rhetoric. As Jean-François Courtine has brilliantly succeeded in showing in his study of the early Latin translations of *ousia* and "la comprehension romano-stoïcienne de l'être," the Roman orators, having appropriated the Hellenistic vocabulary of the *staseis* ("issues" to be dealt with in a speech), grafted onto this stock their transformed understanding of the Aristotelian categories of being. Accordingly, when Quintilian takes over *substantia* as the proper rendition of *ousia* he understands it as a matter of solid fact ("De substantia est conjectura; quaestio enim tractatur rei, an facta sit, an fiat, an futura sit" [*Instit. Orat.* 3.6.39]), fact, that is, supported by a *res certa et solida* corresponding to the Stoics' individual bodies and their palpable actions. The aim of the speaker is to leave his hearer with the convincing impression that his words are or may be "substantiated."[114]

What might have seemed at the start a straightforward philological matter—Stoic *hyparchein* as the semantic precursor of the Latin *existere*—has turned out to be a more convoluted affair, engaging in the course of inquiry many of the cardinal assumptions of Stoic ontology, epistemology, and rhetoric. As I suggested earlier, the question of "existence" cannot be posed in a void; it comes to life, if at all, *in vitro*, in a philosophical culture with its own peculiar roots. Once deracinated, it is not immediately clear that it can be transplanted and made to survive in alien soil. (One additional sign of the Stoics' estrangement from the Euclidean style is, of course, the fact that for them the canonical form of inference is hypothetical implication, in geometry as well as in the natural sciences (see Cicero, *De fato* 15); no explicit instance of that inference-form can be found in Euclid, or, for that matter, in any classical geometrical text.)

If the final home of Stoicism is the sub-uranian domain of the law courts and the forum, the region colonized by pagan neo-Platonism and its early Christian legatees is the hyper-uranian sphere of the pure, beyond affirmative speech: neo-Platonic *apophasis*, the endlessly iterated silence of negative theology, is the counterpole to Stoic *emphasis*. While it would only be fair to try to do at least roughly proportionate justice to the textures of this second domain in which late antiquity found temporary residence, I must pass it by after merely calling attention to two fairly obvious features relevant to the present question.

The idiom of "existence" or "true being" is clearly of great importance to the early Christian apologists and polemicists; thus, Origen challenges the pagan Celsus to prove that the Greek gods are not "mere figments" (*anaplasmata*) but "subsist/exist substantially" (*kat' ousian hyphetēkenai*) (*Contra Celsum* i.23).[115] As in early Greek poetry, the question of divinity once again presides over the issue of what truly is. And it is under this same theological aegis that patristic writers went about elaborating an exquisitely detailed vocabulary for distinguishing the being of a numerically distinct (divine) person from the identical substance (*ousia; hypostasis*) of the Divine. If, for the Stoics, corporeal singularity was the hallmark of a genuine entity, incorporeality and creative (or emanative) omnipotence now become the signatures of what truly is/exists, measured by which all other entities are deficient. What "exists" in a more familiar workaday sense always bears the traces of having been made; divine *facere* takes over the reins from Stoic *fatum*.[116]

iv   Essence and "Existence" in the Medieval Arabic Tradition

A leading controversy within the medieval Arabic tradition set the theme of "existence" in still another arrangement, one which proved of lasting, if soon anonymous, influence, at least until Kant. I have in mind the debate between Ibn Sina and Ibn Rushd concerning the thesis that "existence is an accident of essence" (in the formulation that soon became customary in Latin scholasticism). The obscurities besetting efforts at exact elucidation of Ibn Sina's own intended meaning as well as of Ibn Rushd's fairness or deliberate unfairness in criticizing his predecessor in the way he does are close to intractable in view of the complicated and often elusive overtones each Arabic term has on its own and, more strikingly, when taken in concert first with its presumed Greek models and then with its preferred Latin "equivalents."

Despite efforts by Arabists to unravel these convoluted meanings, nothing that could be called scholarly consensus has so far been in evidence. This is, I think, due first to the different levels on which the Avicennian theme requires consideration, namely, the "logic" of concepts or intentions (*taṣawwurāt* or *maʿānī*) versus the "metaphysics" of actual things (*aʿyān al-ashyāʾ*) and, due second, in partial independence of this first contrast, to the politico-theological setting in which Ibn Sina, in company with the Islamic and Jewish *falāsifā* generally, was obliged to think through the harmony, real or rhetorical, between absolute creation and one or more pagan "proofs" of the eternality of the world. (Indeed, it is often a mark of attempts to vindicate Ibn Sina's "modernity" that the politics of theology is almost wholly left out of account.)[117]

When the Avicennian teachings are assigned to their respective levels and positioned within their apologetic or rhetorical setting, the following basic results seem to emerge with a fair degree of clarity.

(1) With respect to logic, or more precisely, in regard to concepts or intentions signifying essences taken as essences, the question of existence amounts to the question of *instantiation*: Are there instances falling under or answering to (the intention of) some essence? Two considerations are straightaway pertinent. First, Ibn Sina wants to emphasize that an essence's logical architecture, so to speak (such as the genus and specific difference figuring in its definition), does not of itself determine whether there are any real things (*fīl aʿyān*, "in reality") possessing that essence. Accordingly, that there *are* such things in the case of any particular essence or, in Ibn Sina's idiom, that the essence has existence, is an additional truth not derivable by inspection of the logical architecture.[118] Second, it is of the greatest importance for the future of the notion of existence that Ibn Sina, quite possibly under Stoic influence, introduces a mental domain populated by meanings or intentions (corresponding to the Stoic *lekta*) which can themselves be said to *exist* just insofar as they enjoy intelligibility, or, most broadly, insofar as they designate a "something" whatsoever, even if that something fails to exist *in rebus* [*fīl aʿyān*]. Of any such *ens rationis*—that is, the item as it is explicitly entertained in the soul—the subject of a proposition for instance, we can say: "The intention of existence [*maʿna al-wujūd*] never ceases to be concomitant [*luzūm*] to it."[119] It is necessary to keep in mind that the same Arabic term *maʿna* is used to render *both* the Stoic *lekton* and the *eidos* as understood by Plato and Aristotle; in other words, *maʿna* can designate either noematic or eidetic being. Avicennian contexts such as the one from which I have just quoted make it plain that noematic being has the upper hand as far as the *existence* [*wujūd*] of an essence or intelligible nature is concerned.

(2) With respect to metaphysics, the question of existence amounts to the question of actuation or eventuation. I use these rather unwieldy terms to approximate the following outcome of Ibn Sina's intricate version of the Aristotelian teachings concerning form, matter, and their composition: That this one individual entity (the man Socrates, say) actually has some purchase on the world for some time is not fully explained either by its form alone or by its matter or, indeed, by their union. Additional details apart, we could say that Ibn Sina diminishes the "causal" relevance of natural forms to the point that some efficient agent other than the individual's form (such as humanity) is required both to effect union with the appropriate matter (which, according to him, is a nonbeing prior to this union) and to enliven, as it were, that union. In other terms, an artisan's bringing a certain composite into being becomes the model for natural coming-to-be as well. As he puts it

in a passage from his *al-Najāt*: "'Know that two things belong to the effect produced by the agent when the latter gives existence to another thing after this has not been, *viz.*, a non-being (*ʿadam*), which has preceded, and an existence (*wujūd*), which is actually present (*fī ʾal-ḥāl*)."[120]

(3) With respect both to logic *and* to metaphysics, Ibn Sina displays a complex and nuanced vocabulary to capture the status of existence vis-à-vis an essence and the elements of an essence. It must be said that most of the relevant nuances tend to disappear when his teaching is transposed from Arabic into Latin; one result is a rather homogeneous and thus convenient polemical target usually known nowadays as "Avicennian essentialism" and identified with the thesis that "existence is an accident of essence." Ibn Sina distinguishes *three* classes of predicates which can be truly said of a subject: constitutive (*mūqawimāt*), concomitant (*lawāzim*), and accidential or adventitious (*ʾaʿrad*). "Figure" is an example of a predicate constitutive of the essence of the subject "triangle" of which it is predicated; that is, "figure" signifies the genus which is an essential ingredient in the definition of a triangle. "Being young" or "being old" are accidental predicates of, say, a man, inasmuch as either is "that by which the thing is sometimes described, provided that it is not necessary that the thing is always described by it."[121] The second class, that of "nonconstitutive concomitants," is here the most interesting. It is subdivided into "intrinsic" and "extrinsic," which are illustrated by the oddness of the number three, on the one hand, and *the existence of the world*, on the other. "Existence" (*wujūd*), then, is strictly speaking not an accident at all in the sense of a variable and adventitious feature, but is tied much more intimately to the entity or item of which it is truly predicated. The examples of an intrinsic concomitant (the oddness of three or the angle-sum of a triangle; see *Kitāb al-ʾIshārāt*, trad. Goichon, p. 89) make it reasonably clear that Ibn Sina is thinking of the Aristotelian *symbebēkota kath' hauta* or *per se accidents* and that he is bent on assimilating "existence" to these while also preserving the all-important difference between concomitants following from the essence of a thing and those somehow brought into league with an essence by some extrinsic agency. Consider the following passage from the *Kitāb al-ʾIshārāt*:

A thing can be caused in relation to its quiddity [*māhiyya*] and its nature [*ḥaqīqa*] and it can be so [caused] in its existence. It is for you to conceive this by way of example in the case of the triangle: its essence depends on the surface and the line which forms its side; both constitute it insofar as it is [a] triangle and as it possesses the truth of triangularity, since they are its two causes, material and formal. But, from the point of view of its existence, it certainly depends on still another cause which is not these. This cause is not a cause constituting its triangularity and forming a part of its definition, but it is the efficient cause or the final cause, which is the efficient cause of the efficient cause.[122]

Here it becomes manifest that the issue of "existence" is not referential but causal; the extrinsicality in question is the sign that a primordial insufficiency marks an essence (or a form) as such; that is, neither instantiation nor actuation [*ḥuṣul*] is "natural" to essences or forms in their own right. Or, to put it differently, where for Aristotle natural generation (together with extrinsic factors such as the sun's movement along the ecliptic) adequately explains the presence of some individual of such-and-such a form or kind (see, for example, *Metaph* 12.5.1071a11ff.), for Ibn Sina such a "natural" process of coming-into-being *via* the *actual* being of the progenitor's form is no longer simply obvious or patent. We need to recall that both he and Ibn Rushd chastized Aristotle for assuming the existence of "nature" without giving a demonstrative proof![123]

What looms on the horizon at which these lines of reflection converge is, of course, the grand enigma of the world's supposed createdness, the knottiest *aporia* for all medieval philosophers, but especially for those born into the Moslem and Jewish communities, who, to all appearance, tried to "straddle the fence" between the compulsions of scriptural revelation and the reasoned conclusions of the pagan cosmologists and metaphysicians (while also striving to hold at bay or, alternatively, to accommodate, the rationalizing blandishments of the "dialectical theologians," the purveyors of apologetic speech [*kalām*] known under the name of the *mutakallimūn* [Latin: *loquentes*]). Two primary strands can be distinguished in the nodal composition of this *aporia*: first, the question of the creation *in time* of the world as a whole and, thus, the radically contingent status of the world's existence (compared to which the issue of the existence of any one singular entity fades in importance); and, second, the question of the standing of those intelligible essences, the instantiations of which appear to make up the population of that radically contingent world. In at least one tradition of reflection on this complex *aporia*, a tradition to which Ibn Sina seems prepared to give his public allegiance, the second strand is more baffling than, and therefore prior in importance to, the first.[124]

In broadest outline, this more baffling issue has to do with the equiprimordiality or coeternality of intelligible essences "alongside" the Divine agent of creation. Understandably enough, the popular or orthodox interpretation of *creatio ex nihilo* is, at least at first blush, incompatible with the thesis of the coeternality of essences and God, implying as it does that the contingent act of creation is ontologically bound to those necessities which find expression in self-evidently true definitions or in demonstrably true reasonings. To state the issue with greater exactness:

The foremost concern in this setting is *not* with the eternal pre-existence or the generation in time of the *matter* out of which God fashioned the heavens and the earth—that is, the tension between two positions Maimonides distinguishes in the *Guide* as the thesis of the creation of the world *ex nihilo* (*min al-ʿadam*, where the *nihil* is prime matter) and the thesis of the world's creation "after absolute pure non-existence" *baʿd al-ʿadam al-maḥḍ al-muṭlaq*)—but, rather, with the relation between the act of creation and the *forms* or essences possessed or exhibited by created entities.[125]

Ibn Sina follows Al-Farabi in discriminating between two "grades" of necessity, between what is uniquely necessary of itself or *per se* (God) and what is necessary, but nonetheless dependent for its being (*ab alio*) on the necessary *per se*. Both grades or modes of necessity—absolute and derivative or dependent—stand in nuanced contrast with possibility. In Ibn Sina's words:

> Everything that is necessarily existent *ab alio* is possibly existent *per se*. . . . Considered in its essence it is possible; considered in actual relation to that other being [on which it depends] it is necessary, and, when its relation to that other being is considered as removed, it is impossible.[126]

If essences/immaterial beings are possible *per se*—that is, if they have the (non-self-contradictory) character they do have eternally *and* independently of Divine will or intellect, while, on the other hand, their existence (that is [?], their instantiation or actuation *in rebus*) is derivative from the *per se* necessary existence of God—then an exquisite and tenuous compromise is in sight between the Koranic declaration "Everything goes to destruction except His face" [*Sura* xxvii, 88], with the possibility of annihilation it entails, and the philosophers' commitment to the position that intelligibility, necessity, and eternality (*a parte ante* and *a parte post*) mutually entail one another. Disentangling the exoteric from the esoteric elements from this line of solution to the nodal problem of the independence versus the dependence of essences on the Divine Being is exceedingly difficult; at all events, the very publicity of this attempt created the format within which medieval controversy over the contingency of the world's existence is subsumed or by which it is pervasively affected. Hence, we would have to keep in mind in any fuller discussion of this controversy three separable versions or levels of the existence/essence distinction as it has come to sight primarily in Ibn Sina's texts: (1) existence as a nonconstituent concomitant of some particular essence (as in the singular instance or actuation of humanity); (2) the contingent existence of the (sublunary) world as the effect of Divine causation; and (3) the necessary, although dependent or derivative (*ab alio*), existence of imper-ishable, eternal essences.[127]

Reculons pour mieux sauter. As in the case of the Stoics, so here in the case of Ibn Sina I have taken some pains to rehearse at least some highly significant details both to avoid the "flatness" of the respective positions when those details are omitted and, even more importantly, to flesh out with these two particular instances the environmental or domestic character I ascribed to the very notion of "existence." Arabic treatments of *wujūd,* as much as Stoic discussions of *hyparxis* (and *hypostasis*), are, so I have tried to show, at home in determinate settings from which they can be made to emigrate only at the cost of deracination. Moreover, we can detect within each setting the traces of a sundering or diremption which seems to be *anterior* to the posing of the question of "existence." Thus, for the Stoics, the sensible is sundered from the intelligible in such a manner that the latter is subordinated to, and even overshadowed by, the impressiveness of the former. Correspondingly, individual bodies (and their qualities) win the status of (existent) entities (*ta onta*), while the noematic surrogates for intelligible forms (*ta lekta*) reside in the domain of *ta mē onta*. The sundering through which the question of "existence" is situated in early medieval thinking is most in evidence in the Avicennian and post-Avicennian separation of nature from the grounds of the being of nature. As I mentioned above, the natural forms of Aristotelian meta-physics are considered impotent to account for the coming-into-being of natural composites; this is part of the reason why necessity and contin-gency (or possibility) replace actuality and potentiality as the pivotal terms around which the analysis of "natural" coming-into-being turns. In a similar vein we could say that while forms or essences are, in Aristotle, the primordial "causes" of the being-ness (*tou einai*) of what is (see *Metaph.* *Zeta* 17 *in toto,*) in Ibn Sina and others the dominant question concerns the causes of these causes, and, hence, of the very possibility that some-thing like "nature" can be given or be at work at all.

v   Kant's Relation to Pre-modern Understandings of Existence

Let me try to place these results in close connection with the prevailing modern understanding of the relation between constructibility and the alleged postulates of "existence" in Euclidean geometry:

(1) In Kant, as much as Stoicism, sensibility is pried apart from intelligi-bility, but with this crucial difference, occasioned or signalled by the Copernican divorce of explanatory persuasiveness from "aesthetic" or manifest experience: If the sensible, for Kant, retains its impressiveness only as matter, the sensible informed by intuition still has the upper hand over intelligibility in respect to validation and authentication. A concept

has its *Sinn* and *Bedeutung* only in reference to intuitable formations in which the externally presented manifold is first rendered comprehensibly one; in the absence of this *datum* of sensation, however, the concept ceases to have purchase on the "world."

(2) Accordingly, the logical possibility of a concept (or a noematic essence, corresponding to the Stoic *lekton* or Arabic *maʿna*) does nothing to guarantee its "real possibility." Absence of internal contradiction is the "logical mark of possibility," by which the object (*Gegenstand*) of the concept is distinguished from the *nihil negativum*. "However, the concept can nevertheless be an empty one, if the objective reality of the synthesis through which the concept is produced is not specifically exhibited/ proved [*dargetan*]" (*KdrV*. B624 Anm.). However weighty a role "principles of experience" play in this setting forth of the "objective reality of the synthesis," the gravamen of proof always resides in what is accessible to intuitive experience apart from the principles by which it is included in the lawful continuum of all possible experience. It is, once again, noteworthy that the examples brought into play in this passage are the triangle and the concept of an absolutely necessary being (B622–23). Neither the one nor the other imposes itself upon the understanding as a subject the existence of which cannot be rejected without violating the principle of contradiction. Conceptual necessity (something still akin to Ibn Sina's class of the necessary *ab alio*, with transcendental apperception taking the part of the *aliud*) is set apart from "real" possibility, that is, the possibility (but never the necessity) that there may be a "something" external to the understanding which answers to its judgmental predicaments.

(3) Kant draws the lesson of these sunderings with optimal clarity in the first note to the preface to his *Metaphysische Anfangsgründe der Naturwissenschaft* (cited above):

> Essence is the first inner principle of everything that belongs to the possibility of a thing. Hence we can ascribe to geometrical figures (since nothing is thought in their concept which expresses an existence [*Dasein*]) only an *essence*, but not a *nature* (my emphases).

"Essence" points to the intelligibility of a concept (or noematic intention), no longer to the actuality of a form or nature as these seem to have been understood by Aristotle. And with this transposition of meaning, Kant announces his definitive loyalty to that side of the medieval tradition (most evident in Ockham) for which both the necessity of thinkable essences and the *fact* of nature are equally problematic, or somehow optional. That we have the concepts we do and that there is anything actual at all which can be experienced as conforming to the principles and categories of understanding are ultimately opaque signs of the radical contingency, the ontological insufficiency, of the world—opaque, because Kant repudiates any testimony from the world to its ontological, cosmological, or physico-theological ground. As we have already had occasion to observe, construction (in Kant), by effecting the transition from the mere possibility of a concept to the exhibited existence of some relevant item meeting the conditions of sensibility, substitutes in a timely way for the unachievable spontaneities of an *intellectus archetypus*. However, since no theoretical proof can be given of the existence of a god as the absolute ground of a conditioned world, construction is the imitation of a *Deus absconditus*.

IX   The Enigma of the Postulates

This excursus into the distinctive theoretical (and theological) domains in which the notions of "existence" have had their roots has now returned us to its starting point, to the question: Does Euclid, after all, assume a notion of primordial geometrical constructions which amount to postulates or proofs that the constructed entities "exist"? The fate of my main thesis hinges on the answer to the question, since if the answer were "Yes," then any claim that early-modern mathematics begins as a radical break with ancient mathematics would have to be mitigated, perhaps to the point of banality. Despite conspicuous differences in emphasis given to the role of mathematical constructions, their ontological implication would, at bottom, be the same for Greek and early-modern geometers, namely, the "being" of the mathematicals would depend on the human capacity to *produce* the mathematicals, to bring them, or at least their vicarious instances, into being in a sensuously recognizable way. Zeuthen would turn out to be essentially correct: "In the Platonic theory [at the root of Euclid's *Elements*] the primitive geometrical notions are our own creations: we give them an existence either by our definitions or by postulates, which complete the enumeration of the properties we wish to attribute to them without demonstration."[128]

If the answer to this question is "No," then it is clearly incumbent upon me to furnish something in the way of an alternative account of the "being" of the mathematicals in Euclid and their relation to (primitive) constructions. If, as I have suggested, "existence" is always a domesticated notion, what counterpart to Stoic *hyparxis,* Judeo-Islamic *wujūd* or Kantian *Existenz/objective Realität* is native to Euclid, taking on the character of its homely environment? Answering these questions is made all the more difficult by the fact that Euclid, as I have had noted many times before, never stands apart from his work to interpret its ontological or epistemological underpinnings or to comment on its philosophical provenance;

the *Elements* come to us as a *fait accompli*. Thus we are faced with the infinitely delicate task of detecting intentions and understandings presumably governing that achievement, but not discernible apart from its inner architecture, its modes of procedure, and its discursive idioms.

Let me begin, then, with a brief recital of the less controversial results to which my inquiry has so far seemed to lead, as a way of approaching circumspectly the all-important and much-argued issued of the meaning of the postulates in *Elements*, Book 1.

(1) The orthodox identification of "existential" assumptions in Euclid with declarations of the possibility of constructing certain primitive items has long rested (as in Heath) on the interpretation of *hoti estin* judgments in Aristotle as assertions that such-and-such (*a* circle or the "class" of circles) exists. By parity of reasoning, the weaknesses in the case for the latter interpretation—weaknesses brought to light in the studies of Kahn, Mignucci, and Gomez-Lobo—debilitate the former. At a minimum, to the extent that *einai* in the Aristotelian discussion of *ei estin/hoti estin* cannot be given an univocally "existential" or "referential" reading, independent arguments for the assimilation of Euclidean postulates to existence-claims would have to be brought forward. To the best of my knowledge, no such arguments have been produced, nor would they be easy to come by since the only occurrence of *einai* in any form, in the fourth Postulate, is uncontroversially *copulative* ("all right angles are equal to one another"). This philological point does not, of course, settle the philosophical matter, since one could still argue that the *intent* behind the postulates is to present assertions of "existence" by means of undemonstrated postulates of constructibility. (It is significant, however, that when Proclus revives the Aristotelian distinction *ti estin/ei estin* in his *Euclid Commentary*, the latter is associated, not with postulates, but with the *diorismoi*, the specifications of the conditions under which a problem is solvable [202.1–5 Friedlein]; the issue *ei estin auto kath' hauto*—"if it *is* just as it has specified"—is not, on Proclus' view, whether an equilateral triangle "exists" *überhaupt*, but whether the construction of a triangle from three straight lines equal in length to three given straight lines can succeed only under the condition that two of the lines taken together are greater in length than the third [cf. *Elements* Bk. 1, Prop. 22 and Heath's note *ad loc.*].)

(2) Apart from the received term for the axioms, namely, "common notions" (*koinai ennoiai*), which may or may not be a Stoic formation, the vocabulary of the Early Stoa is not audible in Euclid. Again, this by *itself* does not decide the question whether Euclid might have endorsed the Stoic conception of "being" in the case of the geometricals. What does seem close to decisive is the near-total divergence from the spirit of Euclidean demonstration (a) of the Stoic logic of hypothetical inference (a logic, as we might say, of contingent states of affairs rather than a logic

of the constitution of geometrical forms) as well as (b) of the circuit traced by the Stoics from the impressiveness of individual material bodies (and their states) to the formation of general notions in which the classes of those impressions are significatively registered and then back to those individual bodies via the patent act of pointing, of *deixis*, rather than *apodeixis*. Sambursky, in this *Physics of the Stoics*, was eager to show that the Stoics' "physicalization of geometry" via the endowment of "geometrical figures with the elastic properties of material bodies" was a step on the way to the modern conceptions of function and the dynamic continuum (and, therewith, the calculus of variations).[129] The tensions by which the cosmos is held together as an array of ever-varying bodily shapes is mimicked by the process of concept-formation, as in the example given by Sextus Empiricus (*Adv. Math.* 3.51ff.) of trying to conceive a breadthless straight line by "continually intensifying [*aei . . . epiteinontes*]" the narrowness of an actually thick line. The question of geometrical being and its knowability thus becomes first cosmological and then psychological. Euclid, the contemporary of Zeno of Citium, is said by Proclus to have arranged the *Elements* so that the demonstration of the uniqueness of the five regular solids required by the cosmology of the *Timaeus* be its finale (68.21–23). (Consequently, we could conjecture—but only conjecture— that a physics of the world's structure and motions follows upon geometrical science, while in Stoicism geometrical art is embedded in a physics of tensional variations.) Furthermore, even though Euclid says nothing of how we arrive at the notions or concepts set to work in geometry— indeed, the very concept of a "concept" (*ennoia* or *ennoēma*) *qua* mental formation is pregnant with Stoic preconceptions—we cannot resist remembering that his vocabulary of prescribed operations pointed to some kind of pre-acquaintance with the stable character of the figures placed under consideration and that the instruction *nenoēthōsan*—"let them have been attended to by the intellect"—appears not in the statement of a theorem, but in the so-called *ekthesis*, the setting-out in suitably adapted form of what has already been given (and understood) in the theorem itself (see, for example *Elements* Bk. 12, Prop. 18: "Let the spheres *ABC*, *DEF* have been attended to by the intellect").

(3) "Ibn Sina" was introduced above as a synecdoche, standing for the whole class of the post-pagan thinkers to whom the question of the world's being given at all had been made problematic by the insistent and authoritative theological teachings of their native communities. However remote from Euclid such a matter might seem, nothing in fact is more intimately and more perplexingly bound up with any attempt to understand the style of thought his work exhibits. With the medievals, the question of "existence" loses any innocence it might conceivably have had, since from now on, in both letter and in spirit, it is inevitably raised

in the conviction or suspicion that no "worldly" entity, including essences and the world itself as an entity, suffices as the ground of its own being-ness, of its presence or occurrence in a manner to which the intellect can be wholly and confidently responsive. The radical moderns, before and after Kant, cut this Gordian knot by making the intellect wholly *responsible* for worldly intelligibility, if not in every sense for the world's presence. Mathematics, so I have been suggesting since the first chapter, is, for these radical moderns, *the* proleptic and promissory *exemplum* in which intelligibility and presence (existence) can be seen or made to coincide.

Heidegger, in the early *Grundprobleme der Phänomenologie,* allows us to measure what is most deeply at stake here, even if the confines of the present chapter will not permit more thorough probing. His principal claim is that Kant's thesis "Being is no real predicate" has to be understood as a continuation of ancient-medieval ontology or, more precisely, that "ancient ontology interpreted what is [*das Seiende*] in its Being on the basis of establishing/producing [*Herstellen*] or perceiving and that, to the extent Kant, too, interpreted actuality with a view to perception, a direct line of connected tradition here reveals itself."[130]

Crudely put, Heidegger's argument is that no fundamental difference separates Aristotelian from Scholastic ontology, which is in turn the immediate source of Kant's thesis about Being; or, in other words, the doctrine of *creation* is a version of ancient pagan ontology, rather than the most dramatic alternative to it. Athens alone, not Athens in perennial combat with Jerusalem, is the matrix of the Western thinking of Being! "Ancient ontology, despite its different origins, was in its foundations and its basic concepts cut from the same cloth, as it were, as the Christian conception of the world and the Christian conception of what is as *ens creatum.*"[131] Heidegger is brought to this conclusion by his analysis of the "intentional structure of productive comportment [*des herstellenden Verhaltens*] and the understanding of Being" it implies. Far from being antipodal to one another, *Herstellen* and *Anschauen* (*theōrein*), as Heidegger construes them, manifest exactly the same character when grasped as modes of *Dasein*; both come to sight as ways of "releasing" an entity and "giving it its freedom" to be in its own right, as what is already "ready-made" and "complete" (*fertgig*). The ideal of disinterested contemplation of what is there on its own emerges from *Dasein's* putting itself at a distance from any occupation with what is, but only in virtue of having already interpreted the Being of what is as *das Ansichsein des Fertigen*— the Being-on-its-own of what is ready-made. Consequently, the "being-present-at-hand of an entity belongs to it on the basis of its having-been-produced [*aufgrund seiner Hergestelltheit*]."[132]

Much as it would be to my governing purpose to take up the gauntlet thrown down by Heidegger in this challenging account, I must restrict myself to a single consideration, one to which I shall return when discussing, for the last time in this chapter, the style of Euclidean *mathesis*. At least in this early work, Heidegger isolates the orientation of Greek ontology to "this everyday and close at hand comportment [of producing]" from the Greek experience of teaching and learning. Might it be that the Greek *Seinsverständnis* arises primordially from that experience and from reflection upon the grounds of its possibility? That the horizon of ancient ontology, as articulated in Plato and Aristotle, is the *artfulness* of teaching what can be learned and learning what can be taught (beings as *mathēmata*), rather than the *craft* of letting what has been fashioned come to stand on its own? That humans encounter *logos* always in the shape of *dialogos,* the exchange of telling speeches by which souls are turned? Would that horizon and that exchange primarily inform the distinctive comportment, the ethos, of the ancient mathematician? Must what is distinctive in every instance stem from the everyday and close at hand?

My path thus far has been a *via negativa.* The starting point seemed to be a narrowly defined and therefore promising question concerning an interpretation of Euclid given its initial impetus by Kant and his followers and enjoying orthodox standing among most contemporary scholars: For Euclid, the "existence" of a geometrical entity means that its constructibility is either postulated or shown to follow effectively from postulated constructions. To assess the soundness of this interpretation, it became necessary to confront, however obliquely and episodically, the question of "existence" as in each case belonging to a historical and theory-laden domain governed by specific preoccupations. To the extent that none of these preoccupations could be readily matched with Euclidean procedures, the twin issues of "existence" in the *Elements* and their alliance with constructibility remain as open and as baffling as ever. Is there, then, a *via affirmativa? Do* points, say, or straight-lines and circles "exist" for Euclid in some so far uncanvassed sense? Will the postulates reveal that sense or do they invite us to place ourselves in a quite different habitat of thought, one in which "existence," by that name, could never be more than an unfamiliar, if diverting, guest? I want to devote the remaining pages of this chapter to exploring, necessarily in a speculative and abbreviated way, what it could mean to take up that invitation.

Euclid's five postulates have caused inordinate perplexity since antiquity, not least, of course, because of the fifth, parallel, postulate and the long sequence of attempts to demonstrate it until the later nineteenth century. However, perplexity has also arisen over the meaning of the term (*aitēmata*), the relation of the definitions to the postulates and the inclusion of the fourth ("all right angles are equal to one another") amidst three which seem unambiguously to be concerned with possible constructions and the fifth with its own exceptional status.

When interpreters have turned to Aristotle for guidance for the meaning of *aitēma* in Euclid, more darkness than light seems to have resulted. The key passage in *Posterior Analytics* (1.10.76b24–34) distinguishes a hypothesis (*hypothesis*) from an *aitēma* in terms of the acceptance or indifference of a student to the proposition identically formulated in both cases. When the proposition is accepted it becomes an hypothesis; when it runs contrary to the student's opinion or when he has no opinion, it is a postulate. How much of this helps us to become clear about the postulates in Euclid?[133]

First, we are reminded that both in Aristotelian demonstration and in Euclid's *Elements* someone (the student) is being "shown" something by another in a position to give explanations; *epistēmē*, science, Aristotle tells us in the *Ethics*, is the habitual facility in giving explanations. Already we are required to situate the unfolding of any proof in a dialogue, not in the solitary monologue of an *ego cogitans*. Accordingly, before anything else might be said about the respective views of Aristotle and Euclid, it appears that consistency (or non-inconsistency) with the opinions of a student is desired in the case of hypotheses and postulates, rather than the mutual consistency of the postulates on their own, apart from the dialogic enterprise of teaching and learning.

It is simply not clear how Aristotle's own stress on the contrareity or indifference of a student to a postulate would fit in with any of the Euclidean postulates except the fifth. I do not want to plunge into these turbulent waters, especially in regard to the possibility that both Aristotle and Euclid countenanced the intelligibility of "non-Euclidean" geometries when the fifth postulate is negated or omitted.[134] Since Aristotle gives no examples of postulates in the present passage, we are left to conjecture that his remark is an *obiter dictum* putting his listener in mind of the prescientific exercise of topical reasoning in which someone bent on building up or tearing down an argument seeks his antagonist's agreement to a premise in order to discover what will follow. Aristotle nowhere suggests that *aitēmata* as so defined belong to the premises of a demonstrative *science*. In Euclid, there is no hint that the student might be of a contrary or indifferent mind in face of the first four postulates; instead, he introduces, for the first time in the *Elements*, that verbal idiom, the perfect passive imperative (here, *ēitēsthō*) which we detected again and again in the body of his work. As in those occurrences, so here the syntax carries the sense that the "action" demanded has already been carried out and that it has been done so impersonally—that is, not by this one student or reader in his present individuality, but by any one student experiencing what it is to be caught up in a course of studying which somehow has always been under way from a timeless start. Moreover, the nominal rubric *aitēmata* seems to be used here with an ear attuned to its likely derivation from the same root as *aitia*. An *aitēma* would therefore be something more and even other than the object of a request or injunction; it would be what is called upon—or has already been called upon—to bear some measure of responsibility for the dialogue of teacher and student about to unfold once again. A "postulate," on this understanding, would be a condition of learnability.[135]

Can we become more specific about these conditions? The now-familiar standard answer, that at least the first three postulates are meant to win the assent of the student to the possibility of constructing straight lines and circles and thus to their existence, has in its favor the verbal language employed, namely, "to lead [*anagein*] a straight line. . . ," "to extend a limited straight line continuously. . . ," "to draw a circle. . ." It is when these instructions are put in conjunction with the preceding definitions that all of the old difficulties reappear. Consider, once again, the testimony of Heath, who on this matter follows Trendelenburg: "The transition from the subjective definition of names to the objective definition of things is made, in geometry, by means of *constructions* (the first principles of which are postulated), as in other sciences it is made by means of experience."[136] What is meant, of course, is that the definition of straight-line (Def. 4) is merely "nominal," whereas the first postulate tells us that "real" straight lines can be constructed between "any two points." I won't linger on the division of subjective/objective mobilized by this account since it simply repeats the very issue at the heart of my discussion. We do have to notice, however, that if such a transition *is* what is occurring, then something is lost in the translation; that is, nothing in the postulated "objective" constructions either refers to or directly satisfies the relevant "subjective" definitions. This emerges in two different ways: First, the points, straight-lines and circles actually inscribed on any occasion will not be without parts or breadthless, respectively; hence, to "read" these inscriptions as imaging or otherwise conveying to sense or mind what is intended by the nominal definition requires more than an inspection of the "successful" result, as we saw earlier. Second, even if we set aside this first source of uncertainty, it remains the case that the first three postulates presuppose something like the "truth" of the definitions, rather than asserting their "objective reality" in Kant's sense. I mean by this that the second postulate, for instance, which is equivalent to the claim that two distinct straight lines cannot have a common segment, depends on our having already grasped, from Definition 4, that a straight line "lies evenly with its points" and thus cannot share two or more points with another line. Unless this has been acknowledged, there is nothing in the postulate to guarantee that the continuous prolongation of a given line-segment does not yield a new segment not lying evenly with the points on the first. Some other task must, then, be assigned to these conditions responsible for learnability, conditions, as already noted, that are to be acknowledged as having already been met when any one student makes his debut.

Another initially plausible candidate for the intent of the postulates would be individuation, together with the accompanying claim that the constructions licensed by them carry us from the generality of a definition to the singularity of an instance. One might suppose that the testimony of a "realized" definition in any one case would outweigh any doubts of its intelligibility or internal consistency. Standing against this claim is one noteworthy lexical fact, as well as one principal philosophical obstacle. In the formulation of Postulates One, Three and Four Euclid already speaks of "every" point, "every" center and distance and "all" right angles; that is, the standard English versions, e.g., "from any point to any point," misleadingly suggest that some one instance (here, a pair of points) is selected at random and made to carry the force of "existential instantiation." Furthermore, even if we were to substitute "any" for Euclid's "every," we would still have to face, as so often before, the enigma of the 'backward' inference from what *did* succeed in one instance to what *must* succeed in every instance.

Nonetheless, this enigma and the notion of individuation from which it stems do, I think, point us in a new and promising direction along which the identity of the conditions of learnability might be sought.

We need to recall Aristotle's frequent mention of one of the "greater mysteries" of Platonic thinking, the unlimited manyness of *each* of the mathematicals (as in *Metaph.* A6.987b15ff.). Whatever the "official" standing of this so-called "doctrine of intermediates," the relevance of its main thesis (namely, the mathematicals differ from the forms inasmuch as there are many "similar" [*homoia*] squares, say, while there is only one unique form) to geometrical practice is considerable. No definition and no theorem specifies the "absolute" size (length, area, volume) of any of the configurations which it intends; indeed, if they did, they would cease to refer us to *the* square or to every square. As the theme of definition or demonstration *each* square is as good as any other precisely because all manifest inalterably the definite shapeliness to which our learning is addressed. Moreover, each sensible image offered by the "working" ge-ometer is the image of some one of these squares, and yet any image must somehow be as good as any other, lest learning become impeded or brought to a halt by something peculiar to a single image. The manyness intrinsic to each "kind" of figure as well as the manyness displayed by the infinitely various images of each kind must somehow be a multiplicity indifferent to itself, a manyness of differences that make no fundamental difference, while nonetheless never collapsing into indiscriminate sameness or identity with one another. (This, I take it, is the form of Aristotle's description of the mathematicals as *poll' atta homoia*—"many *similars*.")[138]

Viewed against this Platonic backdrop—which is *in all likelihood* the "historical" matrix from which Euclidean geometry arose—"construction" is a lesson in negotiating the terms of this unaccustomed alliance between indeterminate manyness (of the "kind" as well as of its images) and invariant similarity. Each construction must yield—must have already yielded (the perfect passive imperative)—an image which both captures some one member of a kind in its uniqueness (*this* triangle with sides of *these* lengths) and conveys the fundamental similarity or kinship of this member with all the (infinitely many) others of this kind. A construction, then, does not bring into being a nominally or otherwise defined *concept;* nor does it mimic the steps taken by the mind in putting together, synthesizing, a concept (as construction will be seen to do in Descartes). Rather, a Euclidean construction puts us in mind of the stable being of what it images or depicts, not by way of *individuating* it, since the constructed figure is always an image of this *one*, uniquely determinate, specimen of the kind, but by way of exhibiting this being in its indifference to its own intrinsic multiplicity. There is no *one* perfect square, but *every* square has to be perfect of its kind, not *sui generis*.

What do the *postulates* have to do with this complex field on which manyness and kinship, variation and stability, play out their game of intricately mutual implications? After all, Euclid, even if by training a Platonist, never explicitly announces this relationship. Its traces must be discovered in the workings of Euclidean *mathēsis* or, as I have indicated earlier, in the dialogic exchange through which a teacher calls upon and edifies the preunderstanding of his student, thereby setting his own knowing to work, giving it actuality, in the completed insight of the learner (see Aristotle, *Physics* 3.3.202b7: "esti gar hē didaxis energeia toū didaskalikoū, en tini mentoi"). We can perhaps begin to hear something of this dialogic appeal in the postulates since we need no longer hesitate to associate them with constructions, once we have grasped the possibility that what is at issue is not "existential assumptions" meant to allay or forestall doubt about the "objective reality" of our subjective concepts, but the self-preserving and self-perpetuating fit between a constructed image and that one original which nonetheless has infinitely many kindred.

The inclusion of Postulate 4 (the equality of all right angles) among the canonical five has almost always given pause to post-Kantian interpreters convinced that a postulate, as Kant writes in the *First Critique,* "means the practical proposition which contains nothing other than the synthesis through which we first give ourselves an object and generate [*erzeugen*] its concept" (B287). This postulate, unlike the first three, so these interpreters assume, does not refer explicitly to the constructibility/existence of right angles, but to their equality. Heath, for example, construes Postulate 4 as, implicitly, the assumption of the *"invariability of figures* or its equivalent, the *homogeneity of space*."[139] While the second, allegedly equivalent, interpretation begs the question I alluded to above in connection with the modern notion of a three-dimensional, isotropic "space" in which a mathematical *physics* can be embedded, Heath's first suggestion gives a helpful clue.

Suppose that each of the first four postulates is meant to secure acknowledgment on the student's part that geometrical learning can only proceed if any illustrative depiction *and* its referent are unaffected by the parochial circumstances in which they are graphically presented or imaged, respectively. That all right angles are equal to one another means that all rectangular figures of the same kind (squares, for example) will lose nothing of their kindness owing to any local (or temporal) variations in circumstance (such as means and place of sensible inscription). Analogously, no circle is compromised in its intelligible or dianoetic integrity by its particular center and radius or by the graphic image into which *it* is transcribed (Postulate 3): The straightness of any one straight line-segment is not distorted by its particular end-points (Postulate 1) or by any extension changing its length (Postulate 2). What the postulates summon the student to acknowledge are conditions of invariability set into play prior to any one episode of graphic illustration *because* in their absence or merely possible non-fulfillment, learning can never *have been begun* at all, but would have always to be on the way to beginning all over again.

Indirect corroboration of this reading comes from the recent work of Robert Wagner, which takes up in a helpfully novel manner the vexed questions provoked by the apparent assumption (in Euclid, Bk. 1, Props. 4 and 8) that figures can be rigidly transported in "space." Details apart, Wagner's principal negative conclusion is that spatial displacement is *not* the core of Euclid's procedures here; his positive argument is intended to show that figures (such as triangles) constructed by two or more different "algorithms" and occupying different positions are equal, even though not positionally coincident. (As Wagner emphasizes, for Euclid the equality of figures is *not* the same as the equality of the point-sets with which they are identified in contemporary geometry.) Consequently, Euclid requires that different (token) constructions "reproduce" or "copy" one another—to the point of equality, in the cases of Book 1, Propositions 4 and 8, or more generally in the case of the postulates, to the point that each (token) construction permits the learner to "see" his way through to the original, one of whose kindred will be graphically imaged by the next construction and the next, *ad infinitum.*[140]

This corroboration, welcome as it is, leaves us with one final riddle: How can the stable and invariant figure (or figural component, such as a straight-line) be made accessible by the overt motions carried out in authorized constructions? In the Academic debate on which Proclus reported (see above 2, V), the pivotal point was the admissibility of motion or *genesis* into the geometrical domain. If Euclidean (primitive) constructions are now to be interpreted as the conditions of invariance required for learning to have begun, does this imply that they must somehow cease to involve movement and genesis? Or, is another understanding of the tie between motion and *mathēsis* available to us from sources in the same environment in which Euclidean geometric teaching appears to have had its governing roots?

The pseudo-Euclidean *Sectio Canonis,* to which I have referred before, begins with a preface setting forth the conditions under which a strictly mathematical science of musical intervals can be inaugurated:

> If there were rest and motionlessness, there would be silence; if there were silence and nothing moved, nothing would be heard. Therefore if something is to be heard, there must be beforehand a blow and motion. (148.1ff., ed. Jan)

The motions we deliberately bring about (here, the "blows" on strings) are the necessary conditions for the audible presence of the sound-intervals studied by harmonics; but the theme of that study is the pure numerical (in the case of this text, commensurable) ratios made evident in virtue of this audible presence, *although not identical with it.* The phenomena making their appearance only because certain motions have been studiously performed elicit or evoke nonphenomenal ratios whose being consists wholly in their immovable presence for *mathēsis.* It scarcely seems forced to say that the phenomena of audible consonance and dissonance move us on each occasion to recollect those standing ratios.

I have cited the opening passage from the *Sectio Canonis* as an invitation to explore in much wider-ranging fashion the role of sensible or phenomenal motions in the inception of *mathēsis.* If Euclid drew on this understanding, preserved and transmitted by the Pythagorean-Platonic circle in which he found collaborators and students, then we might venture to see in the postulates (and in all subsequent constructions) in the *Elements* the counterpart to the "blows" trained upon the string of the Pythagorean monochord. To draw, on papyrus or on the sand, a "circle" with any center and radius, is to enact certain studious movements that evoke the genuine theme of one's study—here, this one mathematical and intelligible circle—and that permit us to attend to it as both singular and exemplary, a one within the manyness of its own kind. Neither this exemplar nor its kind comes into being or confirms its "existence" by grace of the constructive motions we effect; rather it is the former that somehow hold sway motionlessly in those evocative motions, however often and by whatever means they are carried through. This hypothesis seems to be exactly what must assume responsibility for the discursive stabilities by which the movements of learning are held on their course. And *this* hypothesis, as it were the *aitēma* holding responsibility for all other requisite *aitēmata*, might throw one last ray of light on Euclid's pervasive use of the perfect passive imperative whenever time-consuming operations would seem to be demanded of the student: It is as though student and teacher alike are being invited to regard themselves as reenacting in time what has already been done all along and thus never for the very first time.

X   Conclusions

Let me bring this chapter to a close by repeating three of the major suggestions made in the preceding sections, beginning with the last and moving back to the first.

(1) The Kantian equation of constructibility with the existence or objective reality of mathematical concepts, from which the orthodox interpretation of the Euclidean postulates is taken, is not at home in the theoretical setting of the *Elements*.[141] Indeed, the very notion of "existence" is so thoroughly colored by the different historical environments to which it (or its lexical analogues) belong that we must be wary of injecting it into an inhospitable setting. The alternative reading of the postulates sketched here brings Euclid into close affiliation with the Platonic doctrine of the mathematicals, suggesting that he is principally concerned with having the student acknowledge, as a condition of learning, that the indefinitely *many* intelligible instances of each geometrical *kind* are sufficiently akin so that no circumstances or details of any graphic construction will vary or distort in fundamental ways the nature they share. The movements performed in these constructions do not "create" or "realize that nature," but instead evoke or allow it to make its intelligible presence "felt."

(2) The internal details of the Euclidean constructions inspected over the course of the *Elements* lend their weight to these conjectures. As we saw, the language for operations used by Euclid is almost always sensitive to the specific nature of the figure to be constructed, thereby reminding us that we must somehow be acquainted with that nature prior to any operations we perform. Correspondingly, the properties and relations of a figure or group of figures become manifest and learnable *by means*

of certain constructions (including those introducing new lines, and so forth, into an initially given configuration), but only because they belong all along to the nature(s) of the relevant figure or figures. A line from Archimedes' preface to his *De sphaera et cylindro* captures what is implied by this: "These properties were by nature there in advance all along" (*tēi physei proupērchen*—note the imperfect tense). This is precisely the claim that will be subverted and turned around by the radical moderns, for whom a figure has such and such properties and relations *because* it has been constructed in such and such a way (see Hobbes, *English Works*, vol. 7, p. 183).

(3) Conspicuous in the working "methods" of Euclid is the geometrical version of practical prudence, *phronēsis*. Euclid as teacher strives to find means fitted to the choiceworthy end of allowing a student to learn what is intrinsically learnable. Proclus praises him for including "reasonings of all sorts" as well as "all the dialectical methods" in the *Elements* (*In Euclid.* 69.10–13) and with these latter words reminds us that the context in which Euclidean *phronēsis* is put to work is shaped by the dialogue between teacher and student. This means that the voice of the teacher must be able to elicit and enhance the student's pre-understanding at least of the very basic items to which their dialogue will continue to be addressed. It also means that the prudential rhetoric suited to this dialogue must resist, as far as that is possible, the seductive blandishments of any technique unharnessed from the directing control of that shared pre-understanding. (We witnessed an outstanding instance of what abandoning this resistance entails, namely, the introduction of compound ratios.)

This reminder of the ever-present tension between didactic *phronēsis* and technique returns us to the starting point of this chapter and to the central theme of this work as a whole: Technique in its primordial alliance with *poiēsis* aims at bringing into being what previously was not, while didactic *phronēsis* intends to preserve and enliven what has all along been understood or at least already understood by the teacher. Much earlier I quoted Rousseau's proud claim that the founders of radical modernity had and needed no teachers; each, in his own fashion, understood himself as a "self-made man," as I shall show in some detail in the case of Descartes in the first section of the next chapter. Forswearing the need for teachers seems to go hand-in-hand with reliance on technique or method as something to be set into play in solitude, something looked to for self-certification, and, ultimately, something obedient to the dictates of the personal will.

The Eleatic Stranger in Plato's *Sophist* is especially attentive to the apparently "clean" division of the arts into productive (*poiētikoi*) and acquisitive (*ktētikai*), setting "the entire look of the learnable as well as that of knowing" under the seal of acquisition (219c2]. Whatever else would need to be said about the unarticulated ambiguities of this division in that dialogue, it does serve to remind us of what is finally at stake in the discussion of "mathematical construction/production." If we were to take the Stranger's words at their face value, this phrase would be an oxymoron.[142] Euclid saves it from becoming oxymoronic by circumspectly, if not always successfully, subordinating construction (and technique) to the acquisition or reacquisition of learning.

Proclus tells us of another, long-lost book by Euclid entitled *Pseudaria*, in which he taught beginners how to discover paralogisms and to avoid being misled by specious reasonings; hence its aim was "cathartic and gymnastic" (70.15–16). It is pleasant to imagine that something of these latter goals survives in the *Elements* as well and that it not only sets out "an irrefutable and complete exposition of the very science of geometrical things" (70.15–16), but simultaneously exercises us in the prudential arts of teaching and learning. Seen from this angle of vision, Euclidean mathematics would also be a *mathēsis mathēseōs*. It remains for us to see how this vision comes to be refracted in the medium of Euclid's radically modern rivals.

3

Descartes' Revolutionary Paternity

Mad Mathesis alone was unconfined.
Pope, *The Dunciad*

I Preface

Descartes has more than once been called "the father of the modern mind." Perhaps like the modernity he helped to spawn, he is almost ostentatiously Odyssean in the variety of masks and disguises he was prepared to assume as he made his own way towards his goal or goals; his Ovidian motto "Bene vixit, bene qui latuit" (He lived well who hid well) epitomizes this sense of the need for dissimulation if victory is to be won in the battle against the ancients. In Michel Serres' words, "Like many philosophers, Descartes pursued his military calling in metaphysics"—to which I would only add: not in metaphysics alone and almost always in camouflage, behind enemy lines.[1]

Two immediate consequences follow from this Cartesian gift and need for dissimulation: (1) Descartes' relations to his own roots or "sources" are unsurprisingly *opaque*, and (2) the arguments or positions by which he sets about fathering the modern mind are not in every case *patent* in what have come to be regarded as the official Cartesian texts that gave birth to modern philosophy. Inasmuch as I shall be trying throughout this chapter to bring some of these inaugural or generative arguments to light, especially as they may be found in Descartes' way of doing mathematics, let me say just a few words here about the first consequence.

Since, as I attempt to show in section II, Descartes is very much concerned to erase the traces of any "influences" his predecessors or teachers (such as Beeckmann) may have had on his thinking and since, in any case, the search for "sources" in the accepted scholarly sense of *Quellenforschung* amounts in the end to what Lichtenberg more candidly called an *Ideenjagd* (an idea-hunt), the question of Descartes' roots has to be treated with unusual delicacy. I would pose it, *not* as an issue of specific individuals or schools or movements having any sort of formative influence on the genesis of Cartesian philosophy, but as a matter of the horizons or preconditions thanks to which the central Cartesian enterprises could come to make sense at all. If I had to designate, however loosely, the connection between those horizons and his enterprises, I would certainly eschew suggestions of causality as normally understood in favor of something akin to the relation between ground and figure. Happily, this is not the place to explore and certainly not to defend this choice.

The preconditions most germane to the possibility of Cartesian thinking are two: first, the collapse of rigid distinctions between (productive) arts and (theoretical) sciences and, second, the flourishing (not to say hypertrophy) of the practice of symbolization. The first horizon gives visibility to that domain in which mechanics and the operative arts come to enjoy the prestige of intelligibility hitherto withheld from them in Scholastic as well as Humanistic classifications of the received arts and sciences. The second allows a shifting of attention from "natural" to "artificial" languages and, even more crucially, gives to artfully devised images or contrived signs the freedom to participate both in the interior order of the mind's operations and in the exterior order of public communication. The juncture or fusion of these two horizons, I believe, provides the ground against which Descartes' understanding of science as essentially operative (constructive and productive) *and* as essentially symbolic could emerge. Indeed, it may have been Descartes himself who brought about that juncture, since it is he who most conspicuously exploited the power inherent in the view that symbolization frees us to work ingeniously beyond the boundaries apparently fixed by nature as it is sensuously, premethodically given (see my discussion of dimensional homogeneity below, III, vi), while at the same time serving to direct those mechanical operations or movements from which outwardly manifest configurations artfully issue. He might easily have taken over as his own the Sophist Antiphon's aphorism quoted in the first paragraph of the pseudo-Aristotelian *Mechanical Questions: technēi gar kratousin hōn physei nikōmetha:* "By art we master things in which we are conquered by nature" (847a21).

These comments are meant simply to delineate future lines of research and recollection, most of which would first have to be stretched back into the Hellenistic and Patristic traditions (especially as regards the prestige of signs or symbols) before being brought forward, through the Arabic metamorphoses of Greek conceptuality, into the world of the later Renaissance in which Descartes' thinking had its *factual* roots. If my initial suggestions should prove fertile, then the advent of radical modernity, at least in its Cartesian figure, might be characterized by this pairing of two tactics for outwitting "Nature," mechanization and symbolization. From now on the "natural" will be measured by its accessibility to artifice or, as Descartes himself says more invitingly, "All the things which are artificial are natural as well [*avec cela*]" (*Prin. Phil.* 4.203).

In the three sections which follow, I want to sketch some of the principal aspects of that Cartesian figure. In section II, I consider the autobio-graphical fable devised as a prelude to the new sciences (*Discourse on Method*, Parts 1–3).[2] In this work, Descartes applies to himself (or his fictive persona) the strategem of self-origination or self-production which we shall come to see as pivotal to his most basic understanding of how we can and should go about the project of learning anything with certainty.

Section III is devoted to the main procedures of Descartes' *Geometry* and to their justification in the *Rules for the Direction of the Mind*. I hope to bring into clear relief the central importance of the *Geometry*, both in technique (or method) and results (constructive problem-solving), for Cartesian thinking in its entirety.

In section IV, this last claim is given more substance through a reading of *Meditations* 5 and 6 in the light of Descartes' geometrical style and achievement.[3] At the close of this section, I shall indicate the problematic legacy left by Descartes to his most attentive contemporaries or immedi-ate heirs.

Throughout this chapter I shall be implicitly defending the description of Descartes offered by Paul Valéry:

> You will observe one thing: that in every question to which he could reply by an act of his Self, he triumphed. His Self was a geometer. Without insisting on the point, I shall say, with certain reservations, that the basic idea of his geometry was thoroughly characteristic of his whole personality. It would seem that, in everything, he took his Self, of which he was so powerfully aware, as the point of origin of the axes of his thought.[4]

II   The Art of Origins: Descartes' Fabulous History

It makes no sense to doubt that Descartes understood himself as a modern in the most radical and aggressive—shall we say, the most "modern"—sense of "modernity." While *moderni* in medieval usage was a term of versatile and labile reference—denoting one's contemporaries or those who wrote after the Fathers of the Church; those who studied the "new logic" introduced with the recovery of the whole of Aristotle's *Organon* or those who followed the *via moderna* of Ockhamist nominalism; while to be "modern" meant, for Cassidorus in the sixth century, to imitate or emulate the ancients and, for some later writers, carried the somewhat disdainful sense of latching on to what was up-to-date, *à la mode*—for Descartes and the majority of his contemporaries the modern age was experienced as something decisively and *irreversibly* novel in comparison with the past in its entirety, and in particular, with the ancients and the traditions they spawned, as I have already noted in chapter 1.

This shared sense of "modernity" takes on particular nuances and modulations in Descartes' self-understanding, as becomes apparent from three early, marginal texts preserved among the Stockholm manuscripts and originally published by Baillet. Assessed with care, they can tell us a great deal about Descartes' peculiar version of *his* modernity.

The first passage operates on an ironic inversion of the relation between antiquity and modernity as that relation is commonly or traditionally understood:

> We shouldn't give great credit to the Ancients on account of their antiquity; it is rather we who should be called the more ancient ones. For the world is older now than it was then and we have a greater experience of things.[5]

The trope is not original with Descartes—the incandescent Bruno in his *Ash-Wednesday Supper* and then again Bacon in his *Novum Organum* of 1620 speak in nearly identical terms: "The opinion which men cherish of antiquity is altogether idle and scarcely accords with the term. For the old age and increasing years of the world should in reality be considered as antiquity . . . [the ancients], with respect to ourselves, are ancient and elder, with respect to the world modern and younger," in Bacon's phrasing, from which he then draws the pointed lesson "Reverence for antiquity has been a retarding force in science."[6]

It is noteworthy that this figure—"it is we who are truly the Ancients"— occurs in Bruno, Bacon, and other late sixteenth- and early seventeenth-century writers, but *not* in Descartes, as one element in a network of complementary and mutually qualifying tropes, some inherited from the older "Ancients," others of more recent coinage.

Other prominent elements in this network include: *Veritas filia temporis* (Truth is the daughter of time); the motif of dwarfs who stand on shoulders of giants and are thus enabled to see further; and, third, the theme of nature's inexhaustibility—"Nature has certainly not become so sterile that she cannot breed Platos and Aristotles in our own day," as Joachim du Bellay wrote in his defense of the vernacular tongue.[7] Although each of the elements has a complex literary history, what is common to them and to the whole they compose is the twin sense that (1) the modern age is continuous with, and built upon, what preceded it, and (2) this continuity, far from undermining confidence in the quality of present and future achievements, bolsters the moderns' sense that they can see and go further than their predecessors, the ancients in particular.

An *argumentum ex silentio* usually lacks convincing amplitude; in Descartes' case, however, it is at least striking that, to the best of my knowledge, he nowhere cites or alludes to any of these other commonplaces of his literary and cultural "climate," with one ironic exception. Rather than emphasizing continuity with, and decisive improvement upon, the past, Descartes, we might be allowed to infer, eschews any compromising link with the past; modernity comes into being thanks to a radical and unbridgeable break with all that is anterior. This inference is lent support by the second passage preserved by Baillet:

> Dii male perdeant
> Antiquos, mea qui praeripuere mihi.

> Let the Gods cruelly destroy
> the Ancients, who snatched my things
> away from me beforehand.[8]

At the same time as it gives evidence of Descartes' erudition—the verse may well be an original variation on a line cited by St. Jerome in his *Commentary on Ecclesiastes*—this text, with its compressed violence, aims at eliminating any possible continuity with the ancients, despite, or perhaps because of, every trace of evidence that they anticipated what he himself has said or wants to say. The self-instituted cleavage with the ancients, and hence with the past, is complete and critical.

This posture has several crucial implications which I shall mention briefly here and return to again in greater detail.

(1) Descartes refuses to connect his "modernity" with the enterprise we know under the name of the "Renaissance" and which its participants understood both as the advent of a "new age" (*acetas nova*) *and* as the renovation, that is, the renewal and repristination of ancient letters. He spoke to his early mentor Isaac Beeckman and to others with contempt for the so-called *novatores*—men such as Campanella, Bruno and Telesio—who fell short of making a "clean-break" with ancient philosophers.[9] Moreover, in place of the vacuous "new miracles" promised "in all the sciences" by the Renaissance authors on natural magic, Descartes would provide "the true means for resolving all the difficulties" of a science which must, in principle and in execution, be "entirely new."[10] If, as is sometimes said, the Renaissance provided the matrix from which early modern science and philosophy derive, in Descartes' case it acted as midwife to a matricide.

(2) Radical novelty is an essential, not merely a parenthetical, concern throughout Descartes' works, both public and private. To be wholly original means, for him, to originate the whole, the comprehensive totality of knowledge, from foundations of one's own finding or devising. Radical or complete novelty in the foundations is the necessary and,

perhaps, the sufficient condition for attaining to universality in the finished structure. Beeckman reports that when Descartes brought him specimens of his new algebra in 1628, the latter claimed that he had not only reached by means of it the perfected science of geometry but was also enabled to attain to "all human cognition."[11] The title originally intended for the eventual *Discourse on Method* was: "The Project of a Universal Science which Can Elevate Our Nature to its Highest Degree of Perfection."[12]

Universality and the "elevation of our nature to its highest perfection" go hand in hand and equally depend on fundamental originality. In the last of the texts furnished by Baillet, Descartes compares his situation to that of a writer who must of necessity take his words from an alphabet and a lexicon accessible to all; nonetheless, if the things he says are "so coherent and connected with one another that some are the consequences of others," he can no more be accused of borrowing his thoughts from others than a writer would be reproached for stealing his words from a dictionary.[13]

(3) Accordingly, Descartes rejects not only more remote "predecessors" who might seem to have anticipated him—the case of St. Augustine is the best known; he also repudiates, sometimes tacitly, sometimes stridently, his contemporaries, along with others standing close to him in time and in intention. He keeps Viète, the founder of algebra, Galileo, Beeckman, and Fermat at a distance, often with a disingenuous denial that he has seen or learned from their works. To originate requires that the space of one's own thinking be cleared of all foreign matter. Descartes sees himself, and desires that he be seen, as the very first "self-made man."

(4) Finally, it is in deep harmony with this understanding of modernity as radical originality that Descartes' project is *intended* to be his work and his alone. As Leibniz wrote, with insightful truculence, "Descartes had the vanity of wishing to be a solipsist."[14] The undertaking of an "entirely new science" is, at least at its inception, solipsistic or egocentric, not collective or social. For the "new age" celebrated by Renaissance and early seventeenth-century writers as a shared or collaborative production, Descartes would substitute his singular deeds accomplished in his own time—a time, that is, of his own making.

No one, other than Descartes himself, captured his intention better than his admirer Constantin Huygens in a series of poems and epitaphs. These lines are among them:

> Natus vocari, non Renatus debuit,
> Natura nasci vidit hactenus nihil
> Ex que renato surgeret Cartesius.

"New-born," not "re-born," he ought to be called.
Up till now nature has not seen anything
From which Cartesius could arise "re-born."[15]

Descartes gave birth to himself.

Why, then, does Descartes choose to make his public debut with an anonymous record of the events and lessons of his career? If, as Henri Gouhier wrote, "The first word of Descartes' philosophy, like the last, is a 'No' said to history," how are we to understand his purposes in the *Discourse on Method*, the inaugural document of the Cartesian Revolution?[16]

Unlike Bacon who wrote a *History of Henry VII* to exemplify what he called the "Georgics of the Mind" and who otherwise wrote extensively on the subject of civil history, unlike Hobbes, who made his debut with a translation of Thucydides, and to say nothing of their common godfather, Machiavelli, Descartes never wrote a study of any historical personage or period. To be sure, early in his career he borrowed the name "Polybius" as a pseudonym for a projected "Thesaurus Mathematicus," playfully inflating the historian's epithet from "Megalopolitanus" to "Cosmopolitanus," perhaps to suggest that only a mathematician can answer to the Stoic ideal of making the *cosmos* his *polis*.[17] For the rest, Descartes uses the terms *histoire* and *historia* sporadically, sometimes in the Baconian sense of a compendious catalogue of natural phenomena devoid of "any explanations or hypotheses," sometimes in the sense of already-written historical narratives. This latter usage is, on occasion, ample enough to embrace the writings of past philosophers and mathematicians as well as those of historians in the strict sense.

A key example occurs in the third rule of the *Rules for the Direction of the Mind*. Having begun by recommending that one read the "works of the ancients," Descartes proceeds to undermine his own recommendation by pointing out the unavoidable dangers and irremediable defects of such studies. Not even agreement among all authors on a particular teaching would be a sufficient condition for certainty since "we shall never become mathematicians, for example, even though we hold in our memory all the demonstrations made by others, unless we are fitted by our inborn wit, our *ingenium*, to resolve problems of any kind whatsoever; nor will we become philosophers if we have read all the arguments of Plato and Aristotle but are unable to make a stable judgment concerning matters proposed to us. In that way we shall appear to have learned, not sciences, but histories."[18]

Two points should be noted in this passage. First, the only mode of response to the works of others is, according to Descartes, passive, not active or dialogic. There can be no useful "fusion of horizons," for

Descartes. Second, committing others' works and demonstrations to memory stands opposed to generating demonstrations, not only by oneself, but also according to a method applicable to *any* problem whatever. Histories, unlike science, give us no access to the intuitive principles or elements from which self-generated deductions arise.

This same contrast between inventing sciences oneself and studying what others have said and written is at work in the dialogue *Recherche de la vérité*. Study merely furnishes "simple instances of knowledge [*connaissances*] which are acquired without any discourse of reason, such as languages, history, geography and, in general, everything that depends on experience alone."[19] Whatever "comes to our intellect from elsewhere," rather than from its "reflexive contemplation of itself," to use the language of Rule 12, necessarily lacks utility, that is, it fails to provoke any discourse of reason; it fails to produce any new inventions.

However, it is only in the first part of the *Discourse* that Descartes turns his full attention to the status of historical studies. His remarks have to be examined with greater care than I can devote to them here; I shall have to content myself with sketching the conclusions to which I think a close reading will lead before turning to some of the relevant evidence.

(1) Descartes does *not* repudiate the study of history in the post-Cartesian sense according to which events themselves are "historical," that is, are taken to make up an ontological category *sui generis*—history in the sense of *die Geschichte* in the singular, history which may be said to flow, move, and even come to an end. Rather, he rejects historical writings or narratives; more precisely and importantly, he rejects the written histories belonging to the exemplarist or pragmatic tradition initiated by the Romans and revivified by the Humanists of the fifteenth and sixteenth centuries. What is at issue for Descartes is not the "historical sense" cultivated by the Romantics and Historicists, but the reliability and personal utility of reading books of history.

(2) This rejection of exemplary historiography is not an abstract negation; instead, Descartes, in subverting the genre as it was known to him from his early studies and later reading, simultaneously appropriates its conventions and purposes for himself. The anonymous narrative of a life in Parts 1 and 2 of the *Discourse* is itself an exemplary history; the earlier impression that Descartes never wrote about a "historical" personage proves faulty, since this text corresponds to the "Histoire de mon esprit" demanded of him by his correspondent Guez de Balzac.[20]

(3) This account of Descartes' *Aufhebung* of Humanistic histories is still inadequate; close attention has to be paid to his deliberate confounding of history and fable. The fabulous history he chooses to tell portrays a fictive self, a self-produced *persona*, not at the expense of factual accuracy, but rather by way of showing what it is to become on *unparalleled exemplum*.

Ingredient in this self-display are several of the essential features of the Cartesian project: liberation not simply from tradition and overbearing authority, but from opinions as such; method as an art of inventing new and certain truths each of which is subject to the control of a self-transparent mind; suppression (at least temporarily) of the *polis* in favor of individual autarchy, which in turn is meant to yield a fundamentally new model of the political and the moral domain. Taken together, these features amount to the program of commanding time so as to conquer nature. If Descartes lacked the "historical sense" it is because he was busy inventing history in what becomes at least one of its primary modern senses: history, namely, as something of man's making and, thus, potentially at his disposal. No less than for Vico, for Descartes, too, in human affairs *verum et factum convertuntur*.

Let me try to spell out and to sustain these claims a bit more fully.

Descartes' inaugural narrative traces the course of his various studies and brings out the grounds for his disappointment and disillusionment with each of the disciplines, with the exception of mathematics. The reading of ancient histories (and fables) is of qualified utility; that is, in the final account, it is futile for two reasons: too much "exploration" of what is not one's own leads to self-estrangement (I shall return to this point), and, secondly, even the most faithful histories, those that neither "change nor augment the value of things" leave out the meanest and least illustrious circumstances so as to make the things they narrate "more worthy of being read." This gives the matters retained in such histories a false appearance and misleads those "who regulate their own moral habits by examples drawn from them. They are subject to falling into the extravagances of the paladins of our romances and to conceive designs which surpass their powers."[21] Cartesian aspiration should fit self-guaranteeing powers.

It is, above all, the phrase "those who regulate their moral habits by examples drawn from these histories" that identifies the tradition to which Descartes means to refer. The canons of pragmatic or exemplary history were established by Polybius, Cicero, and Quintilian and reinstalled with the Renaissance recovery and elaboration of their works. Cicero's *De oratore* contains what became the governing maxim of this genre: *Historia Magistra vitae* (History is the mistress and guide of life).[22] In the spirit of that maxim, historical writing, whether a part of rhetoric, as for Cicero, or an independent *ars*, as for many of the Cinquecento Humanists, was concerned not with "simply showing how things actually were" in the past, but with the eloquent and decorous exhibition of deeds and speeches in which matters of nobility and baseness, political success and ill-fortune might be brought to light for the reader's contemplation.

Exemplary history was thus inseparable from moral and political edifica-
tion and presupposed the *recurrent* character of those civic affairs and
challenges to which men respond nobly or meanly. No *exemplum* is without
parallels. According to Polybius, "Plato tells us that human affairs will
then go well when either philosophers become kings or kings study
philosophy, and I would say it will be well with history either when men
of practical action [*pragmatikoi*] undertake to write history . . . or regard
training in practical affairs as necessary for writing history."[23] According
to Jean Bodin, one of the most distinguished premodern advocates of
exemplary history, the prime virtue of "human history—depicting the
deeds of man while leading his life in the midst of society—[is] prudence,
which distinguishes the base from the honorable."[24]

Descartes challenges this genre on the grounds that the incompleteness of historical accounts, even when it is not the result of malice or
vanity, obscures the genuine ratio between the high and the low, the
noble and the base in human affairs. This reminds us of the comment
he makes soon afterwards about pagan writings on moral habits; comparable to "very proud and magnificent palaces built on sand and mud . . .
They raise the virtues too high and make them appear estimable beyond
all the things in the world."[25] The historical *exempla* of the moral virtues
will trap the unwary and lead them to "conceive designs beyond their
forces." Cartesian ingenuity, regulated by methodical precept, not by
moral *exempla*, will guard against the extravagances of the heroes of
romances, such as the ingenious hidalgo Don Quixote, even while matching its forces to deeds undreamed of by those fictive heroes.

Descartes substitutes the tale of his own life for the exemplary histories
and fables of the past. This substitution makes the frequently disputed
question of the "historicity" or factual accuracy of Descartes' account
otiose to the degree that the conventions of his narrative are, in the main,
those of the genre he both subverts and takes over for his own ends.
Nevertheless, one class of deliberate omissions is highly revealing: Descartes eliminates any mention of his teachers Beeckman and Kepler, who
did not simply pass on to him the traditional disciplines of La Flèche but
made him literate, so to speak, in the "great book of Nature." These
omissions confirm our earlier impression that Descartes, or the *persona*
revealed in the *Discourse,* means to appear self-generated.

At all events, Descartes had been planning to appropriate exemplary
history for some time before composing the final text of the *Discourse.* A
letter from Balzac in 1628 reminds him of his promise to send along a
work apparently entitled "L'Histoire de mon esprit." Balzac continues:
"It will be a pleasure to read your diverse adventures in the mid- and
high-region of the air; to consider your deeds of prowess against the

Giants of the School, the path that you have taken, the progress that you made into the truth of things, etc."[26] In his exemplary history, Descartes will thus take on the shape of "the paladins of our romances," with the difference, once again, that he will battle victoriously against real Giants, not tilt at windmills.

This comparison, as suggested by Balzac's comments, brings us to the next significant feature of Descartes' *Aufhebung* of ancient and Humanistic histories: He sets forth the *Discourse* as "a history or, if you prefer, a fable." This is not so much a signal to the reader not to expect from Descartes complete fidelity in reporting of the facts of his life, as it is an ironically subversive blow aimed at one of the foundations of traditional histories. These drew their legitimately persuasive force, at least in part, from the preservation of the distinction between historical and mythical accounts. According to Cicero, the three species of narration over which the rhetorical art (except in legal cases) extends are *fabula*, which is neither true nor similar to the true; *historia*, which gives an account of a deed done remote from the memory of our day; and *argumentum*, defined as "ficta res, quae tamen potuit," namely, a "feigned thing which nonetheless could happen."[27] History must be purged of fables and directed upon either the true or the verisimilar. "The rhetorician," adds Quintilian, "who makes his start with history, will be stronger, the more truthful he is."[28]

Why does Descartes blur or transgress the line dividing history from fable? One reason, in all likelihood, is the challenge to the authority of authors, or, for that matter, to the reliability of "what others think," mounted and sustained in the *Regulae*. In dismantling the authoritative teachings of schools, Descartes will not *straightforwardly* install his own "authority." His relation to the reader necessarily blends allurement and detachment. To imitate Descartes' example one will need to practice and apply it, not memorize or passively receive it.

But there is, I think, a still deeper reason for Descartes' tactic: to write a fabulous history of one's life is to create or recreate that life on a pattern unlimited by the constraints of "fact" narrowly construed. (Strictly speaking, of course, no one is to know that René Descartes is the author of the *Discourse*.) In other words, the gesture of fictive autobiography can unfold in "imaginary spaces" like those in which God is made to make a wholly new world in the fable of *Le Monde*. One can, so to speak, apply the "method of annihilation" to the history of one's life and then recommence it *ab ovo*, this time with rational surety and self-transparency; Descartes can, in retrospect, see his knowledge advancing *by degrees*, just as the third precept of method will demand.

The gestures of fictive self-creation are of a piece with the epistemic gestures of Cartesian thinking in general; their ensemble points us close to the heart of the Cartesian "attitude."

History, in the following crucial respect, is like pre-Cartesian philosophy (and theology): it is a matter of *opinion*, more exactly, of *many, diverse opinions*. "Opinion" is one of the most pervasive words in the first three parts of *The Discourse* (it appears twenty-four times); it designates the antithesis to Cartesian *science*, for we are passively *exposed*, and can succumb, to opinions, while science is always a matter of the mind's active invention. We can usefully recall the argument of Rule 3 in the *Regulae:* There is no way to make active, spontaneous contact with what others think or say (see Rule 12; "quidquid ex aliis audimus"). (". . . whatever we hear from others.") Simply the fact that we gain or inherit our opinions by way of what others think, write, and say is sufficient, in Descartes' eyes, to disqualify them *en bloc* from the body of genuine science.

History, then, in the narrower sense of *ars historica*, as well as tradition taken more amply to include past *and* present opinions, is or ought to be, *alien* to the singular mind's endeavor to take its bearings by itself alone.

This, more fundamental, basis for the indictment of traditional studies carries several implications which I can only mention here.

For the Aristotelian "tradition" which Descartes means to subvert, *dialectic* is the art of faculty of passing from reputable opinions—*ta endoxa*—to the first principles of all the sciences, theoretical, and, one presumes, practical alike. Needless to say, for Aristotle the *endoxa* are no more "probable opinions" than the *endoxoi*—men who stand in good repute in the city—are "probable persons." Rather, they are the record, as it were, of the looks beings initially turn upon sensing and speaking men. They need to be sorted through and, ultimately, refined or purged; nonetheless, *ta phainomena* are, in the main, equivalent to the trustworthy *legomena*, the things responsible men say about the *phaimonea*.[29]

Cartesian "dialectic," if I can use that term, builds no bridges from *doxai* (opinions) or *ta endoxa* (reputable opinions) to the governing principles of a demonstrative science. Descartes discredits the *endoxoi* of the remote past and what was recently the present, both because of the disparate multiplicity of their opinions—"considering how many diverse opinions there can be touching on one and the same subject . . . I deemed almost nearly false all that was only probable," he says in the *Discourse*[30]—and because our sights are inevitably lowered from *scientia* to *memoria* when what others say or have said is passively engaged. Furthermore, Descartes makes it plain that an active or *dialogic* engagement with others' opinions is impossible, as well as impractical.

The self is exposed to opinions from birth; in other words, the locus or medium of opinions and the traditions they sustain is the *body*. The world of shared and preserved speech must be suspended if the singular mind is going to assume autonomous command over its thoughts—if, that is, its motions are to be directed along straight lines instead of being left to turn fitfully in the disoriented circles of others' speech.

Consequently, since one's body cannot be literally, but only fictively, put out of play, steps must be taken to make it and the mind impregnable to the solicitations of opinion. One could, for example, having witnessed the coronation of an emperor, retire from the *polis* to a *poêle*, there to rid oneself, definitively, of all prior opinions.

Having mentioned the suspect judgments of "our friends" in the first *Discourse,* Discartes does not use the word "friends" again until the sixth *Discourse,* once again with marked reserve as to the dependability of their judgments.

The Greek term *syneidesis* originally designates conspiratorial knowledge—one is politically "in the know" together with others; in Aristotle's *Ethics,* friendship on the basis of excellence is uniquely the medium in which self-knowing or self-consciousness can occur. Borrowing an image from Plato's *First Alcibiades,* Aristotle, or one of his more astute pupils, argues in the *Magna Moralia* that "just as when we want to see our face we do so by looking into a mirror, so, too, when we want to know ourselves we can obtain that knowledge by looking at our friend. For the friend is, as we say, a second self."[31] At the end of Book 9 of the *Nicomachean Ethics* Aristotle argues that "Whatever it is to be for various men or for the sake of which they choose life, this they wish to pursue in common with their friends." Among these common pursuits Aristotle lists "philosophizing together."[32]

Philia, for Aristotle, in addition to being the enabling occasion for self-consciousness, is the protoform of the city. For Descartes, the mind is its own friend exclusively; it conspires with itself, at least when it is bent on inaugurating science.

To expunge opinions is to detach oneself from the "city," much more resolutely than a Socrates ever did. This inference is borne out by two further aspects of Descartes' self-presentation in Parts 1 and 2 of the *Discourse.* The first half of his critique of the reading of fables and history turns on a complex comparison with journeying; Descartes, in effect, offers us a topology of reading the thoughts of others. It is "worthwhile to know something of the manners of other peoples, so as to judge our own manners more judiciously." This can, however, be overdone: "when one spends too much time in travelling one ends up a stranger in one's own land; and when one is too curious about the things practised in past centuries one ordinarily becomes quite ignorant of the things practised in this age."[33] Excessive attention to what is foreign, in space or in time, leads to estrangement from the manners and practices of one's own land.

The remedy for estrangement, however, as the recommendations of the *morale par provision* in Part 3 show (not only, I submit, a provisional or temporary morality, but also one that equips Descartes with "provisions" in the military sense), is not to reappropriate what is native, nor to strike a balance between the alien and the native, but to see each as equally contingent and opinionated. Neither the indigenous nor the foreign is the product of self-generated ideas. Accordingly, the fictive self of the *Discourse* keeps its distance from both alike; to secure an Archimedean point from which to initiate a wholly new science that self has to remain *atopos*, outlandish. Its natural home is not the *agora*, but the mechanic's private workshop.

Cartesian self-exhibition in the *Discourse* has a second, even more revealing, aspect. The manifold and, hence, intrinsically discrepant, character of traditions and opinions is a sign of the fact that many hands or minds have been at work in their genesis. In his *poêle*, the now-solitary self recollects that "often there is not as much perfection in works composed of several pieces or made by the hand of different masters, as there is in those on which a single hand has worked."[34]

Descartes sees the ancient *ontological* problem of the One-and-the-Many in *anthropological* terms: Cartesian science, like the Cartesian self, ought to be "all of one piece." The discursive manyness of clear and distinct ideas, or of simple natures, rests on a more primordial oneness and homogeneity conferred by the unity of *scientia humana* and the regulated uniformity of the mind's motions. Cartesian science, like the Cartesian self, must be liberated from opinions and, thus, from the manifold contingencies of history, so that it can become *inventive*.

From the first, Cartesian science is conceived as an *ars inveniendi*, not as an *ars demonstrandi* or *judicandi*: syllogistic or synthetic demonstrations presuppose that middle terms have already been found; Cartesian analysis discovers those "means." But Cartesian science is not only "inventive" in this sense of discovering something new rather than reordering what is already supposed to be known; beyond this, the art of invention is an art of production or fabrication in accordance with rule-governed procedures. The enactments of this art testify to the mind's self-control.

The third maxim of the *morale par provision* prescribes that "There is nothing entirely in our power except our thoughts." Unlike the Stoic in Hegel's phenomenological portrait, the Cartesian self is *not* closed off from, and enslaved to, the world by force of its inward thoughts; to the contrary, Descartes construes thoughts or "ideas" on the model or artisanal or demiurgic designs. In his reply to the third set of objections, Descartes defends his choice of the term *idée* on the grounds that "it was already commonly received by the philosophers for signifying the forms of the conceptions of the divine understanding."[35] At the end of his sensitive study of the relationship between the Thomistic and the Cartesian idea of *ideas,* William Carlo concludes that the psychological entity Descartes identifies as an *idea* "was not the principle of knowledge functioning as the principle of knowledge, but the principle of knowledge functioning as the principle of artistic production. Descartes mistook the idea for the concept."[36] This "mistake," I might add, is not the outcome of terminological confusion, but of conscious design, that is, of a consciousness the essence of which is designing. A Cartesian *idea* is a force of production.

These conjectures are, I think, corroborated by Descartes' guileful appropriation of the Scholastic vocabulary of "formal" and "objective reality" especially in the French version of Meditation 3: "we must, in the end, arrive [that is, in pursuing, in reverse, the sequence of antecedent causes of our ideas] at a first *idea,* whose cause is like a pattern or original, in which all the reality or perfection is contained *formaliter* and effectually which is found only objectively or by representation in those ideas [which are produced from it]."[37] Simple ideas give birth to complex sequences of ideas; such sequences, in form, can receive external embodiment in, say, a machine. The functional success of an artifact, its "objective perfection," is proportionate to the eminent and formal reality of its productive source, namely, the inventive, originative ideas in the mind of its maker. Moreover, the products in question bear the marks of their maker—that is, the structures they embody are recognizably the work of the ordered or regulated efficacy of the mind.

The most notable example of this regulated efficacy is Descartes' geometry, as we shall see in section III. For the moment let me simply underscore one motif technically conspicuous there, but pervasively manifest in Descartes' works as the node at which several dominating themes of his "philosophical anthropology" meet and mate. This is the motif of *discretion.* Descartes tells us in the first book of the *Geometry* that the unit-length, by comparison to which all other magnitudes in a problem-complex are to be determined, "can ordinarily be taken *à discrétion.*" "Discretion" here means both arbitrary choice and prudence and in each meaning stands opposed to the artless way of doing things *à tâtons,* "gropingly," or *sans industrie et par hazard,* expressions of opprobrium in Descartes' critical comments on contemporary mathematicians. If we ask how arbitrariness and prudence can be the two faces of a single coin, the answer is supplied by the Cartesian concept of *utility.* The choice of unit-length (where the unit, according to the idiom of the *Rules,* is "presumptive" or "assumed" (*unitatis assumptitiae* in Rule 14), as well as the choice of "principal lines" or axes, is *arbitrary* in the sense that nothing in the nature of the magnitudes or lines in a given problem dictates to us the selection we must make; these same choices are *prudential* in the sense that one is already looking ahead to the fruitfulness of the solutions obtainable once the chosen unit (or pair of axes) has been placed at the head of a sequence of operations through which the "unknowns" (the *quaesita*) are progressively aligned with, and ultimately equated to, the "knowns" (the *data*). Discreet choices take their bearings from the overriding desire that the mind's efforts not be "uselessly consumed" but rather pointed towards the step-wise or continuous augmentation of science (see Rule 4: "nullo mentis conatu inutiliter consumpto, sed gradatim semper augendo scientiam").

Utility as the union of free choice and prudence is by no means confined to mathematical technique narrowly construed; on the contrary, Descartes suggests that the relevance, if not the entire being, of things themselves is gauged by their utility to our proposed tasks (see Rule 6: "res omnes eo sensu quo ad nostrum propositum utiles esse possunt et seq."). This "utilitarian" orientation should not come as a surprise; after all, Descartes emphasizes in the final part of the *Discourse*, as well as in the very first rule, that he aims at reaching "connaissances qui soient fort utiles à la vie." It is only in the light and under the guidance of this utilitarian *telos* that we can prudently map the methodical paths to be followed from simple ideas to the complex executions which they initiate. This same connection is brought out even more readily by the Baconian provenance of Descartes' understanding of *rules* (*regulae/règles*): "that which in contemplation is as the cause [*instar causae*], is in operation as the rule [*instar regulae*]."[38] Prudently regulated discretion is the virtue required for the sake of productive operation.

And yet, the final sense of utility itself still remains enigmatic as long as it is left on its own. Although the *telos* of the new sciences, it stands in need of its own *telos*. When we explore the identity of this second-order *telos*, the wheel turns full circle, returning us to the theme of self-origination made visible in Descartes' autobiographical history or fable.

In a letter to Hogelande dating from 1640 Descartes made reference once again to the contrast between history and science:

> By 'history' I understand all that has already been invented and is contained in books. By 'science' I understand skill at resolving all questions and in inventing by one's own industry everything in that science that can be invented by human ingenuity. Whoever has this science does not desire much else foreign to it and indeed is quite properly called *autarchēs*—self-sufficient.[39]

To be autarchic, one's own commanding source and sole resource, is both to be a principle oneself—that is, the generative or productive source of all that can be invented by science—*and* to be the principle *of* oneself—that is, to have one's life and work fully under command. Later, in the *Passions of the Soul,* Descartes gives this autarchy the name *générosité*—the key and coping-stone of all the other virtues.

Generosity, the more-than-nominal substitute for magnanimity or *megalopsychia* (compare *Passions of the Soul,* article 161), shows itself as a "firm and constant resolve to make good use of the free disposition of our wills, that is to say, never to lack the will to undertake and to execute all the things a man will judge to be the best" (art. 153). The criterion of the best, however, seems to be for Descartes the most feasible and, hence, the most amenable to our self-inculcated and self-developed powers.

Three concluding reflections seem aptly to follow from this.

(1) The desirable or the good must be bounded so as to fit easily within the compass of the feasible. If, for Descartes, infinity of will is the single *analogatum* of the human and the divine (see *A.-T.,* t. 2, p. 628), nevertheless this infinity is not to be recklessly or promiscuously exploited. Its proper use is most visibly displayed in the programmatically anticipated "infinity of artifices" through which the new sciences are to prove their credentials. This discreet use of the will's infinity is, one could say, the Cartesian counterpart to Machiavelli's insistence "to go straight to the effectual truth of the matter [*andare direto alla verità effettuale della cosa*]."[40]

(2) As we have had more than one occasion to observe, Cartesian self-making is simultaneously the repudiation of otherness. Opinions, histories, and teachings, as well as presystematic or uninterpreted perceptual experiences share alike in the quality of coming to one "from elsewhere" (*aliunde*) than oneself. It appears an inevitable consequence of this that utility, too, must be measured, not by the expressed or manifest needs of others, but by the dexterities of invention which are under one's own control. Utility as much as certainty is, at bottom, always a first-person singular affair. Accordingly, the desire for autarchy and the self-referring virtue of generosity combine to fix the true Archimedean point from which Descartes can move the world.

(3) We should now be in a better position to analyze Descartes' intention in producing for the public an *anonymous* autobiography which is deliberately poised between the traditionally antagonistic genres of history and fable. The identity of the author of the *Discourse* is the unknown term of a problem set to the reader. By resolving the complex phenomenon of the author's career and ideas into its simplest terms, thence to construct the image at its root, the reader should succeed not only in unmasking the true visage of Descartes but should also, and more fruitfully, learn the "utility of cultivating [his] mind by exercising it," to borrow the words of the *Geometry.*[41] When one has reflected on what it took to solve the problem of Descartes, one can discard the Cartesian *exemplum* and start becoming an *original.*

III   Mathesis and Mathematics in Descartes' *Geometry*

i   Prelude

One traditional, and initially promising, way of penetrating behind Descartes' masks has been to explore the nature of Cartesian *method*. After all, Descartes first came to notice in Parisian intellectual circles by claiming to have a "natural method" for "refuting the sophisms of the learned and protecting himself against ever committing an error."[42] It was Descartes, too, whose *Discourse on Method,* published in 1637, raised to fever-pitch the mania for method characteristic of the late sixteenth and early seventeenth centuries. (As his contemporary Pascal wrote: "The whole world is in search of a method of reasoning that does not go astray.")[43] Lest it be thought that method is insubstantial "window-dressing" for the less glamorous, but more solid stock of Cartesian arguments and positions, we should keep in mind throughout what follows that, once again, the title he had originally planned to give the *Discourse* was "The Project of a Universal Science that Can Elevate Our Nature to Its Highest Degree of Perfection."

Nevertheless, Descartes proves as elusive in this realm as he does in others. In the *Discourse,* he told Mersenne, he did not intend to teach his method, but only to talk about it. Readers of this work might be relieved to hear this assessment, since the four rules of method stated there are enormously compressed and full of cryptic allusions for which no key is given.[44]

Descartes' contemporary adherents were equally baffled as to the true nature of *the* Cartesian method. Clauberg labelled the *Discourse* an "exoteric and popular work" from which no serious teaching could be drawn, while Daniel Lipstorp, the most talented of Descartes' early disciples, asserted that the true method was to be found in *Geometry,* since geometry is "the mother and source of all the liberal arts." Leibniz, on the other hand, denied that there *was* a Cartesian method at all.[45]

This ambiguity suggests that we might do well to examine in detail one of the three "samples" (*échantillons*) of method Descartes appended to the *Discourse,* rather than to begin by asking after his method in general. The *Geometry* seems a good choice for a variety of reasons, some historical, some reaching deep into the body of Descartes' work.

(1) The *Geometry* undoubtedly contains what came to be appreciated as Descartes' major technical achievement in science. In John Stuart Mill's words: "Analytical Geometry, far more than any of his metaphysical speculations, immortalized the name of Descartes and constitutes the greatest single step ever made in the progress of the exact sciences."[46]

Whether or not Mill was right to set apart geometry and metaphysics in this way, it is surely the case that the history of mathematics and physical science in the seventeenth century and, thus, the formation of the distinctively modern understanding of the world owe an incalculable debt to the initiatives Descartes took in his mathematics.

(2) Furthermore, the goal to which these initiatives was meant to lead was not, for Descartes himself, confined to technical mathematics in a narrow sense. In one of the most famous sentences anywhere in his work, Descartes declares to Mersenne in 1638, "The whole of my physics is nothing other than Geometry." While Descartes distinguishes here between "abstract geometry" and "another sort of geometry which proposes to explicate the phenomena of nature," the former and the latter versions of geometry enter into the most intimate complicity.[47] This is the spirit that informs Descartes' epitaph, perhaps of his own choosing, "Putting together the mysteries of nature with the laws of mathematics, he dared to hope to be able to unlock the secrets of both with the same key."[48]

(3) Moreover, Descartes gave the *Geometry* a distinctive status in respect to the first two "Essays" following the *Discourse*, the *Dioptrics* and *Meteorology*: "In the [*Dioptrics* and *Meteorology*] I tried merely to persuade others that my method is better than the ordinary one; but I claim that I have demonstrated its superiority through my geometry."[49]

(4) Finally, the *Geometry* is linked by multiple ties to the uncompleted *Rules for the Direction of the Mind*, which would have been Descartes' most expansive statement of his methodology (if not of his philosophy *tout court*). It is by inspecting those ties that we might come eventually to grasp how Descartes' *mathesis universalis* (mentioned by name only in Rule 4) found expression in his actual mathematical practice and to understand why the latter was carried out in the form of a *geometry* and only in that form.

ii   *Reading the* Geometry

However persuasive these and similar indications might be, the philosophically minded are still bound to wonder what actually reading the *Geometry*, or, for that matter, any strictly mathematical text, can do to enhance or supplement our comprehension of strictly philosophical arguments. (D'Alembert, in the preface to the French Encyclopedia, lends his endorsement to the nowadays standard attitude towards this matter: "Descartes can be considered on the one hand as a philosopher, on the other, as a geometer.")[50] Most of those interested in questions of Cartesian philosophy are apt to be indifferent to the detailed technical content of the *Geometry*, contenting themselves instead with ritual acknowledgments of its importance and glib summaries of its achievements. On the other hand, historians of mathematics, in the main, practice a kind of reductionism, gauging the merits of a past work exclusively in terms of its contributions to the subsequent course of mathematical development.

More dismaying still, the philosophical reader curious enough to turn to these historians for an account of Descartes' principal contributions will encounter astounding discrepancies among their interpretations. The following are general characterizations of Descartes' aims and achievement to be found in the works of some of the most adept and respected scholars of this field: "The arithmetization of geometry"; "the geometrization of arithmetic", the "algebraization of geometry"; "the geometrization of algebra."[51] Unless one is willing to treat these merely as insubstantial variations in nomenclature, it is clear that even the technical significance of Descartes' mathematical work remains shrouded in obscurity. To penetrate into its genuine meaning requires, then, a suspension of prejudgments or, more positively, an attempt to accommodate ourselves to the text itself, searching for what I would like to call its "epistemic signature," that is, the way or ways certain acts and forms of knowing become incarnate in the contents and results that materially define the work. This signature is as much a matter of rhetoric as it is of logic and syntax, if not more so; it constitutes a mathematical work's "style" in the broadest sense of that word.

Descartes himself provides another and possibly better rubric for the goal of this search. Communicating to Mersenne his reaction to Desargues' sketch of projective geometry—the only work by a contemporary mathematician that commanded his respect—he writes that Desargues' reasoning "seems to be drawn from what I am in the habit of calling the metaphysics of geometry, which is a science I have not noticed anyone else using, except Archimedes. For my part, I always make use of it in order to judge in general what things can be found and in what places I ought to look for them."[52]

Descartes is referring primarily to Archimedes' frequent use of so-called *diorismoi*—statements of the conditions that must obtain if a certain construction is to be carried through—but the resonances of his phrase "the metaphysics of geometry" go beyond this. In the case of Descartes' own work, it can be taken to mean the implicit and, more rarely, explicit commitments that define the range of mathematical possibility and intelligibility. For example, it is the "metaphysics of geometry," as I am construing it, that is in play when Descartes speaks of "the curves that ought to be admitted into geometry." Thus we can see in advance of any further discussion that the "epistemic signature" of a mathematical work will also spell out ontological limits; that is, only what satisfies the conditions of possibility and intelligibility governing the discursive structure of that work can be promoted to the rank of a geometrical entity in good standing.

Thus, it is with a view to the underlying "metaphysics of geometry" that I turn now to Descartes' *Geometry*. I shall begin with a superficial inspection of its compositional structure, turning afterward to the very deep questions this structure raises both for Descartes' understanding of mathematics and for his understanding of the act or process of learning—of *mathesis*—embodied paradigmatically in mathematics.

iii   The Structure of the Geometry

The *Geometry* is composed along two axes, if I may use that term, the problematic and the systematic. In the end, neither the terminology nor the figure is wholly satisfactory, since these "axes" do not intersect once only, but continually interpenetrate.

Let me begin, then, with the problematic axis. The immediate occasion of the *Geometry* was a single, notorious problem that had defied the efforts of the very best ancient and early modern geometers—the so-called "Pappus-locus problem." It was first transmitted to Descartes in 1631 by a correspondent named Golius who seems to have been skeptical of the reports that the former had a marvelous new method for solving such problems. Descartes eagerly picked up the gauntlet Golius threw down and, as he later told Mersenne, worked for six weeks until he reached a solution. In his own eyes, his success placed him beyond both the ancients and his contemporaries alike.[53]

It is crucial to emphasize that Descartes does not treat his solution of the Pappus-problem as an isolated display of his problem-solving skills; on the contrary, as we shall see, in presenting his solution to this one particular problem, Descartes wants to show the reader how to go about solving *all* geometrical problems, or more radically still, how all geometrical problems can be reduced to a single type of problem. To anticipate even further, the route Descartes takes in solving the Pappus-problem furnishes him with the key to the systematic classification of "all the curves that ought to be received into Geometry," as well as of their corresponding equations. In short, the Pappus-problem is the matrix of the *Geometry* as a whole. We would do well, then, to take a careful look at this problem and at Descartes' way of solving it.

The "Pappus-problem" is, as I have said, a locus-problem. (See Figure 1 for an illustration of a simple instance of a locus-problem.) Unfortunately, nearly all of the ancient works on *loci* have disappeared; the surviving discussions of what a *locus* is are often of an occasional character, and the resulting classifications are very far from being transparent. Proclus, in his commentary on Book 1 of Euclid, gives the following definition of a locus, a *topos:* "a locus is the position of a line or a surface producing one and the same property [*symptōma*]."[54] Proclus offers this definition in connection with Proposition 35 of Book 1, that parallelograms on the same base and within the same parallels are equal to one another. What is "topologically" invariant here is the *area* of the parallelograms; the base and the parallel lines fix the parallelograms in position so that their area is always the same. Three observations are in place at this point.

(1) Although Proclus says that this theorem belongs to the "paradoxical *topos*," that is, to a collection of paradoxical locus-theorems, it is not at all clear that the locus-problems we meet with in Apollonius and Pappus are meant to be similarly paradoxical in nature.

(2) The general meaning of *topos* or locus is not transparent in Proclus' definition, or in the classification given by Pappus; for instance, what is meant by saying that "a point is the locus of a point"? (A definition, by the way, that Leibniz repeats, with only slight modification, in his "Metaphysical Principles of Mathematics.")[55] It would, for example, be of great interest to know what relation, if any, the geometrical theory of *loci* bears to Aristotle's concept of enclosing *place* (*topos*). In any case, Proclus goes on to cite the opinion of the Stoic Chrysippus that loci-theorems are like the Platonic forms, since "just as the ideas encompass the genesis of an indefinite number of particulars within definite limits, so, too, in these theorems an indefinite number of cases is encompassed in definite loci."[56] I must leave it an open question whether the mathematical treatment of *loci* owes anything to either of these philosophical traditions.

(3) It is clear from Proclus' example of parallelograms equal in area, as well as from cases that yield the Pappus-problem itself, that locus-properties were not taken by the ancients as *defining* the essential nature of certain kinds of lines, figures or solids. Apollonius in the preface to his "Conic Sections" tells us that Book 1 is devoted to the genesis of the three sections and the opposite branches and to their *archika symptōmata*, their chief or governing properties.[57] The latter are what nowadays are called "planimetric properties," such as the fact that, in a parabola, the square on any ordinate drawn to the diameter is equal in area to the rectangle formed from the *latus rectum* (or parameter) and the segment of the diameter towards the vertex cut off by the ordinate (the abscissa). The locus-properties of the three sections, on contrast, are *not* included in this primary group.

The variations among the conic sections are accounted for by the different relations the square on any ordinate can have to the associated rectangular figures. In the particular cases of the ellipse and the hyperbola, Apollonius brings into evidence a special form or look (*eidos*), corresponding to the defect or excess of the square on the ordinate in comparison to the area contained by the abscissa and the parameter.[58]

That is, for Apollonius, and perhaps for Greek mathematicians generally, the locus-properties are not constitutive of the essence of those sections; although the locus-properties follow necessarily from the nature of the sections, they are secondary in respect to the mode of genesis and to the primary "symptoms" through which the sections display their identity. Descartes, it will soon become apparent, takes a rather different tack.

What Apollonius shows at the end of Book 3 of the *Conics* is that each conic section has a certain three- and four-line locus property, namely, that the area contained by three or four lines drawn from a given point on the section to lines belonging to the section have a constant ratio to one another (see Figures 2 and 3). What Pappus, in Book 7 of his *Mathematical Collection,* and following him, Descartes, are interested in is the *converse* of this: Given three, four, or even more lines fixed in position and three, four or more lines drawn from a point at given angles to the first set of lines so that the areas they contain have a constant ratio, one must show that the point from which they are drawn lies on one of the three conic sections or on a more complex locus (see Figure 4). For Descartes, the conic sections as well as these other curves are essentially *loci,* that is, collections of an infinite number of points, all of which satisfy the "equation" extracted from the conditions of the problem. They are analogous to "point-sets," in modern idiom.[59] Secondly, what challenged Descartes most deeply were two remarks by Pappus, first that when more than four lines are given in the problem the resulting loci "are not known up to the present time but are merely called "lines" (*grammai*) or *linear loci* (Descartes calls them "*supersolid* loci"); and second, when more than six lines are given, the figures contained by these—not being either plane figures or solids—are incomprehensible, since they would be of more than three dimensions. (Pappus then adds, prophetically, as we shall see, that these higher-order problems can be handled by means of "compounding" ratios.)[60]

This passage establishes the twin tasks Descartes will discharge in the *Geometry:* first, to determine with exactitude the nature of the curves generated by the linear or supersolid locus problems and then to extend the method used in doing so to any number of lines whatever, without being inhibited by restrictions on the "dimensions" of the products of the lines involved.

Descartes' *tour de force* is to take two of the lines specified in the problem, one whose length is known, the other whose length is unknown, and then to relate all the other lines in the problem, however many there are, to these two "principal lines," as he calls them, which are, by the way, an ordinate and the related segment of the diameter towards the vertex in Apollonius; this is the origin, needless to say, of the so-called Cartesian coordinates, although Descartes' "principal lines" need not be perpendicular to one another.

The rest of this story is generally familiar. Descartes shows us how to write "equations" for each of the unknown lines so that its length can be determined by the "roots" of these equations. Even more remarkably, Descartes discovers a strict correlation between the number of lines involved in the problem, the degree of the equation of the curve on which the points lie (where the degree of an equation is determined by highest exponent that occurs in it), and the degree of the simplest curve that can be used in actually constructing the locus defined by the corresponding equation.[61]

For example: If there are three, four, or five (non-parallel) lines in the given problem, we can express the value of the unknown line in a quadratic equation (one of the second degree); the locus, all of whose points satisfy the equation, is of the first class (which comprises circles, parabolas, hyperbolas, and ellipses). Furthermore, the locus can be constructed by elementary means, using only a ruler and compass. If there are six, seven, eight, or nine lines in the problem, then the equation will be of the third or fourth degree (Descartes groups these together into what he calls the second genre or class) and the locus will belong to the second class of curves. This locus can be constructed by means of the conic sections—hence, by curves of a simpler class. For ten, eleven, twelve, or thirteen lines, the equation is in the third genre, the locus is in the third class, and we can construct the locus by means of curves of the second class. And so on. Problems, equations, and the curves required to *construct* problems—that is, actually to draw the required locus—thus come together in an orderly array, a harmonious sequence that passes step by step from the most simple to the more and more complex, *ad infinitum*. Accordingly, we can classify curves and equations by their degrees and produce the relevant constructions in a fully determinate way—proceeding "as by degrees," as Descartes recommended in veiled fashion in the third rule of the *Discourse*.

It is especially worth noting that Descartes can reduce any problem of a given class—solid problems, for instance—to a single type of problem—the finding of two mean proportionals, which, in turn, is nothing other than the classic problem of doubling the volume of a given cube.[62] Or, alternatively, any solid problem can be reduced to the one problem of trisecting a given angle. (*Which* reduction depends on the problem at hand.) In both cases the solution can be reached by constructing the curve resulting from the intersection of a circle and a parabola, that is, by the very same means employed in solving the Pappus-problem for the case of solid loci. This pattern of reducing all problems to a single type of problem, with its associated constructive solution, can itself be generalized, so that Descartes can write: "It is only necessary to follow the same course in order to construct all problems, more and more complex, *ad infinitum*."[63]

The final "surprise" comes when Descartes shows that (by his lights) the solutions to the generalized Pappus problem "teach in order all the curved lines which ought to be received into Geometry." These *and only these* curves also satisfy the "kinematic criterion" by which Descartes, speaking in Book 2 of the "nature" of curved lines, discriminates between geometrical and mechanical curves, excluding the latter from the body of geometry. So, in the end, all problems in geometry "reduce to a single type," namely, to the question of finding "the value of the roots of an equation."[64] Thus, everything, or almost everything, the Greek geometers found frustratingly problematic, is meant to yield to Descartes' systematic analysis of the nature of "problems" as such.

It must be kept firmly in mind that the *roots* spoken of here are the line-segments that can actually be drawn. The lengths of these segments determine the distances of points from the previously selected "principal lines" (axes) and thereby determine the "graph" of the curve on which all the relevant points fall. To draw those roots or the curve they determine is to do exactly what Descartes means when he speaks at the very start of the *Geometry*, and throughout, of "constructing problems." (See III, iv, below.) Let me quote the opening sentence of the work: "All the problems of geometry can easily be reduced to such terms that afterwards we need only know the length of some straight lines in order to construct them."[65]

With these words, understood retrospectively in light of the entire *Geometry*, we have finally located the strictly technical sense of "construction" which, so I have been claiming, serves as a leitmotiv for the radically modern enterprise of philosophy and for the self-understanding it spawns. Much more needs to be said, in this and the following section, about the ramifications of this technical sense in Descartes himself. For the moment, it is enough to emphasize that we also encounter in this opening line the clearest proclamation of Descartes' *constructivism*. Whatever else it may appear appropriate to say about the role of Cartesian *intuitions* or clear and distinct ideas grasped reflectively or introspectively by the *ego cogitans*, here, in the quite determinate setting of the *Geometry*, constructions hold the center stage. In truth, the implied discrimination is too harsh and inexact, since, as I shall be arguing later, for Descartes graphic constructions *are* the outward counterparts and sensible signs of the inward activity of mind.[66]

iv   The Style of the Geometry

At all events, this provisional inspection of the plan of the *Geometry* puts us in a position to begin taking the measure of Descartes' distance from ancient mathematics such as it came to view paradigmatically in Euclid's *Elements*. The results of section II should already have hinted at the appropriate scale by which this measuring has to be carried out, for we found in the tactics of Descartes' fabulous self-portraiture compelling evidence that he wants his "samples" of method to display those virtues of absolute originality and self-sufficiency by which genuine learning is distinguished from simple *historia*. Accordingly, Descartes' "borrowings" from near or remote predecessors, often an issue of some controversy in his own correspondence and amply ventilated by contemporaries including Leibniz and Wallis, *should not* mitigate the force or obscure the implications of his own desire to break free of the Others, to be the sole source of his experience (Rule 12). The *Geometry*, too, must be deciphered in the light of that desire. Only in this sense can we give credible meaning to Chasles' famous (or infamous) sobriquet for Descartes' analytical geometry: *proles sine matre creata*—"a child created without a mother."[67]

As soon as we undertake this task, however, we are straightaway confronted by the "sedimented" character of most of Descartes' founding gestures. Their results were so quickly absorbed in the technical repertory of modern mathematics that contemporary historians are customarily impatient in the face of his hesitancies and failures to proceed as far as his own successors. To bring these sedimented evidences or evidentiary claims back to the surface, we have to emulate the innocence we can imagine a reader schooled in the ancients bringing to his first encounter with the auroral strangeness of the *Geometry*. "Innocence" is perhaps ill-chosen here; we would do well to recall Newton's marginal notes in his copy of Descartes' text: "Error, error, non est Geometria."[68]

In this spirit I shall first try to call attention to some of the most conspicuous features of Descartes' mathematical "style," his way of "doing" geometry, with the emphatic inclusions and allusive exclusions which the notion of style, here as elsewhere, implies. With a compendious inventory of these stylistic features in hand, I shall then turn to the *Rules*, whose own inaugural gestures originally call these features of the *Geometry* into being, that is, give them their methodological utility or necessity.

(1) First, the *Geometry* is devoted exclusively to solving *problems* and not at all to proving *theorems*. This contrast straightaway reminds us of the intramural debate in the Academy between the followers of Speusippus and Menaechmus, respectively (see above, chapter 2.V). Descartes' single-minded focus on problems seems to place him squarely in the latter camp; nonetheless, we must be on guard lest this reminiscence be misleading.

Proclus' report of that ancient debate suggested that the axial tension around which it turned was the unequivocal distinction between knowing (*gnōsis*) and making (*poiēsis*); thus, for the Speusippeans, geometry fails to qualify as a genuine science if (the truth of) what it studies is not in every case eternally or timelessly present. If we set aside the important nuances and qualifications to be discovered upon closer examination of the relevant ancient sources (Oenopides, Euclid's *Data,* Proclus, Pappus and Marinus),[69] then we could say that the issue concerns the ontological standing of the "mathematicals" and *not,* or not primarily, the cognitive privileges allotted to two contrasting ways of learning about the mathematicals.

In Descartes' case, the situation is reversed. Happily, he did have occasion to comment on the reasons or motives behind his choice of the problematic over the theorematic approach to geometry, in a letter to Princess Elisabeth from 1643:

> In exploring a geometrical question [the context makes clear that *question de géométrie* is a problem], I always take care that lines which I use for finding the solution are either parallel or intersect at right angles, as far as that is possible; and I do not take into consideration any other theorems besides these, that the sides of similar triangles have to one another a similar [*semblable*] ratio and that, in right-angled triangles, the square of the base is equal to the two squares of the sides. And I am not afraid of supposing several unknown quantities in order to reduce the question to such terms that it depends only on these two theorems; on the contrary I prefer [*aime mieux*] to support more unknowns rather than less. For, by this means, *I see more clearly everything that I am doing* [*je vois plus clairement tout ce que je fais*] and by disentangling [these unknowns] I can better find the shortest paths and spare myself superfluous multiplications. In contrast, if one draws other lines and makes use of other theorems, although it can happen, *by chance* [*par hazard*], that the path one finds is shorter than mine, nevertheless, the contrary almost always happens.[70]

A great deal could be gleaned from this remarkable testament. For the moment it will be enough to emphasize that underlying Descartes' defense of the procedures he followed in the *Geometry* is the desire to exhibit virtuosity—that is, by increasing the number of unknowns (thereby rendering the problem and the corresponding equation[s] apparently more complex) and simultaneously decreasing the number of presupposed theorems, Descartes will succeed in finding a solution by the shortest and therefore most artful or ingenious paths. To embrace such a methodical course is to replace fear ("And I am not afraid . . . ") with love ("On the contrary, I love better/prefer . . . "), a love that is, for better or worse, narcissistic. The key line in the cited passage is, no doubt, "for by this means I see more clearly everything that I am doing." Descartes' way of problem-solving makes his own activity cognitively transparent, or, in other words, the "theoretical" is absorbed into the "problematic" as the optimal lucidity of self-couscious feats. One can see (contemplate) all that one is doing while in the very act of doing it!

This passage impresses us in much the say way as a more famous discussion in the *Rules* in which Descartes accuses the ancient geometers of jealously covering over the traces of their heuristic art through "a sort of pernicious cunning" (*perniciosa quadam astutitia*).[71] The proof of a theorem is, as it were, a *fait accompli* which occludes the ingenuity required to find the means for reaching this end; according to Descartes' suggestion, the old geometers *exercised* their astuteness in *hiding* their astuteness or artfulness, and they did so out of jealousy, a species of blameworthy fear at the prospect of losing what is one's own (see *Passions of the Soul*, arts. 167–169). Descartes, on the contrary, intends to publicize the manner and routes of his doing itself, his *savoir-faire*, not his *faits accomplis*. He is proof against the passion of jealousy since it is his artfulness itself, and *not* the finished products of his art, which should be the focus of admiration or emulation; furthermore, as he is at pains to stress at the end of the *Geometry*, any who do emulate his art will still have infinite scope for their own deeds. Descartes can afford to be generous.

These two passages combine to put us in mind of two more related implications of Descartes' choice of the problematic over the theorematic style. First, this choice seems to be rooted in an ethos quite unlike the Euclidean as it appeared in chapter 2. While the latter is bent on inculcating the appropriate virtues in the learner *qua* learner, the Cartesian ethos, as so far revealed, concentrates on exhibiting the virtuosity of the artisan *qua* inventor. The inventor deliberately multiplies what will strike his audience as obstacles to success ("j'aime mieux en supposer [*quantités inconnues*] plus que moins"). So overcoming them will be a *tour d'adresse*, to use the phrase Fermat turns against Descartes' *Dioptrics*. Second, these passages begin to make us aware that for Descartes, as for other radical moderns, the *topos* of wonder finds a new home: No longer nature and no longer the truth of a mathematical theorem, but from now on the artistry of the technician evokes wonder or admiration in those to whom both the effects and the causes of his artistry are unfamiliar (compare *Passions of the Soul*, art. 75, and *Meteors*, Discourse 8, final paragraph). Descartes' predecessor Stevin captures the installation of this new attitude perfectly in two statements on the *topos* of wonder: "Wonder en is gheen wonder" (What appears a wonder is not a wonder, from *De Beghinselen der Weeghconst* [Principles of Statics]) and "alderwonderlicste Reghel der Reghelen Algebra, de toetse vande subtijheyt des menschelicken verstandts" (algebra, the most wonderful rule of rules, the greatest of the subtleties of the human understanding, from his *Dialecktike*).[72]

This is not yet the whole story, however. It will not suffice to say that what is at stake in the contest between *gnōsis* and *poiēsis* is ontological standing for the ancients and cognitive privilege for Descartes. As we shall be seeing with increasing clarity in the remainder of this section (starting with the discussion of kinematic continuity, in section III, v. below), for Descartes it no longer makes sense to drive a wedge between these two options in imitation of Speusippus and Menaechmus since (a) making is already a privileged form of knowing and (b) having-been-made in a suitably artful way *is* the index of a thing's ontological standing, in the relevant sense of being a fit *object* for mathesis.[73]

Two preliminary observations on this second, crucial point. First, in Aristotle's canonical classification of sciences into theoretical, practical, and productive, the basis of distinction between the first and the other two is the locale of the governing source (the *archē*) of motion and rest. In the case of the theoretical sciences this principle is "in" the beingness of what they study, while in both practical and productive sciences the principle of motion and rest is "in" the actor or the maker (*Metaph.* El. 1025b 20–25). As will become patent in the *Rules,* Cartesian science is by this criterion productive through and through. Secondly, we shall have to struggle to become as precise as we can in respect to the "ontological" implications of Cartesian productive or poiētic science; in particular, the medieval question of divine *creatio ex nihilo,* which was explicitly taken as a model for human constructions and productions by post-Cartesians such as Fichte (see above, chapter 1, II), will continue to haunt any attempt to reckon with Descartes' express or implied position. It may be that re-creation, re-generation and re-production, in each case through the designs and within the constraints of human technique, are the Cartesian surrogates for creation, and so on, *simpliciter.* In any case, it is difficult to resist remembering here these well-known lines from his suppressed work *Le Monde, ou Traité de la lumière:*

> For a short time, therefore, allow your thought to leave this world in order to come to see a wholly new one, which I shall cause to be born [*naistre*] in the presence of your thought in imaginary spaces.[74]

(2) Descartes' emphasis on problems, rather than theorems, is closely linked with his promotion of what he and his contemporaries called "analysis" or the "art of invention" (*ars inveniendi*), an art celebrated at the expense of the Aristotelian syllogistic logic taught in the schools and latterly used as the regulative model of Euclidean proofs. In Descartes' judgment, syllogistic demonstrations or "synthesis" do not lead to the discovery or invention of anything *new,* but merely rearrange knowledge already acquired by other means into a didactically convincing sequence. The art of invention, by contrast, produces genuinely novel knowledge; when the process of invention is reenacted for the benefit of others, they

can participate in the discovery as though it were their own. Two classes of readers are discriminated according to their preference or need for synthesis rather than analysis, or conversely, "obstinate and inattentive" readers, in the first case, those with a flair for invention, in the second.[75]

Although this characterization merely touches the surface of the meanings and functions of analysis in Descartes, the recent work of Hans-Jürgen Engfer, *Philosophie als Analyse,* relieves much of the obligation one would otherwise feel to explore its historical logic to a greater depth. Engfer succeeds admirably in identifying separately the various strands of usage and tradition which were to be so fatefully entangled in the seventeenth-century notion of analysis (for example, analytical method in Greek geometry, *regressus* in the empirical natural sciences, and *analysis* in modern mathematics). His contribution allows me to confine myself to a few points of significant detail.[76]

Descartes' critique of the traditional syllogism is not a simple repudiation but amounts to a veritable *Aufhebung,* at least by his lights. Two logical moments can be usefully distinguished within this process. (a) First, Descartes replaces *per se* predication with *comparison* (see Rule 14 on *comparatio,* which is one of the standard terms for ratio). The attention of an analytically minded searcher must be quite deliberately shifted from predicates or attributes either belonging to the essence of a "subject" or in whose definitions the "subject" is present (*Post. An.* 14. 73a35ff.), to *quantitative relations* between subject and predicate, or, more precisely, between terms in a series. Rule 14 makes the quantitative character of these comparative relations unmistakably clear: only when the "sought" [*quaesitum*] and the "given" participate in a common nature which admits of the more and the less, as well as equality, can they be set together with one another in a way that should finally yield knowledge of the unknown being sought. This "common nature," then, is neither a species nor a genus, each with its *per se* attributes, but quantitative being or magnitude in general.[77]

(b) Having replaced predication with quantitative comparison, Descartes goes on to remove the remaining linch pin from the apparatus of syllogisms, namely, the middle-term. In the early *Cogitationes Privatae* he still retains something close to the traditional conception when he writes, "In every question there ought to be some mean [*medium*] between two extremes, through which they are linked together either explicitly or implicitly: as, for example, the circle and the parabola, by means of the cone."[78] He may also have at least flirted, in this same early period, with the notion of mapping geometrical analysis onto syllogistic structure, just as his then-mentor Beeckman attempted to reduce Euclidean proofs to syllogisms.[79] By the time of the *Rules,* however, the place of the syllogistic middle-term has been usurped by the mean-proportional (*medium* being, of course, the Latin expression for both). This second substitution goes hand in hand with the first inasmuch as Aristotle requires that the middle-term in a strictly demonstrative syllogism be (or denote) the essential definition of the major (*Post An.* 2. 8, and 9.94a20ff.), while Descartes is intent on banishing essences, at least essences in the Aristotelian sense, from the rule-governed domain of inquiry. (See IV, iv.)

The discovery of one or more mean-proportionals through the appropriate association of known with unknown terms is *the* paradigm of ingenious discovery in the *Rules* and is at the heart of both the algebraic and the geometrical techniques deployed in the *Geometry*. So, for example, extracting a root is a matter of establishing an appropriate series (a continuous proportion) in which one or more mean-proportionals are inserted between a chosen unit and the term whose root is being sought (for example, $1:x::x:x^2::x^2:y$; therefore $\sqrt[3]{y} = x$. When questions or problems are "involuted" or "complicated" (*involutae*), that is, when we are *given* the extremes and must discover certain *intermediates* in an inverted order (*turbato ordine*), then "the whole artifice of this topic consists in this, that by supposing the unknowns to be knowns we can prepare for ourselves an easy and direct route of investigation, no matter how intricate the difficulties are" (Rule 17).[80] (To take a very simple instance, if 3 and 24 are the known or given extremes and x and y, the unknown intermediates, then we must set up the continuous proportion $3:x::x:y::y:24$.) Examples need not be multiplied for us to see how central a place this artifice occupies in the format of Descartes' method. In the Latin translation of the *Discourse*, the following phrase was added to the statement of Rule 4: "Ut tum in quaerendis mediis, tum in difficultatum partibus percurrendis" (both in seeking means as well as in running through the parts of the problems), which, like the opaque allusion "comme par degrés" in Rule 3, allies these more exoteric maxims with the precise techniques of *pura mathesis*.[81] Nicolas Poisson had seen an alternative text of the *Rules* which he used in his *Commentaire ou Remarques sur la Méthode de Rene Descartes*, published in 1670; his comment on the last in a series of five very general rules ("One must set all these parts [of a problem] in relation, by comparing them to one another") is worth citing in full:

> The most difficult article of these rules to put into practice is the last: as much because one does not know well enough the terms one has to compare as because one has need of a means [*moyen*], which is called *Medium* in the schools, and this is not easy to find.[82]

(c) So far I have been pointing to the replacement of middle-terms by mean-proportionals in *algebra;* it is when we consider the effect of this same replacement in *geometry* that we begin to experience the full force, not only of Descartes' dismantling of the traditional syllogism in favor of a genuinely *inventive* art, but also of his mobilization of *analysis* in the service of a thoroughgoing *constructional* programme. Since this latter claim goes to the heart of my interpretation, I shall need to proceed circumspectly.

Engfer comes to the conclusion that when, with a and b as the "knowns," we have established the appropriate equations for the unknown mean-proportionals x and y, namely, $x = \sqrt[3]{a^2b}$ and $y = \sqrt[3]{ab^2}$, these "display the goal of [Cartesian] analysis," since by inserting the known values of a and b, we can compute the values of x and y (that is, 6 and 12). And the lesson he draws from this: "the synthesis attached to this analysis in the Pappus-model is totally missing here."[83]

Engfer's conclusions would be in order if it were *not* for the fact that Descartes does *not* stop with the algebraic solution of the equations, but *always* points beyond this to the construction or inscription of the line-segments and the loci their lengths determine in a given context (as set by the choice of "principal lines" and the unit-measure). Moreover, Descartes *does* follow the Pappus-model, not by retracing the steps of the analysis in a *deductive* order to secure a proof from first principles or already-proven *theorems*, but by accompanying the analysis or the actual construction it makes possible with a "demonstration" (in his idiom) showing that the construction is indubitably what was sought. Without delving into the details of the much-debated and beleagured passage from Pappus' *Collectio* on which all interpretations of ancient analysis ultimately depend, we can nevertheless say that Descartes, faithful in this to his fundamental concern, opts for the second and *only* the second of the two kinds of analysis distinguished there as the "theoretical" and the "problematic" (the latter defined as *to poristikon tou protathentos:* "the provision of what has been set before one as a problem").[84]

Book 3 of the *Geometry* offers sufficient evidence that this is indeed the canonical sequence Descartes intends to follow even when some of its steps are omitted: "analysis" (preparation and reduction of the relevant equations) → "construction" (production of the lines whose lengths are the roots of the equations) → "demonstration" (that is, a showing *ad oculos* that the distances determined by the points on a curve defined by the construction are indeed the roots satisfying the equation). This third step, the "demonstration," can itself have a constructive character; so, for example, the "very easy (*assez facile*) demonstration" begins: "Applying the ruler *AE* and the parabola *ED* on the point *C* . . . " (original edition, p. 408). If any doubts on this issue were to linger, they would be quickly dispelled by two additional sources of evidence: Descartes' epistolary responses to the Parisian critics of the *Geometry* and the presentation of his procedure for finding mean-proportionals preserved in Beeckman's *Journal*. In the first of these sources Descartes is primarily intent on rebutting the charge that his method of resolving problems failed to provide "the construction and demonstration *in addition to the analysis*."[85] Thus he writes to Mersenne in March 1638: "But the good part, concerning this question of Pappus, is that I have put only the entire construction and demonstration, without including the whole analysis which they [that is, "vos Analystes," including Fermat] imagine is the only thing I included; in this they testify that they understand very little of what I have done."[86] Or, to Florimond De Beaune, in February, 1639: "I have not given the analysis of these loci, but only their construction, as I also did in the majority of the rules in the Third Book."[87] With these and other, nearly-identical, comments to his partisans, Descartes clearly and distinctly emphasizes that his general method in the *Geometry* is aimed at the exhibition of successful constructions (or constructional procedures) *based* upon the formal results of algebraic analysis; the constructions, not the analyses, are meant to be the touchstone of his dexterity and ingenuity.

Beeckman's report of an earlier (ca. 1628–1629) success (finding two mean-proportionals between two given lines through the use of a parabola) is especially illuminating since Descartes, who is being quoted verbatim (*quod ad verbum descripsi*), is still using the idioms both of the syllogism *and* of analysis followed by synthesis.

> Let the two given lines be the minor *gb*, the major, *bh;* then it is necessary to find two means in continuous proportion between them.
>
> Ἀναλυτικῶς [Analytically]
> . . .
> Συνθετικῶς [Synthetically]
>
> [The figure] therefore is composed [*componetur*] in this way. . . . And thus it will come about that as *gb* is to *de*, so *de* is to *ae* and *ae* to *bh*.[88]

Descartes does not understand *synthesis* as the deductive legitimation of analytically achieved results; he uses the adverb "synthetically" in its root sense of "putting together," that is, the recomposition of the figure dictated by the terms of a problem so as to show that the relevant ratios being sought are manifest in this structure. For Descartes, "Q.E.D." is always a pendant to "Q.E.F."

(d) Several noteworthy consequences follow from this inspection of the place of analysis in what I have called Descartes' "canonical sequence" (analysis → construction → demonstration). As a proof-procedure, *algebraic* analysis is self-sufficient and irreproachable, since all of the "propositions" included in it are "equations" and therefore straightforwardly reversible. We need only recall Aristotle's remark in *Posterior Analytics*, "If it were impossible to prove the true from the false, analysis [*to analuein*] would be easy; for [conclusion and premises] would reciprocate necessarily" (1.12. 78a6–7), a *contrary-to-fact conditional*, to grasp the magnitude of the change wrought by the introduction of *equations* (see III, v)! In the account of analysis furnished by Pappus, p (the proposition being "sought") is shown to imply $p_1 \rightarrow p_2 \rightarrow p_n \rightarrow K$, where K is some *theorem* which has already been (synthetically) demonstrated; however, it is not necessarily the case that $K \rightarrow p$, since the converse of any link in the chain $p_K \rightarrow p_K$ may not hold. Once equations are substituted for propositions, reciprocal implication follows without further ado, since equations have the logical form of bi-conditional entailments $(p \rightarrow K)$ if and only if $(K \rightarrow p)$. Consequently, while Archimedes, as we saw in chapter 2, held that analysis yielded only "a certain impression that the conclusion is true" and Geminus (first century B.C.) defined analysis as "the discovery of a demonstration" which is not itself a demonstration, for the contemporaries of Descartes, analysis, in André Robert's words, "is at one and the same time a method of invention and of demonstration."[89]

Why, then, is Descartes not content to provide the "whole analysis" in the case of each family of problems since he would have thereby, at a single stroke, also provided all of the steps required for a synthetic demonstration in Euclid's sense? The answer indicated by the evidence of Cartesian *practice* is twofold. First, as we have already noted, Descartes is interested exclusively in what Pappus called *problematic*, as distinct from *theoretical*, analysis, and problematic analysis is devoted to discovering the possibility or impossibility of certain constructions. Second, and even more revealingly, *not every formally possible solution* of an algebraic equation is in fact geometrically constructible for Descartes. This observation ushers in the complex issue of "true" (that is, positive), "false" (negative), and "imaginary" roots, about which only this much need be said in this setting: Only "true," and "false" roots are admissible as authentic solutions to geometrical problems since there is no "space" in the local expanse determined by the principal lines in which "imaginary" roots can be inscribed. As Descartes states the point: "For the rest, the true roots as much as the false are not always real [not in the technical sense of "real" numbers, but in the sense that a *res* corresponds to their value in each instance], but sometimes they are also imaginary; that is to say, one can indeed always imagine as many roots as I have said in each equation, but sometimes there is no quantity which corresponds to those roots which one imagines."[90] Accordingly, Descartes rejects what came to be called "complex numbers" even though the algebraic techniques he has on hand for extracting roots do nothing to militate against them. (This is one respect in which Descartes exhibits something akin to the *phronesis* I attributed to Euclid, although the *rationale* underlying their refusal to reify the products of available technique differs importantly. It is also instructive to note that as late as 1867 the famous mathematician Hankel was drawing attention to continuing confusion about the referents of complex-number expressions: "in a word, the true metaphysics of imaginary numbers is in a very bad way in most presentations hitherto.").[91] Therefore, the logical reciprocation (mutual entailment) of the equations worked out by *analysis* is a necessary, but *not* a sufficient condition for the "existence" of quantities corresponding to the values determined by those equations. The "abstract" formulae derived through the solution of equations remain "altogether inexplicable" (*ominino inexplicabiles*), as Beeckman reports Descartes' position, and this means that "explication" is the display or inscription either of an integral numerical value, a measurable geometrical, or some other physical magnitude.[92]

These four reflections should suffice to persuade us that Cartesian algebraic analysis does not stand on its own but, rather, is employed as an instrument designed to facilitate and (partially) authenticate geometrical constructions. Later, in section IV, I shall have more to say about the roots and the implications of this conjugation; prior to this discussion, I want to offer one final suggestion extremely relevant to the constructive aim of Cartesian analysis.

Descartes was surely conversant with the intense discussion of the putative functions of *synthesis* and *analysis* which was wide-spread since the mid-sixteenth century. It is therefore more than merely likely that his own express pronouncement on the difference between these two is couched in knowingly heterodox terms. In the main line of tradition stemming from the *Posterior Analytics* (compare 78a20ff.) and consolidated in Galen's *Ars medica*, synthesis (or *compositio*) is a proof of the "cause" (*tou dioti*), while analysis (or *resolutio*) is a proof of the "fact" (*tou hoti*). Zabarella, in the most fastidious account of this line of tradition as it was understood by many authors in the Renaissance, assigns the label *compositio* to the discursive movement "from cause to effect" (*a causa ad effectum*), the label *resolutio,* to the movement "from effect to cause."[93] Descartes, prodded into recasting the *Meditations* "synthetically," transforms and subverts this latter conception of the distinction; for him, *analysis,* as a manner of demonstrating, that is, bringing us to see something, shows how a thing was "methodically and as it were a priori invented/found," while synthesis "clearly demonstrates what it concludes by the opposite way, a way sought as it were *a posteriori* (although often the proof itself [*ipsa probatio*] is more *a priori* in this case than in the former)." The authorized French translation restores the idiom of cause and effect alongside the a priori/a posteriori contrast; thus, analysis "makes us see how effects depend on causes" whereas synthesis "examines causes by their effects."[94] Only the parenthetical clause in the definition of synthesis matches the Galenic description; otherwise, it is *analysis* that moves from cause to effect! It seems fair to interpret this to mean either that the possessor of the *ars inveniendi* acts as the cause from which effects (discoveries) methodically follow or, more narrowly, that the terms to which algebraic equations have been reduced are the causes from which constructions effectively ensue. In truth, as I hope to show, the narrower interpretation is simply a conspicuous specimen of the former.

In any case, by reworking the traditional distinction such that analysis, *not* synthesis, is an explanation *tou dioti* or *a causa ad effectum*, Descartes places himself in near-perfect harmony with still another characterization in Zabarella:

> The resolutive order is a logical instrument by which, from the notion of an end *which can be produced and generated by man operating freely [qui ab homine libere operante produci et generari queat]*, we progress to the principles to be found and known; after beginning our operation from these principles we can produce and generate that end.[95]

Synthesis is suited to the theoretical sciences, analysis, to the practical and productive sciences, according to Zabarella, who hereby unpremeditatedly anticipates Descartes' own declared desire to find *une pratique* to replace "that speculative philosophy people teach in the schools."

(3) I have already mentioned what is possibly the most consequential feature of Cartesian mathematical style when set in contrast with the traditional syllogism; Descartes' *ars inveniendi*, as its name promises, yields novel knowledge, while the syllogism produces "nothing new" (*nihil novi*, Rule 10). The third and final aspect of style to which I want to pay heed concerns the conditions under which novelty may be authenticated not merely as an unprecedented, albeit coincidental discovery, but as a self-sufficiently originated invention. Descartes' remarks on autarchy as an index of mathesis in his letter to Hogeland (see above, II) make a fitting preamble to his specific practices in the *Geometry*.

The conditions germane to inventiveness reduce to two: economy, or relative simplicity, of means and orderliness of sequence. We have a splendid example of the first in Descartes' discussions of how to go about discovering and constructing mean proportionals between two given quantities. He tells us in the opening paragraph of Book 3:

> Although all the curved lines which can be described by some regular movement ought to be received into geometry, this is not to say that we are permitted to make use indifferently [*indifférement*—"at random" in

the Smith-Latham translation] of the first one encountered, for the construction of each problem.[96]

He goes on to illustrate what he means by invoking the "compasses" introduced in Book 2: A mean proportional between two lengths measured along the base of that instrument can be, so to speak, automatically determined by taking the longer of these as the radius of a circle and then opening the compass; the point of intersection at that circle with the top arm gives one of the mean proportionals we are seeking. However—and this is the crucial point—the curve produced by the act of opening the instrument belongs to the *second* class, that is, the equation is $x^4 = a^2 (x^2 + y^2)$, while two mean-proportionals between two quantities (where $x^2 = ay$ and $y^2 = bx$) can be determined by the intersection of two parabolas or a parabola and a hyperbola, that is, curves of the *first* class. Accordingly if we use the Cartesian compass to solve this problem, we are employing means more complex than the "nature" of the problem itself, as shown by the class/degree of its equations. While it is *easier*—manually or mechanically—to solve the specific problem by using the designated instrument, facility is here at odds with fertility. Complexity must emerge from (relative) simplicity and not vice versa. Or, in other words, nothing *new* is to be expected if what is required for solving a given problem (alternatively, constructing a specific locus) already involves more "industry" than the "nature" of the problem. (Conversely, as Descartes adds, "it is a fault on the other side to exert oneself uselessly in trying to construct some problem by a class of lines more simple than its nature permits.")[97] In general, the means for constructing a curve of class K ought to be drawn from elements of class K-1. To do otherwise is to betray ignorance of the "natural" relation between simple and complex. And, as we are told in Book 3, "Anything testifying to ignorance is called an error."[98]

This summary account of economy of means already implies its companion condition, orderliness or, more exactly, seriality. Authentic inventiveness is not a one-time affair which happens to meet the criterion of economy. Instead, matters should be so arranged that a sequence of means → solution can be constantly and consistently iterated; more complex curves constructed by simpler means ought in turn to be usable as means for the construction of even more complex curves. Descartes' solution to the Pappus-problem, as I noted above, "teaches in order [*par ordre*] all the curved lines which ought to be received into geometry." The exact correlation among the number of lines in the problem, the degree of the relevant equation, and the class of the curves for determining points on the resulting locus means that *we always know where we are.* Hence we never run the risk of committing the fault Descartes most often castigates when he is criticizing contemporary mathematicians such as Fermat, proceeding *à tâtons*, gropingly, or *par hazard*, at random. He turns aside an objection to his own work with these words: "As for his third objection, which is that this rule [for the reduction of cubic equations] proceeds *à tâtons*, I reply that it is in no way to proceed *à tâtons* to examine in order diverse things when one knows them all, as one does here, and when their number is determined, as it is here, even though there were a thousand of them."[99] We should not, however, mistake orderliness or seriality for an *ex post facto* virtue alone; more importantly, it is a generative power inasmuch as knowledge of where we are and of the principle by which we have arrived there allows us to make continuous headway, just as in the simple case of a continuous proportion. Most of all, we can advance "without uselessly consuming any effort [*conatu*] of mind, but always increasing knowledge by degrees [*gradatim*]," as Rule 4 promises. It is no wonder, then, that when Descartes divulges "the principal secret" of his art, this secret teaches us that "all things can be disposed in certain series."

At this point, the conditions promoting novelty and the critique of the syllogism join forces. "Middle terms" are not only infertile, allowing us only the rhetorical advantage of arranging "old" knowledge, they also seem to elude systematic or methodical discovery. Aristotle calls proficiency in coming up with middle terms without further ado *anchinoia*— "quickness of wit" or "acumen" in standard renditions, "having one's intellect [*noūs*] close by," more literally.[100] Descartes substitutes the orderly directions of *ingenium* for this hazardous knack, since Cartesian means are generated by the terms of the problem itself. For instance, the class of a locus we are seeking, as indicated by the degree of its correlative equation, informs us that the inventive means for constructing it are lines from the next-lowest class.

Inventiveness is the fruit of discreet beginnings (compare "l'unité . . . qui peut ordinairement être prise à discrétion" "the unit . . . that can ordinarily be chosen at one's discretion") mated to discrete series.

v Liberty and Constraints: Why Can Problems Be Solved?

The collaboration of discretion with discreteness is, I have just suggested, at the base of Descartes' technical program for "finding something new," indeed, for continuing this inventive process *ad infinitum*. This technical program, however, is by no means exempt from the need for guarantees of success; if Descartes refuses to proceed *à tâtons*, neither is his confidence in orderly progression a matter of "blind faith" in our power freely to execute our intentions and designs. What we need, then, is some assurance in advance that we can look forward to a regular passage from the starting point of our inquiry to its (infinitely projectible) goal. Such an assurance would mean that we can locate and partially identify what we are seeking before we actually find it. Cartesian method requires that nothing be so unknown (or unknowable) that it cannot be at least partially assimilated to what is already known (or knowable). Hence Descartes must undercut "Meno's paradox of inquiry" by denying its key premise, that we have no way of identifying our supposedly unknown goal at the start of our search.

This is the force of Descartes' initial move in his solution of the Pappus-problem: namely, to assign names to all the relevant lines, "to those that are unknown as well as those that are known. Then, making no distinction between known and unknown lines, we must unravel the difficulty in any way that shows most naturally in what way they depend mutually on one another."[101]

Consequently, unhindered access to the unknown demands the nominal and more than nominal assimilation of the unknown to the known. But we must then ask: What conditions or restraints underlie their assimilability? What, in other words, gives the domain of our inquiry its uniformity and the "objects" in that domain their determinateness, in regard both to their respective identities and to their interrelations?

The text of the *Geometry* allows us to observe the constraints responsible for uniformity and determinateness at work in two of Descartes' technical policies for "constructing problems": the establishment of *equations* and the *kinematic criterion* by which admissible "geometrical" curves are distinguished from inadmissible, "mechanical" curves. After having investigated these two policies in some detail, I shall look to the *Rules* in the hope of uncovering their ultimate roots.

Equations. So long a familiar and fundamental part of modern mathematics and physics, equations initially come into play more problematically as the culmination, or, alternatively, the subversion of the Greek understanding of ratio and proportion. Or, to be more exact, the technique of forming equations is made possible and feasible thanks to two "faults"—in the geological sense—which we have already detected in the seemingly intact corpus of Euclidean geometry: first, the violation of the principle of homogeneity expressed in the definition of ratio (Euclid, Bk. 5, Def. 3) and, second, the "reification" of ratios implied by the operation of compounding (see chapter 2, II, ii). It is only through the simultaneous exploitation of these two "faults" that an equation becomes available not only as a way of naming or expressing a quantity in terms of its (often complex) connections with other quantities, but also as something like a quantity in its own right, insofar, that is, as it itself become subject *en bloc* to arithmetical operations on quantities (such as multiplication and division).

Unabashed appreciation of the opportunities this exploitation makes possible was not immediate. This is most readily seen in the insistence of the major sixteenth-century algebraists, including Stevin and Viète, on keeping the tie between *equalities* and *proportions* explicit; indeed, it can be argued that the latter retain the "upper hand" over the former in the inaugural stages of the development of early modern algebra. Thus, in his *General trattato di numeri et misure* (Venice, 1556–1560), Tartaglia states that "the fundamental principle of the rule of algebra is the proportion of equality," where the last phrase stands in contrast with "proportion of inequality," that is, cases of greater or lesser ratios ($a:b > c:d$; $a:b < c:d$).[102] For the lesser author, but influential translator of Tartaglia, Guillaume Gosselin (?–1590), "the whole object [*ratio*] of this science of algebra is comprised in proportion."[103] Stevin, to the annoyance of his twentieth-century biographer and editor Djikterhuis, persists in interpreting the solution of an *equation* as the finding of a fourth *proportional* to three given quantities. In the second book of his *L'Arithmétique* (1585), he takes pains to explain why he calls "the rule of three, or invention of the fourth proportional of quantities, what is commonly [*vulgairement*] called 'equation of quantities.'" Thereby he intends to disabuse "apprentices" of any notion that "this word 'equation'" refers to "something special [*quelque matrière singulière*]," rather than to the familiar operation of determining a fourth proportional.[104] The "equation" is nothing other than the equality of the product of extremes and the product of means in a proportion where the fourth-term is as yet unknown. (For instance, $4x = 112$ is just a transcription of $4:16::7:x$). With Viète, a threshold is reached, beyond which proportions (or ratios) are on the way to being conceptually and operationally absorbed into equations (and hence into the generalized notion of a function of one or more variables). For him "a proportion can be called the composition [*constitutio*] of an equation [*aequalitas*]; an equation, the resolution of a proportion."[105] Viète maintains this symmetrical equilibrium by pairing proportion and equation with the "methodological" notions of *constitutio* and *resolutio*, that is, synthesis and analysis, respectively. So, to resolve a proportion is to set out its ingredient terms in such a way that the unknown term is compared with (made equivalent to) the known terms. Leibniz will still speak of the *conversio aequationis in analogiam vel contra* as one of operations included under the general rubric of *syllogismi algebraici*.[106] An equation can now be understood as an instrument or template for the construction or exhibition (as in Viète's *exegetic* art) of the quantity denoted by the term for the unknown.

One additional facet of "le grand mistère de proportion en quantitez" (the great mystery of proportion in quantities), as Stevin called it, ought to be briefly indicated before the logic of Cartesian equations can be directly examined. This is the practice, apparently initiated by Michael Stifel, of setting all the terms of an equation (known and unknown) on one side and then setting them equal to zero. The venerable utility of this practice for solving systems of simultaneous equations should not mask its exotic novelty for those still habituated to the Greek conviction that ratios are always relations between instances of *manyness* (numerical or geometrical, as the case may be). Nothing testifies more patently to contemporary sensitivity to the exotic novelty involved here than Thomas Harriot's contention that setting all the terms equal to zero signals or symbolizes *creatio ex nihilo;* the quantity or manyness proleptically denoted on the left-hand side of an equation springs into being from the *nothingness* vacantly pictured on the right.[107] This interpretation, extravagant as it may seem, could be said to reproduce in miniature the grand conflict between pagan and biblical cosmologies or onto-theologies. It is difficult to forebear remembering at this point Solomon Maimon's celebration of mathematical construction a century and a half later: "We are in this similar to God."[108]

For Descartes, the practical autonomy of the equation seems a *fait accompli,* not requiring legitimation or even extensive clarification. (Compare his terse remark on the "nature" of equations in Book 3 of the *Geometry.*) Nonetheless, it is important for us to observe in precisely what way his easy manipulation of equations rests upon that twofold subversion or exploitation of the Euclidean theory of ratios to which I called attention above.

Since the operation of compounding treats a two-term ratio of magnitudes as though it were a magnitude or quantity in its own right, it was, I would think, a fairly easy step to treat a 4- or n-term proportion as a "quantity" in *its* own right. In other words, if $m:n$ can be multiplied by an integer k, then the floodgates have been opened, allowing us to transform the (Euclidean) proportion $a:b::c:d$ into the (Cartesian) magnitude $ac/bd = 1$, which can in turn be subjected to the entire array of arithmetical operations. Moreover, as Descartes' own practice testifies, equations understood as names for quantities can themselves be added, subtracted, multiplied and divided by other equations. Although I am still uncertain of the exact semantic steps involved in the passage from *proportio ex proportionibus,* such as we found it in the sixteenth-century translation of Eutochius, where *proportio* means *logos*/ratio, to the full-blown treatment of equations as compositions of two or more proportions (sets of ratios), nonetheless, the significance of this last stage is patent: Membership in the class of *quantities* is now to be understood as a function of *operations* which yield further quantities (such as that quantity which is the measure of the root of an equation).

We can observe in more particular terms the utility of compounding ratios to Descartes' practice if we turn our attention to continuous proportions and to their role in fixing the *new meaning* of algebraic exponents. Starting from a discreetly chosen unit-magnitude, we can institute the sequence $1:a::a:a^2::a^2:b;$ then, $1:b$ is the ratio compounded of the three constituent ratios or, in other words, $b = a^3$, where the exponent simply denotes the number of ratios between the two quantities 1 and b. No geometrical or figural references is (here) assigned to the expression $a^3;$ that is, it does not refer to the *cube* on the line whose length is measured by a.

The result is of the greatest importance to Descartes' entire procedure in the *Geometry* as well as in the *Rules*. Stated baldly, this reinterpretation of the exponent as an index of the place of a term in a sequence means that, at the level of algebraic notation, the mind is liberated from enthrallment by the imagined or actually visible shapes of Euclidean geometry. Each particular term, whether in a continuous proportion or in the equation into which that proportion can be transformed, makes sense only in virtue of its inclusion in a network of terms related to one another in virtue of their respective positions in a clearly formulated sequence. Hence, no limits are placed on the "value" of an exponent inasmuch as *both* the sequence in which it figures *and* the operation of compounding by which its value is determined are iterable *ad infinitum*.

Just how crucially fitting this reinterpretation of the exponents proves for Descartes' problematic endeavor can also be inferred from Pappus' remark quoted directly in Book 1 of the *Geometry* and alluded to above (III, iii):

> If there be more than six straight lines [in a locus problem], they can no longer say 'if the ratio be given between some figure contained by four of these lines to what is contained by the remaining lines, 'since there is nothing contained by more than three dimensions. . . . Some rather recent interpreters of these matters have acquiesced in using these expressions; but without in any way signifying something comprehensible which is contained by these lines [more than six]. They might, however, have said these things by means of compound ratios [*per conjunctas proportiones*].[109]

Where Pappus, on the one hand, is clearly hesitant about the willingness of "some rather recent interpreters" to speak as though there were figures contained by more than six lines (hence, figures of more than three dimensions), but, on the other, is amenable to translating such manners of speech into the idiom of compound ratios, Descartes dissolves the quandary by identifying *dimensionality as such* with the place of a term in the sequence of ratios and thus with its "value" when those ratios are appropriately compounded. (Compare Rule 7, *A.-T.*, t. 10, p. 387, 22–388, 9.) Equations codify these sequences and their arithmetical interrelations, thereby giving body to Descartes' bold disclosure of the "principal secret of the art," namely, "that all things can be disposed by certain series . . . insofar as some can be known from others" (Rule 4).

These last considerations also serve to bring into view Descartes' way of exploiting the second of the two "faults" in the Euclidean terrain; the circumvention of the principle of dimensional homogeneity for the sake of lifting restrictions on the operation of *alternando* ($a:b::c:d$, therefore, $a:c::b:d$). It scarcely needs to be said that the very formation of an equation, for instance, by cross-multiplication ($a:b::c:d$, therefore, $ad = bc$, or $ad/c = b$) presupposes that no such restrictions are in force.

Descartes, however, is not content simply to display *in praxi* the useful results of abandoning homogeneity requirements. More subtly and, one might say, impishly, he pays "lip-service" to the notion of homogeneity while in speech and in deed completely undermining the ontological pre-understanding on which that notion rests.

In brief, his "impish" strategem, set forth in Book 1, is to insist that "all the parts of a single line (as denoted by the equation of the line) should ordinarily be expressed by the same number of dimensions, when unity is not determined by the question, as, for example, here a^3 contains as many dimensions as ab^2 or b^3, of which the line I have named $\sqrt[3]{a^3 - b^3 + ab^2}$ are composed. But, it is not the same thing when unity is determined, because unity can be implicitly understood [*sousentendue*] everywhere, whether there are too many or too few dimensions. Thus, if we have to extract the cubic root of $a^2b^2 - b$, we have to think of the quantity a^2b^2 as divided once by unity and the other quantity b as multiplied twice by the same unity." In the first case considered, a^3 has just as many "dimensions" as the quantity ab^2 as can readily be seen by taking the sum of the latter's exponents ($1 + 2 = 3$). In the second case Descartes' ruse consists in recommending that a^2b^2, which as it stands has four dimensions, can be reduced to three dimensions if we *divide* it once by the chosen unit, while b, with one dimension at the start, can be raised to three dimensions if we *multiply* it two times by the same chosen unit. In fact, what Descartes does is to subtract 1 from the sum of exponents in the first quantity ($4 - 1 = 3$) and to add 2 to the exponents of the second quantity ($1 + 2 = 3$). That the resulting sum of the exponents is the same in both cases is taken to insure that they have the same dimensions. However, his misstating the relevant operations as division ($a^2b^2/1$) and multiplication ($b^1 \times 1 \times 1$) seems to show that these prescriptions are meant only to allay the anxieties of traditionalists. (Much the same "dodge," as Michael Mahoney calls it, was already put to use by the sixteenth-century algebraists; compare the explanation given by the anonymous "Calcul de Mons. Des Cartes," *A.-T.,* t. 10, pp. 672–73.)[110]

Incalculably more important than this computational strategem is Descartes' explicit subversion of the pretheoretical or "natural" understanding of dimensionality as intrinsically characterizing distinct genera of magnitudes (one-dimensional lines; two-dimensional plane figures; three-dimensional "solids"). All Cartesian magnitudes are homogeneous *because* dimension is no longer a feature belonging to (premodern) geometrical "shapes," but a benchmark indicating the sequential order and the relative measurement of the terms ingredient in an algebraic equation. Moreover, homogeneity now refers to the closure of a field under the permissible operations of addition, multiplication, exponentiation, and so forth; it is a property of all magnitudes just insofar as they are all equally accessible to that set of *operations.*

Two clarifications and one amplification are needed before I can pass on to the second liberating constraint in play in the *Geometry.*

(1) First, Descartes divorces dimensionality and homogeneity/heterogeneity alike from any figural attachments such as they had in the Euclidean tradition. At the same time, he retains the traditional nomenclature of "squares," "cubes," and so on. Together these procedures yield a twofold effect: The reader is simultaneously drawn toward the *new* conception of dimensionality (and of a quantitative field closed under permissible operations) and held by the old visual (or spatial) understanding for which the operation of, say, multiplying a square by line-segment has no sense. Descartes, so it seems, intends that as the student becomes increasingly accustomed to that new, operational conception, he will also learn to reinterpret the geometrical lines, figures, and solids putatively encountered as "naturally" different in kind as in truth owing their *intelligibility* to the artifices of symbolic or specious comparison.

(2) However, this is decidedly *not* to say that the old-style geometricals utterly lose their *spatial or corporeal being* (that is, their extendedness). Since I shall have more to say on this distinction between specious intelligibility and corporeal being in section III, let me pave the way by concentrating here on the remarkable ambivalence of straight line-segments in the *Geometry.*

As we know from Part 2 of the *Discourse,* Descartes chose to represent particular ratios and proportions by straight lines on the ground that "I could not find anything more simple or that I could more distinctly represent to my imagination and to my senses."[111] This marks a departure from his policy in the *Rules* (18), where lines, squares, and rectangles are put separately into play to represent the results of distinct arithmetical operations. The shift from the multiple devices of Rule 18 to the single device of the *Discourse* and the *Geometry* does not, however, alter the essentially operational and symbolic bearing of these devices as such. That is, in this aspect, the line-segments in Book 1 of the *Geometry* serve only to record in the simplest or most economical way those relations among magnitudes which are discerned or produced *via* the manipulations of algebraic letters and numerical coefficients. To the virtues of simplicity and imaginative or sensory distinctiveness we can also add their ready susceptibility to the full range of arithmetical operations in such a way that from straight-lines we get only other straight-lines. (For instance, multiplying line a by line b does not yield a rectangle with sides a and b, but a new line c, by means of similar triangles and the finding of a fourth proportional—i.e., $1:a::b:c$, so $c = ab$, that side of a triangle with base b which is parallel to the side a with a unit-base; see Euclid Bk. 6, Prop. 12.) In this application, therefore, these line-segments have ceased to be geometrical, that is, *linear* magnitudes in their own right or "by nature"; instead, they symbolize or explicate the terms in a sequence of ratios among any magnitudes whatever. And it is only with respect to this application of straight line-segments that we can talk of "symbolic or abstract magnitude in general" as *the* subject-matter of Cartesian *mathesis*.[112]

In their *second* aspect, line-segments as actually constructed—that is, the roots whose measures are the values of the pertinent equation or equations—are restored to their standing as geometrical, linear extensions in something akin to the pre-Cartesian understanding of what a straight-line naturally is. (*Mutatis mutandis*, the same point holds true of the curves resulting from the "construction" of the problems in which their equations figure.) Accordingly, we have to be alive to the difference between the line-segment in its symbolic, abbreviatory function, in virtue of which it can most usefully be deployed in specious calculations, and the (apparently identical) line-segment recognized as the product of constructive genesis. The homogeneity made possible by the first function is, as I have stressed, a matter of the unhindered performance of the same set of arithmetical operations no matter what "naturally" different genera of quantity are in question. With the turn to the latter, constructed presence of line-segments and other geometrical formations, something like the heterogeneity of magnitudes reappears—a straight-line and a parabola, for example, have no determinate ratio to one another. Nonetheless, the reappearance of this version of heterogeneity must still be interpreted in the light of the preceding operational homogeneity. Or, in other words, differences in corporal or extended kind are never wholly regrounded *in rerum natura* or in the dianoetic shapes of these natural beings, even while some degree of correspondence between the former and the latter has to be acknowledged. In the absence of any basis for this acknowledgment, the Cartesian project of a mathematical physics of extended bodies ("toute ma physique n'est autre chose que Géométrie") would never "get off the ground."

(3) Unlike the late seventeenth- and eighteenth-century algebraists who took their start from the results and difficulties of the *Geometry*, Descartes was not principally or even persistently concerned with the theory of equations as such—with, say, techniques and algorithms for determining classes of solutions to equations by inspecting the combinational properties of their signs, coefficients, and exponents— any more than he was with *proofs* of such purely "algebraic" matters as the theorem that an equation has as many roots (real or imaginary) as its unknown term has dimensions. (The latter is presupposed as a result in Book 3; Descartes' apparent qualms at probing the issue further quite probably have to do with the "merely imaginary roots" which do indeed satisfy the equation computationally, but for which there is, so to speak, no "room" in the Cartesian field where "real" roots are to be constructed.) He was, to be sure, aware of some of the combinational procedures later given prominence by Leibniz and his successors; thus, in a letter of December 18, 1648, he takes up a question posed by Fermat concerning the number of combinations of so-called "asymmetrical terms" (*A.-T.*, t. 5, pp. 256–57). Nevertheless, Descartes gives sufficient evidence of having regarded such issues as at best marginal to his fundamental concerns, concerns I have throughout identified as "constructive." This subsidiary status of equations, when taken in their own right, is further borne out by Descartes' recommendation to Desargues to simplify his demonstrations for the many by "using the terms and the calculus of arithmetic, just as I did in my *Geometry;* for there are many more people who know what multiplication is than know what the compounding of ratios is, etc."[113] (See below, III, vi on the fuller import of this remark.) At all events, Descartes' lack of preoccupation with equations *in the abstract* does nothing to undercut his attachment to them as devices both for calculation and for representation (see section IV, iii on the equation as a transcription of mental operations); the common final aim of both being the construction or exhibition of these admissible curves and real roots which solve geometrical problems in the most orderly and fruitful way. This brings us quite naturally to the topic of admissibility itself.

Kinematic Determinateness. I have more than once alluded to Descartes' decision to admit only some curves into geometrical science. This decision is not an arbitrary one, but rather appeals to a criterion by which one is meant to judge the geometrical *and* epistemic credentials of any candidate-curve. Descartes' criterion is in fact a kinematic criterion. That is, it turns on the admissible and the inadmissible patterns of motion through which various curves can be produced. Let us look at this somewhat more closely.

In Book 2 of the *Geometry,* devoted, so its title tells us, to the *nature* of curves, Descartes divides the class of all curves into the "geometrical" and "mechanical" and tells us that only the former are to be received into geometry; this division goes hand in hand with a critique of the ancient geometers who used the same terminology but came to a different decision.[114]

If a "mechanical" curve is one that can only be produced by a machine, a physical device or instrument of some sort, then the Greeks were inconsistent in denying that, say, a circle is a mechanical curve since it can only be produced "on paper" by the use of compasses. Having silently rejected the Platonic notion of "ideal" circles, and so forth, Descartes proceeds to make his main point: the use of instruments is not the decisive factor in classifying the two kinds of curves—rather, everything depends on the nature of the movements consequent on the operation of instruments. If there is only one continuous movement or if several motions succeed one another in such a way that the later motions are "completely regulated" by those that precede them, then and only then are the resulting curves to be called "geometrical." Otherwise, they are termed "mechanical." The ancients, through failure to recognize this principle, wrongly classified some legitimate geometrical curves—for example, the cissoid and conchoid—as mechanical.

This summary of Descartes' text requires a number of brief comments:

(1) Descartes' distinction corresponds roughly to the more modern distinction between algebraic and transcendental curves, first introduced by Leibniz. Algebraic curves are those in whose defining equations only rational numbers can appear as exponents. The equations of transcendental curves can have irrational or indeterminate exponents, as in $y^{\sqrt{2}} + y = x$

(2) Descartes was himself thoroughly familiar with at least some of these "transcendental" curves; he studied the logarithmic curve in particular, contributing a great deal to its mathematical analysis. Nonetheless, he rejects such curves from the body of geometrical knowledge. He does so both because, as we shall see in a moment, these curves violate the kinematic criterion and because, in any case, the instrument used to generate the relevant curve is more complex, algebraically, than the curve it generates. "So that this does not give us anything new in Geometry," as Descartes says in a letter of 1629.[115]

(3) To see what the kinematic criterion means, let us take the example of the so-called quadratrix, the curve devised by an ancient geometer Hippias, for solving the problem of squaring the circle:

Take a square

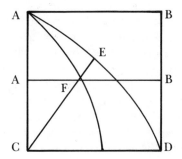

and let the side *AB* move downwards to *CD* at the same time as the radius *AC* is moving along the arc of the quadrant *AD*. The point of intersection (*F*) of the lines *AB* and *AC* defines the locus of the quadratrix. Two objections come readily to hand:

First, the quadratrix is to be constructed for determining, in effect, the value of π; yet, unless one knows the value of π in advance, one cannot guarantee that the two lines move with velocities in a ratio such that they will end up simultaneously at *CD*.

Secondly, even if the relative velocities were known in advance, the radius *AC* and the side *AB* both coincide simultaneously with *CD* and so will no longer intersect one another.

This diagnosis, known to Descartes via Pappus, brings home the importance of linking the motions produced by an instrument in a determinate way, so that the exact ratio of their magnitudes can be known. Only under this condition is the constitution of the resultant curve transparent to the rational mind.

(4) Descartes furnishes an example of the right sort of mechanical instrument in the guise of what have come to be called the "Cartesian Compasses." This is a set of wooden or metal rulers connected with one another in such a way that, when opened, the motion of one causes the motion of another; the curves that result from their joint motions are "geometrical"; in fact, they are the very curves needed for the solutions of all cases of the Pappus-problem. (It is perhaps a minor irony that the total set of these curves is a logarithmic curve, inadmissible on Descartes' criterion).

(5) Descartes summarizes this discussion as follows:

> All points of those curves which we call geometric, that is, those which admit of precise and exact measurement, must bear a definite relation

> to all points of a straight line, and . . . this relation must be expressed
> by means of a single equation.[116]

Since the ratio between straight lines and curved lines cannot be discovered by human minds, Descartes tells us, any geometrical curve is to be made *determinate* by measuring the rectilinear distance of each of its points to some pre-selected straight line or lines, namely, coordinate axes that first came on the scene in his resolution of the Pappus-problem.

We can now begin to examine some of the major implications of this second constraint thanks to which problems become solvable.

Descartes' dismissal of ancient objections to organic or instrumental constructions, except in a limited class of cases (construction by ruler and compass), should not be read simply as a gesture in support of mechanism or mechanization. Even though he would most probably have endorsed something close to Newton's reverberant thesis that "geometry is founded in mechanical *praxis,* and is nothing but that part of universal mechanics which exactly proposes and demonstrates the art of measuring,"[117] his defense of the relevance of kinematics turns centrally on the insight that it is by continuous movements deliberately executed (by the hand or by its instrumental counterparts) that the legitimate curves *come into being.* The kinematic criterion is thus the Cartesian analogue of what Hobbes and Spinoza understand as genetic definitions. In the former case as well as in the latter, an "object" owes its intelligibility to its mode of genesis, *not* to its being somehow accessible to the intellect as ungenerated and ungenerable (compare chapter 2, V). And to this we must add that, for Descartes, the crucial emphasis falls on the mind's having control of this genesis throughout its course, on its being master of the determinate ratios by which each point of a locus is strictly and precisely identified. That *genesis* can be accomplished by means of mechanical instruments is not sufficient to secure the full-dress intelligibility of the generated curve; the instruments must be so chosen and crafted that the requirements of continuity, point-wise determinateness, *and* inventive fertility can be met. At the same time, however, free recourse to devices which do meet these requirements provides us with a most palpable illustration of the way Cartesian *mathesis* cancels the distinction between the "artificial" and the "natural," or, more exactly, invites us to regard the natural as a consequence of the artificial. As Descartes (or, possibly, his translator Abbé Picot) later wrote in the French version of the *Principles of Philosophy,* "so that all the things which are artificial are natural as well [*avec cela*]" (4. 203).

Earlier, I spoke of Cartesian loci as the *analogue* to the contemporary topologists' point-sets (III, v). This qualification was not necessary for the sake of avoiding anachronism but rather because the program of genesis to which the kinematic constraint has made us more acutely alive harbors an especially intransigent problem. On the one hand, the generative motion producing an admissible curve must be continuous; on the other hand, it is the infinity of solutions to the relevant equation(s) which yields an infinity of points each of which is on that curve. Would this entail that the Cartesian geometer, *per impossible,* must single out each one of this infinite number of points before he can take full responsibility for the intelligible being of the whole curve? In Book 1 of *Geometry* Descartes asserts: "Taking in succession infinite different values [*grandeurs*] for the line *y* we shall also find infinite different values for the line *x*, and in this way we shall have an infinity of different points . . . *by means of which* we shall describe the curved line required" (my emphasis).[118] The seemingly unqualified language of this passage notwithstanding, Descartes elsewhere is expressly disdainful of arguments which assume that a line "is composed of an infinity of points *in actu,* which is simply a pure fantasy" (to Mersenne, October 11, 1638).[119] His position, then, would appear to be that the locus does not consist *ab initio* of an infinite set of points, but brings those points into being as it itself is *in statu fiendi,* on the way to becoming. A graph which can only be "constructed" between discrete points is, in fact, a nonconstructible, a mechanical or transcendental, curve, according to Descartes. Indeed, having "to trace all the points only by means of the encounter of two independent movements" is, by his argument, the same as not being able "to find geometrically *any* of the points which are necessary for the effects desired" from such nonalgebraic curves (here, the helix and the quadratrix; see *A.-T.,* t. 1, p. 69).

Narrowly focussed though it may seem, this consideration of the interplay between continuity and punctual discreteness will prove to be on a reduced scale a technically perspicuous version of perhaps *the* great riddle of Descartes' theory of thinking; deduction, understood as the "continuous and nowhere interrupted motion of thinking" (Rule 7), produces pointlike instants of intuition which exist *in actu* only as "values" of that motion on some particular course. Cartesian minds, just as Cartesian bodies, are always on the move. (See III, vi on the tie between the kinematic criterion and Descartes' conception of mental activity.)

I have spoken of equations and the kinematic criterion for admissibility as liberating constraints through which it becomes possible for us to solve problems. Taken together, they teach us a supremely revealing lesson: In Cartesian geometry, if not in Cartesian philosophy as a whole, the being of magnitudes (discrete or continuous, homogeneous or heterogeneous) and of curves (naturally occurring or only instrumentally producible) is bracketed or made tributary to the facilities and limits of technical genesis. The appearance of a certain kind of curve (such as a spiral) in the "world" is not a decisive reason for accepting it as a geometrical entity in good standing; we must first determine the structure of its equation and the possibility of generating or regenerating it by continuous and interdependent movements in virtue of which its reappearance is a human feat. So, for example, every algebraic (nontranscendental) equation can be constructed as a curve or locus, but the inverse is not true. Cartesian prudence amounts essentially to being loyal to the regulations of technical exactitude, where Euclidean prudence showed itself, for the most part, as responsiveness to nontechnical shapes and the pre-understanding which they evoke. In the final portion of this chapter I shall try to probe more deeply into the basis of this contrast.

vi   *The Roots of the* Geometry: *Unity, Order, and Measure in the* Rules

The *de facto* operation of these two liberating constraints—the homogeneity of the dimensions of the terms in any equation and the kinematic determinateness of the relations or ratios among the lines to be constructed in accord with an equation or a set of equations—leaves hanging the question of their rightful or justifiable employment. When we shift our focus from the question of fact to the question of right or entitlement we are brought into the ambit of Descartes' earlier, incomplete work, the *Rules for the Direction of the Mind.*

I have already suggested that the *Rules* and the *Geometry* enjoy significant affinities with one another. To be as brief as possible: The whole of the second book of the *Rules* seems to have been conceived as a prolegomenon to a presentation of Descartes' new geometry, even if not precisely to the *Géométrie* published in 1637. In particular, the final three rules—19–21—are concerned directly with the formation of equations. It is also worth mentioning that the most probable dating of the *Rules,* to the years 1628–1629—places it in close proximity to Descartes' visit to his former teacher and confidant, Isaac Beeckman, who reports in his *Journal* that when Descartes arrived, he claimed to have perfected in Paris his algebra, "by which he attained to the whole of human cognition."[120]

Let me leave questions of chronology and textual stratification to one side. In the *Rules,* Descartes addresses in the most direct way the two questions that are centrally germane to his work in *Geometry*—if not to all of his thinking without exception. The two questions are: What is a science, an organized and unimpeachable body of knowledge? And, what is it to come into possession of a body of knowledge? That is, what is it to come to understand something, to learn it? What, in short, goes into the achievement of what the Greeks, whom Descartes follows here, called *mathēsis?*

The interplay of these two questions and of the answers Descartes furnishes defines the space in which all of his thinking takes place—or so I would want to argue. Since I cannot give the full argument here, let me suggest some of the directions it would take:

(1) First, *mathēsis*, the process of learning, is, for Descartes, the measure of *scientia*, of *epistēmē*. That is, how it is we can come to know something, how we learn or ought to learn, determines the character and the claims of the science we come to know. The logic of discovery is one and the same with the logic of justification, as we might say; this in prominent contrast to Aristotle, for whom *epistēmē* in its most exact sense does not reproduce or reenact the actual (or even ideal) course of learning.

(2) Secondly, the nature of *mathēsis*, the completed activity of coming to know, furnishes us with criteria for discriminating between authentic and inauthentic sciences. More particularly, it is the nature of *mathēsis* that allows us to understand why the pre-Cartesian tradition, perhaps without full self-consciousness, singled out some discipline as "mathematical." Descartes tries to articulate the reasons behind this selection; at the same time, he tries to show why all the disciplines, that is, the total range of what can be scientifically known, *must* conform to the format *elicited* from mathematics in the traditional, or, as he says, "vulgar" sense, and exhibited in its pure and unique form in something he calls *mathesis universalis*—comprehensive or universal learning.

(3) Descartes scholars often insist that Cartesian *method* ought to be distinguished from this *mathesis universalis*, perhaps in the way form is distinguished from content, or, more precisely, in the way rules of procedure are distinguished from the domains of their application.[121]

If my suggestions to this point are right, then no such distinction can be maintained: Method, in Descartes, not only codifies rules of procedure; it constrains those "objects" to which it is applied to such an extent that their very intelligibility becomes identical with their susceptibility to methodical treatment. In other words, all and only those beings apt for inclusion in the one comprehensive *mathesis* fall under the sway of method and vice versa.

Let me substitute for a detailed defense of these claims a very brief consideration of what Descartes tells us about *science* and about *mathesis* in the *Rules*.

The first rule in this book of *rules* has to do with the scope and the unity of *scientia humana*, "human science or knowing as such." In the boldest of all his images, Descartes compares this "human science" to the light of the sun: The latter illuminates all visible objects equally without being modified by their individual differences; analogously, "human science" remains one and the same no matter what is the nature of the subjects to which it is applied. Descartes has interwoven here, as far as I can tell, two ancient images: the sun as the image of the Good in Plato's *Republic,* where the illuminating power of the Good bestows intelligibility, the capacity for becoming known, on the distinctively knowable things (that is, the forms), and Aristotle's baffling and elliptical reference in *De Anima,* Book 3 to a variety of *nous,* of intellect, which "makes" all things in the same way that light can be said to make potentially colored beings into actually or actively colored beings. In combining these two central images of the ancient theory of mind, Descartes silently suppresses what is crucial to both, namely, that, in Plato, the human soul, although it has the look of the Good, is not the Good itself and thus is not the source of the intelligibility of the Forms, while, in Aristotle, much more ambiguously, this poetic or productive mind is never fully identified with what is elsewhere called "*nous* in the [human] soul." Indeed, a venerable tradition that begins with Alexander of Aphrodisias and reaches the School of Padua, Galileo's school, in the sixteenth century via Averroes, held that this *nous poiētikos* or *intellectus agens,* as it came to be called, can only be identified with the apparently self-reflective and fully actual thinking of the Divine Mind or Prime Mover.

Whatever might be said for or against this reconstruction of the sources of Descartes' initial image, it should be evident that what he achieves by it is nothing less than the unification of the sciences and intellectual disciplines *by fiat:* Since it is one and the same "human science"—or, as he later called it, *mens pura,* pure mind—that is applied to diverse subjects, its unity and singularity override what might otherwise seem to be the *irreducible* diversities or discrepancies among the things which it knows. Thus, with a single stroke, Descartes undermines the Aristotelian and Scholastic distinctions among the theoretical sciences—metaphysics, physics, and mathematics—where these distinctions stem, not from different "methods," but from intrinsic differences among the sorts of entity to which our "methods" are addressed.

The same point can be put in a different way: If Aristotle calls the mind the place of the forms (*topos tōn eidōn*), it is because the mind becomes actually what it is only insofar as it takes on the identities of the intelligible forms it meets with in the world; for Descartes, the mind is, so to speak, the locus or place of the forms which are intelligible only insofar as *they* take on the identity of the mind.

What underlies the condition of homogeneity is, therefore, the unity and uniqueness of "human science." No differences among "objects" will ever be so extreme as to defy or disturb the sameness which the mind by its nature imposes on its fields of inquiry. However, this sameness remains "abstract" or "empty" so long as its nature or more exact content is left indeterminate.

It is at this point that the interplay between *science* and *mathesis,* to which I alluded earlier, becomes all-important: It is by studying what distinguishes success in learning from failure, that we shall, according to Descartes, come to recognize what it is that gives determinate content to the sameness or homogeneity imposed by the mind on everything that can be learned.

Descartes first raised the question of the nature of *mathesis* in the form of an etymological query: Why do the disciplines we commonly call *mathematics* have a seemingly preemptive claim on this designation? The question is a traditional one. Proclus, from whom Descartes is likely to have learned it, answers by saying that the mathematical sciences deserve this title because they are preeminently apt to awaken recollection in the soul. "Learning," so Proclus interprets Socrates as saying in *Meno*, "is nothing other than the mind's recollecting its own reasons, its own *logoi*."[122]

Descartes' etymological exploration brings him to a result extraordinarily close to that of Proclus, although their intentions are, in the final analysis, incommensurable: For Descartes, the mind, by taking account of its own innate or congeneric *logoi* (the "simple natures," in Descartes' vocabulary), comes to understand how any one thing it intends to learn can be placed in a determinate relation or ratio with something else it has already come to know or is antecedently capable of learning. The action or motion the mind performs in setting up such a relation is called *comparatio*—"in every instance of discursive reasoning we know the truth with precision only by way of comparison," as Descartes says in Rule 14.

It is from this generalized notion of comparison that Descartes extracts what he calls, once and once only, in his published and unpublished writings, *mathesis universalis*.

The phrase is not original with Descartes; it is first attested in a text by the Dutch mathematician Adrian van Roomen, *Ideae Mathematicae*, published in 1593. Nevertheless, Descartes gives the notion it was intended to convey its most radical inflection.[123]

The history of this notion of a universal *mathesis* beings with Aristotle, who in *Metaphysics*, Book Epsilon, refers to a "universal or general mathematics which is common to all the branches of mathematics." What he means becomes somewhat clearer in the *Posterior Analytics* when, in Book 1, chapter 5, he discusses the theorem that proportional magnitudes are proportional alternately or *alternando*, that is, if $A:B::C:D$, then $A:C::B:D$.

Previously, Aristotle says, this theorem was proved separately in the case of numbers, lines, solids, and times. Now the theorem is proved for all these cases at once, even though they differ in species and do not fall under one common name denoting a feature in respect to which the theorem is unrestrictedly true. Aristotle's brief discussion immediately evokes the Euclidean theory of ratios and proportions embedded in Euclid, Book 5, where the theorem in question appears as Proposition 16. (We would do well to remember, at the same time, that although Eudoxus is almost always credited with articulating a theory of ratio that applies to commensurable and incommensurable magnitudes alike, there is not a single, unequivocal indication that either he or Euclid understood number as a species of magnitude (*megethos*); this opens up those immeasurably intricate questions of the relationships among Books 5, 7, and 10 of the *Elements*, to which I devoted considerable attention in chapter 2.)

At all events, it was Proclus' commentary on the *Elements*, edited in Greek by Grynaeus in 1533 and translated into Latin by Francesco Barozzi in 1560, that gives the topic of a general, universal, or common mathematics a special prominence and urgency in the sixteenth and early seventeenth centuries.[124] Especially relevant in this context are Proclus' efforts to set "the single science that embraces alike all forms of mathematical knowledge" into relation with Aristotle's science of being *qua* being and with the science of dialectic programmatically discussed in Book 7 of Plato's *Republic*. It is of paramount importance that Proclus and most of the Renaissance scholars who commented on his comments viewed the relationship as one that preserved the hierarchical distinction between ontology and universal mathematics.

Descartes, unlike any of his predecessors, sets out to collapse this distinction into an identity. The result is a *mathesis universalis* in a sense peculiar to him: Namely, that the beingness of objects, what they are *qua* beings, coincides with their mathematical intelligibility, and this means, their comparability—that is, in the strictest of senses—their ability to enter into determinate ratios and proportions with other, homogeneous beings. It is with this in mind that Descartes, in the second half or version of Rule 4, defines the proprietary domain of *mathesis universalis* as that of *order* and *measure*, a domain he claims to have detected beneath the covering—the *integumentum*—of arithmetic and geometry as they have been vulgarly or traditionally taught.

To understand the coordination, indeed the inseparability, of order and measure is to penetrate to the roots of Descartes' decisive innovations in mathematics, and perhaps, in philosophy as a whole. Let me try to capture the spirit of this connection in as short a space as possible:

(1) *Order,* for Descartes, is always sequential or linear order; it consists in "lining up," so to speak, a *series* of items so that their relative positions in the sequence are unambiguous.

(2) *Measure,* again for Descartes, is applicable to *order* when the first element in such a sequence is given a special status, that is, when it is made the unit in terms of which the intervals between all the remaining terms can be precisely expressed.

(3) When Descartes writes, in the climactic Rule 14, "All the relations that can obtain among entities of the same genus are to be referred to two headings, namely to order or, alternatively, to measure," the phrase "of the same genus" is disingenuous, since it is the main business not only of Rule 14 but of the *Rules* as a whole to show that there is only one *genus* of entities. At the start of Rule 6, Descartes had said: "This rule contains the principal secret of this art . . . for it advises us that all things can be disposed in certain series or sequences, not, to be sure, insofar as they are referred to some genus of being, as the Philosophers (i.e., the Aristotelians) have divided them into their categories, but insofar as some things can be known in terms of other things."[125]

It is an immediate consequence of the homogeneity imposed by "human science" on the totality of what can be known that all items in that totality are comparable with one another; comparability means that we can always speak in a determinate manner of one item's being more than, less than, or equal to, another item. Distinctions among the categories of being, in the Aristotelian sense, fly in the face of this requirement—for we can give no sense to the statement that a quality is greater or less than, say, a spatial position. Moreover, as Aristotle says in his book *Categories,* we can never say that one substance (*ousia*) is more of a substance than any other substance. So, Descartes is obliged, by the uniqueness of "human science" and the determinateness of *mathesis,* to reduce all the categories of being to a single category, that of *relations,* which Aristotle had regarded as merely an "offshoot" of being. A text from *Metaphysics,* Book Nu, comes readily to mind: "There is nothing great or small, many or few or, generally, relative, which is many or few, great or small, or relative to something without [also] being something else in its own right."[126] It is precisely this thesis that Descartes repudiates.

The Cartesian category of "relation" can be further specified: It is not, as should already be clear, any and every relation that qualifies as a candidate for inclusion in this *mathesis universalis,* only relations among "magnitudes" participate in the "more and the less" and can thus be "reduced" to equality when equations are appropriately formed from the ratios these magnitudes sustain to one another within an ordered sequence.

It is of fundamental importance here that although these *magnitudes* are said by Descartes to be "magnitudes in general," that is, in abstraction from any distinction between the discrete and the continuous, the numerical and the geometrical, nonetheless, the whole idea of magnitude in the *Rules* is modelled on *continuous,* geometrical quantity. Number, in the sense of a multitude of discrete units, far from having an equal share in this new notion of magnitude in general or "symbolic magnitude" is suppressed in favor of a remarkable derivation of numerical multiplicity and discreteness from continua.

If we divide up a continuous magnitude into several equal parts—where equality is determined by some measurement—and then consider these parts in their order or ordination to the original whole, we can be said to be counting or "numbering." This conception, drawn from Rule 14, is not without its difficulties—as Descartes acknowledges when, after claiming that number can properly be called a species of dimension, a parameter for comparative measurements, he immediately adds "even though there is some diversity in the signification of this name 'dimension.'" This much, however, is clear: The "parts" we count up have no integrity of their own; in other words, they depend wholly on the pregiven continuum from which they are extracted by division. Furthermore, these parts are *units* for counting in a totally arbitrary sense of "unit," since there are innumerable ways in which we can cut a continuum into equal parts. This result is of a piece with one of Descartes' decisive commitments in the *Rules* and elsewhere: Unity and integral units are not to be found in the nature of things; instead, they are in every case the product of an essentially arbitrary choice to regard such-and-such a quantity as the unit relative to which other quantities can be measured. This "unit"—Descartes refers to it in Rule 14 as an "assumed or presumptive unity"—is an *absolute* relative to the magnitudes it is used to measure; "in itself" it is as relative as they are and hence equally subject to the mind's arbitrary or discretionary choice. Descartes, we might say, gives radical sense to Protogoras' maxim: "Man is the measure of all things."

This absorption of discreteness into continuity has enormous consequences for the *Rules* and for the *Geometry*.

Most conspicuously and bafflingly, Descartes appears to gloss over the *aporia* of incommensurability, the problem which, as we saw in chapter 2, inaugurates that line of mathematical thinking which finds its classical expression in the Eudoxian-Theaetetan theory of ratios embedded in Euclid, Books 5 and 10. When, in Rule 15, Descartes refers to cases in which "two magnitudes are incommensurable with unity," he does not pause to take stock of the inconcinnity between their cases and the unrestricted possibility of measurement envisaged in Rule 14, where measuring is reduced to, or reinterpreted as, the inspection of the order of successive multiples of some unit. As Costabel points out in his note to the relevant passage of Rule 14 (A.-T., t. 10, p. 452. 11–12): "To think that, whatever the continuous magnitudes being considered, there exists a magnitude of the same kind taken as unity, such that one or the other of the given magnitudes is the exact multiple of unity, or at least is captured between two successive multiples [of unity], this is to fail to recognize that metric art requires the axiom of Eudoxus-Archimedes."[127] (See Heath ad Euclid Bk. 10, Prop. 1.)

Descartes' haste to pass by these troubling implications of the reduction of *all* measurement to something like the inspection of an ordered sequence of multiples of a unit is illuminating in two connected ways. First, he tacitly undermines the ancients' distinction (examined above, in chapter 2, II, i) between the *naturally* and *conventionally* commensurable/incommensurable magnitudes, as this is made known to us in the first section of Euclid's Book 10. Taken strictly at his word, Descartes countenances only those straight-lines (*depicting* magnitudes of whatever sort) which are *rational* (*rhētai*) relatively to a chosen straight line (Bk. 10, Def. 3); still in Euclidean terms, there are no *alogoi*, unsayable or irrational magnitudes, for Descartes.

In turn, this abandonment of the *physis/thesis* distinction with respect to (in)commensurables and (ir)rationals is of a piece with the subordination of measure to order. In the most summary terms, Descartes' thought seems to be that securing precision of measurement (by whatever scale) is much less consequential than preserving exactness in the sequence of steps or moves the intellect makes as it sets about disposing and comparing the terms ingredient in a given problem. In this respect Descartes does not so much "arithmetize" geometry, as it is often said, as claim that the *operations* of arithmetic are independent both of the nature of arithmetical or quasi-arithmetical multitude and of the strictly metric character of geometrical *magnitude.*

Up till this point, I have been looking upon these operations of comparison, of ordering and measuring magnitudes in general, as occurring in or by means of, the intellect alone, the "pure mind" as Descartes calls it in the *Rules.* In Rule 14, however, Descartes introduces certain representations, namely, line-segments and geometrical figures, for magnitudes in general. These representations bring into play, and appeal to, the imagination, thereby ushering in the most difficult of all the problems Descartes confronts in the *Rules,* if not in his entire philosophy: What *is* the relationship between the intellect and the imagination?

Since a second, book-length, study would be required to unravel this enigma, let me simply say here that the very possibility of geometrical construction, as Descartes understands it, depends on the collaboration of these two "faculties," one "pure," the other, at least in the *Rules,* called "a true body, extended and figured."[128] To be a bit more precise, the imagination has to serve as both the instrument by which and the medium in which the prescribed courses of technical genesis can be both carried out and appreciated for what they produce. This collaboration, in turn, vitally depends on having the intellect place the imagination in its service, not vice versa! That is, images (such as algebraic symbols) must be formed as the records of intellectual motions, not as the by-products of untutored or presystematic perception. On the other hand, the mastery of intellect over imagination is never absolute, as the techniques and limits of the *Geometry* itself have taught us. Not every equation can be "constructed" in accordance with Cartesian regulations, and this means that not *every* intellectual motion or sequence of motions can in fact find its corporeal representative or surrogate. Analogously, any rightly formed equation has as many roots as exponents of its unknown term, but some of these roots will be "imaginary," that is, strictly speaking, *unimageable* and hence without potential purchase on the extended world. Conversely, not every imageable curve answers to an algebraic equation (with only integral exponents) and thus cannot be the product of purely intellectual motion or technical genesis. Even if we are considering the imagination as an inward faculty or power, its "corporeality" is a kind of vicarious worldliness or externality. As such, it acts as a check on the intellect's aspiration to total mastery, although without disrupting or infecting the homogeneity of its operations.

The *Rules,* treated here so hastily and inadequately, thus supplies the justification for the dimensional homogeneity presupposed by the *Geometry:* "Magnitudes in general" are all homogeneous with one another and can be arranged accordingly in all sorts of ratios and proportions— x^3 is not a cube, a three-dimensional solid, but the sign for a general magnitude standing in the relationship $1:x::x^2:a$. Thus, $a = x^3$.

Similarly, the kinematic criterion employed in the distinction between geometrical and mechanical curves rests on the continuity of the motions the mind performs as it passes along a series of relations in quest of new knowledge, that continuous motion Descartes calls "deduction."

Indeed, Descartes at his most radical understands *ideas* as operations and thus as motions (see the identification of *ideae* with *operationes intellectūs* in the *Fourth Replies* [A.-T., t. 7, p. 232]). It is difficult to know with certainty whether "motion" is used both of the mind and the body metaphorically or equivocally. In any case, Descartes can be seen pursuing the correspondence in the notes and marginalia preserved by Leitniz: "Intellectio est ad mentem ut motus ad corpum et voluntas ut figura: deflectimus ex una intellectione ad aliam, ut ex uno motu in alium" (A.-T., t. 11, p. 647; compare p. 650 as well). Gilson captures the force of this radical reconstual when he writes: "Cartesian logic will be such . . . that the content of reasoning will *ipso facto* engender its form, the latter doing nothing more than formulating the very movement achieved by the mind in the analysis of ideas."[129] To which I would want to add, these ideas themselves, taken in their singularity and distinctness, are in essence only so many way stations or resting points along the path of that mind's "continuous and nowhere interrupted movement of thinking" (Rule 7).

Consequently, Descartes' choice of a kinematic criterion for deciding the admissibility of geometric curves stands in the closest harmony with, and rests upon, his understanding of thought as continually restive; the curves actually constructed, imaginatively or instrumentally, but always in a properly determinate, gap-free fashion, depict in a worldly way how productive that restiveness can be.

Mathesis universalis, therefore, is not an axiomatic or formalized system from which an infinity of statements can be derived; it is rather a set of useful prescriptions or laws for the motions the mind ought to make whenever it undertakes to solve a new problem. As one might expect, the legitimate motions are all rectilinear and continuous, reaching its limit in the instantaneous velocity of intuition, the light of the mind. The *Rules* direct the mind, which naturally tends to move in futile circles, along straight paths towards its cognitive goals.

This suggestion allows me to pose one final question, to which I have already referred: What is the relationship between *mathesis universalis* and Descartes' *Geometry?* Put in a different way: Why did Descartes entitle his one and only published work of mathematics *La Géométrie,* rather than "Algèbre" or "Ars analytice" or even "Mathesis Universalis" itself?

Despite the fact that he uses the latter phrase only once in his extant works, we do have more documentary evidence of its familiarity to Cartesian disciples and critics than has usually been noticed. Thus, while Florimond de Beaume continues to use Viète's rubric, *algebra speciosa,* to name the content of Descartes' *Geometry,* Frans van Schooten, in his edition of the Latin translation of the latter published in 1661, did not hesitate to entitle his own contribution *Principles of Mathesis Universalis, or, an Introduction to the Method of Cartesian Geometry (Principia Matheseos Universalis, seu Introductio ad Cartesianae Geometriae Methodum).*

Descartes' English critic John Wallis similarly chose *Mathesis universalis, seu Opus arithmeticum* as the title of his book (London, 1657). Most revealing of all are Leibniz' very frequent discussions of Descartes' pretension to have furnished the *complete* version of a *mathesis universalis;* so, for instance, he writes in his essay "On the Origin, Progress and Nature of Algebra": "Algebra [of the Cartesian sort] cannot be confounded with *Mathesis Universalis.*"[130]

These indications show us that both the letter and the spirit of Descartes' designation remained alive; it is equally likely that at least some of its uses on the part of his own students had his *imprimatur.* In light of this evidence, the *titular* choice of "Géométrie" remains perplexing.

A quite similar perplexity was felt by one of Descartes' correspondents, Ciermans, who wrote in 1638 to ask why he had not called his *Géométrie* "mathematica pura," since it concerns matters common to all the mathematical sciences (*A.-T.,* t. 2, p. 56). Descartes' reply is instructive:

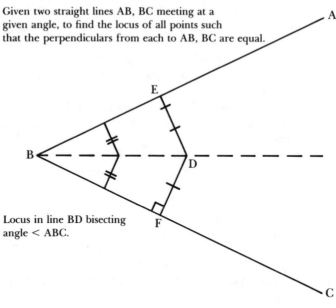

Given two straight lines AB, BC meeting at a given angle, to find the locus of all points such that the perpendiculars from each to AB, BC are equal.

Locus in line BD bisecting angle < ABC.

A Simple Locus Problem

I have not explained any of the things which pertain properly to arithmetic, nor have I solved any of those questions in which order is considered simultaneously with measure, of which we have examples in Diophantus. But even more, I have taught nothing concerning motion, with which pure mathematics, which I have particularly cultivated, is especially concerned. (*A.-T.*, t. 2, pp. 70–71).

This reply was, I think, to be taken *cum grano salis* at least as far as it touches on our present topic (and insofar as we can, here, overlook any differences between *mathesis universalis* and *mathematica pura*). First, Descartes often expresses his disdain for purely arithmetical or number-theoretic problems; his grudging role in the controversy between Stampioen and Waessenaer over the particular *numerical* values of the roots of a certain cubic equation is adequate testimony even by itself.[131] More importantly, though, we have already seen how in the *Rules* arithmetic, the domain of strictly discrete multitudes, was either eliminated or sublated into the continuous. We could have also said that *cardinality* is suppressed in favor of *ordinality*.

As for Descartes' second indication of what is lacking in the *Geometry* (cases where order and measure are considered simultaneously), he is almost surely both referring to Diophantus' work on polygonal numbers and anticipating his own (unpublished) essay "De solidorum elementis"

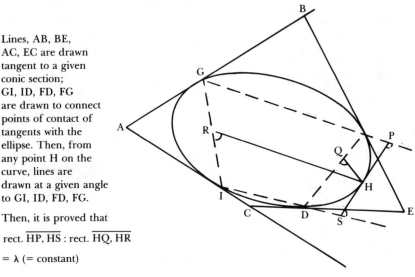

Lines, AB, BE,
AC, EC are drawn
tangent to a given
conic section;
GI, ID, FD, FG
are drawn to connect
points of contact of
tangents with the
ellipse. Then, from
any point H on the
curve, lines are
drawn at a given angle
to GI, ID, FD, FG.

Then, it is proved that

rect. $\overline{\text{HP, HS}}$: rect. $\overline{\text{HQ, HR}}$

$= \lambda$ (= constant)

"Appollonian" 4-Line Locus Property

in which he examines a variety of connections between the sides (or faces) and the angles of polygons and polyhedra. What is important here is that Descartes' "Theorem" (later proved or reproved by Euler), $S - A + F = 2$ (where S is the number of the vertices of a simple polyhedron, A, the number of its angles of intersection, and F, the number of its faces), is tied to the *measure* of the solid angles involved. It is a comparatively minor lacuna of this kind to which he is drawing Ciermans' attention.[132]

As for the absence of attention to motion, we can merely conjecture that Descartes has in mind here what he elsewhere terms "une autre sorte de *géométrie*" in which the mathematical laws of the actual motions of body are to be exhibited. The *Geometry* itself, however, has already put sufficient emphasis on the kinematics of geometrical construction for us to know that motion as such is in no way alien to its constitutive subject matter.

On balance, then, Descartes' reply to Ciermans does nothing to compromise the choice of *Géométrie* as a title for at least a specimen portion of *mathesis universalis*.

This leaves as geometry's only titular and possibly, substantive rival, *algebra*. What has to be most carefully noted is that Descartes himself, unlike most of his modern interpreters, treats algebra as a mere "calculus"; as we noted earlier, in a letter to Desargues, he even goes so far as to suggest that the "arithmetical" calculus—that is, his algebraic symbolization and manipulation of line-lengths in Book 1 of the *Geometry*, is

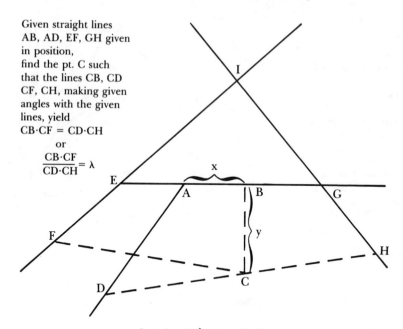

Given straight lines
AB, AD, EF, GH given
in position,
find the pt. C such
that the lines CB, CD
CF, CH, making given
angles with the given
lines, yield
$$CB \cdot CF = CD \cdot CH$$
or
$$\frac{CB \cdot CF}{CD \cdot CH} = \lambda$$

Cartesian 4-Line Locus Problem

largely a concession to the vulgar. (See, too, his slighting references to "ma vielle Algèbre.")[133] It appears, then, that the *Geometry* is not mistitled; it calls into its service a certain technique of calculation that makes possible the formation and transformation of appropriate equations. The "symbolic abstraction" on which this technique rests is, however, only a preparatory first step. The real work of geometry begins when these equations are taken as instructions for the graphical construction of the relevant loci and roots themselves. In this work of graphical construction, the intellect and the imagination must somehow be equal partners. Indeed, we can say that the equations encode those acts of the mind by which order and the possibility of measure are self-reflectively grasped; the *proof* of the validity of these acts lies in the success we enjoy in embodying their claims in the 'imaginative,' but not merely imagined, space of geometrical construction. This must be so if Descartes is to pass from his abstract geometry to that strange and fabulous new physics in which, so he was convinced, he could disarm what he once called "the objection of objections," namely, that mathematical "bodies" are real physical "bodies."[134]

Accordingly, the *Geometry* itself, as it unfolds before the reader the motions the mind makes when it learns and as it embodies or exemplifies the liberating constraints which give these motions canonical direction, can be seen to be, if not the whole, at least the heart, of Descartes' *mathesis universalis.* (See Figures 1, 2 and 3, illustrating the locus-problems in Apollonius and Descartes.)

IV *Objectum Purae Matheseōs:*
Mathematical Construction and the Passage from Essence to Existence in Descartes' *Meditations*

i *Reading Notes*

Reflections on Descartes' exemplary fable of self-making in the *Discourse* and on the style of mathematical practice in the *Geometry* are quite likely to seem, to most orthodox interpreters of Descartes, oblique to the main issues he raises in what as come to be regarded as his central philosophical text, the *Meditations.* These interpreters could be said to agree heartily with d'Alembert's terse discrimination, cited in the preceding section, that "Descartes can be considered on the one hand as a geometer, on the other, as a philosopher."[135] Moreover, these same readers, whether analytical or phenomenological in their allegiance, are customarily suspicious of any suggestion that Descartes would have written so axial a text as the *Meditations* with ulterior or veiled purposes significantly distant from, or even at odds with, his publicly announced intentions. In particular, the implication that one or more subtexts or pretexts are latent behind the official text of the *Meditations* strikes such readers either as impertinent or, at least, as bringing an irreparable loss in epistemological and metaphysical relevance.

In this section I want to allay such suspicions by trying to unveil one of the subtexts which has been in great part obscured by the imposing surfaces of the *Meditations.*

Fortunately, two of Descartes' own private declarations both dissolve any anxiety that such an attempt will inevitably be arbitrary and furnish essential clues to the directions in which this (and other) subtexts should be sought.

On the eve of the publication of the *Meditations,* he wrote to Mersenne:

> And I will tell you, between us, that these six *Meditations* contain the foundations of my physics. But, please, one must not say this; for those who favor Aristotle would perhaps have more difficulty in approving them and I hope that those who read them will become accustomed to my principles unawares and will recognize their truth before noticing that they will destroy Aristotle's. (*A.-T.,* t. 3, p. 298)

We need only remind ourselves of his equally familiar assertion, also to Mersenne: "The whole of my physics is nothing other than geometry" (t. 2, p. 268), to be set straightaway in search of the *geometrical* foundations of Cartesian physics, foundations kept far from visible in the public text, but potentially recoverable from a subtext or pre-text towards which Descartes might have wanted to point some of his contemporary readers.

In the finale of the *Meditations*, the passage from essence to existence, Descartes does indeed seem to be giving such clues when he uses the phrase *objectum purae matheseos* three times in the last two parts and only in those parts.[136] (Readers of the French translation were spared the evocations of the Latin phrase by circumlocutions such as "l'objet des demonstrations de géométrie" with its much more traditional ring.) Accordingly I want to try to see how the last two Meditations might look *sub specie purae matheseos*, once this has been more distinctly identified. At the end of section IV. v, I want to compare some of the features of the sketch which will emerge with a few prominent aspects of the published or official text. One intended consequence of this strategy should be that Descartes' search for firm foundations is not to be understood as a free-floating, "epistemological" enterprise indifferent to the technical achievements and prospects of particular bodies of knowledge; on the contrary, Cartesian foundationalism remains moored, for good or for ill, to the specific format and the warranted procedures of *physico-mathematical* explanation such as he envisioned and elaborated them.

ii   Pura Mathesis *and Descartes'* Geometry

Even a slightly more than cursory inspection of the tortuous paths of doubt laid out in Meditation 1 and pursued until the end of the text will have encountered a number of ambiguous signposts and apparent *culs de sac*. (1) How are the arguments against the veracity of sensory information and inference related to the arguments against the reliability of mathematics? For example, how strongly or how faintly are the premises and prejudices uncovered in perceptual beliefs—the externality of the referent and its (putative) existence, the resemblance of the *sensum* to its "object" and the causal origin of sensation in something other than the pertinent faculty—still at work in the critique of mathematical imagination and, even, mathematical intellection? (2) What takes place in the transition from mathematical simples to imaginative complexes, that is, the reverse of the explicative movement Descartes traces in Meditation 1? Are those simples themselves locatable in, or fabricated by, the same imagination which the painter uses to depict a Siren or do they reside and originate elsewhere, perhaps in the pure intellect? And if the latter, how do we get from these "more simple and universal" items to the complex images fashioned out of them? Is this process essentially one of demonstration or deduction and hence open to doubts of its validity on each occasion or of its reliability over time? Or does Descartes envision another sort of procedure, possibly subject to another variety of doubt? (3) What does it signify that initially the mathematicians are said "not, or barely, to care whether their 'things' are in the nature of things" (*utrum eae* [*res*] *sint in rerum natura necne, parum curant* [A.-T., t. 7, p. 20]), while in the opening paragraph of the last *Meditation* the now-indubitable existence of the object of *pura mathesis* is the bridge from essence to existence generally? What, in other words, is the relation of mathematical objects treated as *res simplicissimae et valde generales* [Med. 1] to the "same" items considered as *res materiales, quatenus sunt purae matheseos objectum* [Med. 4]?

What light, if any, might Descartes' only published work in mathematics, the *Geometry* of 1637, throw on these and cognate questions? These are not questions that he treats *ex professo* in that text, although the near or exact counterparts of several of them *are* on exhibit in the *Rules,* the prelude to something quite akin to the later *Geometry.* It is rather as if the *Geometry* offered us in the mode of realized and felicitous practice what the *theoretical* reflections of both the *Discourse* and *The Meditations* must reckon with and account for. The temptation to allot priority or greater importance to the theoretical must, however, be checked by the recollection that, for Descartes himself, the hallmark of his new science is precisely that it is not unproductively speculative but apt to the making of useful results. Gauged by that explicit measure, a work embodying the products of epistemic inventiveness (such as the *Geometry*) cannot be simply subordinated to a text exploring the "theoretical" foundations of that inventiveness.

Even granted that this is interestingly so, are we entitled to take *pura mathesis* in the *Meditations* as an analeptic allusion to the *Geometry* or, at least, to the project it actively, if partially, displays? Doubts are quickly put to rest by Descartes' own declaration, in a letter to an anonymous correspondent in 1637 setting forth the plan of the *Essays* following the *Discourse:* the *Dioptrics* treats of "a subject mixing philosophy [physics] and mathematics," the *Meteors,* "a whole purely of mathematics"—*un tout pur de Mathématique* (A.-T., t. 1, p. 370).

This documentary evidence does not, however, eliminate further corollary questions provoked by the identification of the *Geometry* as the subtext of *Meditations* 5 and 6: What does *objectum* mean here? What are the ties that bind *pura mathesis* in the style of the *Geometry* to that "other kind of geometry which takes for its questions the explication of the phenomena of nature" (t. 2, p. 268) on which Descartes tells us his attention was trained? Finally, what affiliations are detectable between the "pure *mathesis*" exhibited in the *Geometry* and "*mathesis universalis,*" the notorious phrase used only in Rule 4 and, for some, the matrix of the Cartesian project in its entirety? Let me say a word about these three queries.

Objectum in Cartesian idiom refers to the subject matter of a discipline, a *field* of intellectual concentration, as it were, held in place and shaped by the sustained methodical attentions of the mind. An *objectum,* far from being an entity (or class of entities) standing on its own, is just what method requires in response to its objectives (see Rule 2: [*Arithmetica et Geometria*] *habentque objectum quale requirimus*) (t. 10, p. 365). More determinate items (such as geometrical formations) belong to a specified *objectum* as so many modal versions, multiple fashions in which the general character invested in the *objectum* takes on distinctive particularity. This explains why Descartes can shift from plural nouns back to the singular *objectum* in describing "res materiales . . . quatenus sunt purae matheseos objectum" (Med. 5). The synonymous expression *natura corporea in communi* (Meds. 1 and 5) plays the same role in designating a unified field of attention, but inevitably raises the second corollary question concerning the bearing of the *objectum* of pure mathematics on the extended body or substance with which physics is exercised. Not only does Descartes argue against any ontological distinction between extension and *res extensa* (of a kind that would lend support to an abstractionist view of mathematical entities); he is willing to assert the complete identity of mathematical and physical body, even while denying that, for example, impenetrability is an essentially necessary feature of the being of any body. As Descartes notes, this identification constitutes "the objection of objections" raised against his *mathematical* physics; failure to understand that objection would entail abandoning the very science requiring a *fundamentum inconcussum.*[137]

Thirdly, despite both the long controversy dating at least from Liard and the recent historical clarification brought to the matter by Giovanni Crapulli, the meaning and scope of Descartes' *mathesis universalis* remain quite dark. In particular, it has not been sufficiently noticed, as far as I can judge, that this rubric reappears in the titles of Descartes' mathematical successors and opponents and does so in ways clearly suggesting the most intimate affiliation between what he had actually achieved in the *Geometry* and this encompassing science, method, or mathematics *sensu latissimo.* As I remarked in section III, vi, van Schooten, for example, entitled his contribution to the 1651 Latin edition of the *Geometry,* together with contributions by younger Cartesian experts, *Principia Matheseos Universalis, seu Introductio ad Geometriae Methodum Renati Cartesii.* Leibniz, on the other hand, criticizes Descartes' version of *mathesis universalis* not for its ambitiousness but for its overly constricted range: "Meanwhile, Algebra is not to be confused with *mathesis universalis.* . . . In truth, whatever falls under the imagination, insofar as it is distinctly conceived, falls under *mathesis* and for this reason *mathesis* treats not only quantity [as Algebra does] but also the disposition of things."[138]

I have mentioned these three corollary questions in order to keep firmly in view the limits of the sketch to follow. Even if the *Geometry* provides the best specimen of Cartesian *pura mathesis* at work and hence furnishes indispensable clues to an understanding of what Descartes appears to presuppose in the *Meditations*, on its own the *Geometry* cannot explain to us in full *why* methodical attention should take priority over what there is (or seems to be) in the world in advance of *method* or *how* the identity of mathematical and physical body is to be secured. It may, however, illustrate what it would mean for *order and measure* to serve as the unique and exhaustive topics of Cartesian mathematics and, thus, of his philosophy itself insofar as it aims at rigorous certainty.

iii The Principal Characteristics of Cartesian Geometry: Recapitulation and Extensions

Writing to Mersenne about his reactions to Descartes' *Projet brouillon*—the only work by a contemporary mathematician that inspired his enthusiasm—Descartes, as we saw, refers to what "I am in the habit of calling the metaphysics of geometry" (*A.-T.*, t. 2, p. 490). He had specifically in mind the ancient use of so-called *diorismoi* to determine the limits within which the solution to a geometrical problem is possible; but the phrase can also supply a leitmotiv for Descartes' own text if we take "metaphysics" to mean here the establishment of the most general conditions which must be met by any course of fruitful and hence, legitimate, mathematical procedure. What Descartes both does and says in the *Geometry* has already brought to light for us the following aspects of his implicit "metaphysics of geometry." Let me rehearse these again before turning to their implications for the *Meditations*.

(1) Descartes' mathematics is devoted exclusively to problem-solving, *not* to theorem-proving. The proximate occasion for the composition of the *Geometry*, the challenge to solve the long-outstanding Pappus locus-problem for four or more straight lines, also shapes the work's systematic dimension. The latter is not a chain of theorems deduced from axioms and definitions, but a lesson evolved from the sample Pappus-problem concerning how to go about fruitfully solving problems of any sort in geometry (that is, how one can reduce them all to a single type of problem: "to find the value of the roots of some equation" (*A.-T.*, t. 6, p. 401). Two points are in order:

To solve a problem is, in Descartes' vocabulary, to *construct* a problem, not, of course, in the sense of making up a problem, but as the successful

process of finding and exhibiting the relevant geometrical item(s) satisfying the conditions set out in an algebraic equation. This technical usage,
borrowed from Viète, later passed into the phrase "construction of an
equation" in Leibniz and Wolff, whence it became truncated into Kant's
simple word "construction."

Descartes thereby shifts the center of emphasis from the *mos geometricus*
to a dramatically revised conception of mathematical procedure as the
production of an infinity of more and more complex solutions, always
moving from a lower to the next-higher degree of the pertinent equa-
tion(s); this is the technically solid counterpart to the more "atmospheric"
recommendation of Rule 3 in the *Discourse* ("comme par degrés"). Correspondingly, when Descartes talks of *synthesis* in the strictly mathematical
context, he means the *constructions* of the required loci and/or the roots;
these he is content to leave to readers who can enjoy "the pleasure of
inventing them" on their own (*A.-T.*, t. 6, p. 413). In every sense, he wants
to substitute the inventive *activity* of constructing solutions at first hand
for the passivity both of traditional mathematical education via authors
and of the inspection of given figures or complexes of figures in the hope
of catching sight of relations already proved by theorems.

(2) The "metaphysics of geometry" demands a set of uniformly applicable operations so that all the "items" singled out as ingredient in any
problem can be treated on a par and, hence, be subjected to manipulations (multiplication, division, etcetera) guaranteed to yield a new item
or group of items belonging equally to the original format of activity.
The mutually discrepant labels "the arithmetization of geometry," "the
geometrization of arithmetic" capture only superficially the force of Descartes' principle of homogeneity invoked at the debut of the *Geometry* and
sustained throughout. Most fundamentally, this principle cancels the
seemingly natural heterogeneity between discrete multitudes and continuous magnitudes, on the one hand, and, within the latter category, between continua of different dimensions. Descartes invites us to reinterpret the dimensional differences expressed by exponents (x^2, x^3, x^n) as
marking positions in a *continuous ordering of ratios*, not as signifying perceptible (or, in cases where $n > 3$, imperceptible) distinctions (as, say,
between a 2-dimensional square and a 3-dimensional cube). Moreover,
arithmetical integrity is suppressed in favor of the arbitrary choice of a
line-segment to be the unit-measure within each problem; this selection
à discrétion, as Descartes terms it, seems to allow him to circumvent the
issue of incommensurability, the crux of the Euclidean tradition.

(3) Along with the principle of homogeneity, the metaphysics of geometry imposes another strong constraint on full-dress mathematical intelligibility. This is the principle of kinematic determinateness, as I have called
it, according to which only those curves are to be admitted into geometry

which are producible either "by a single continuous motion or by several successive motions, the later ones being completely ruled by those which precede; for by this means one can always have an exact knowledge of their measure."[139] Descartes' decision to widen the class of geometrical curves to include mechanically produced curves such as the conchoid or cissoid, while barring those which cannot be described, by hand or machine, through appropriately conjoined successive movements (that is, to restrict geometry to curves whose equations have only rational exponents), brings to sight several essential commitments.

First, the ancients' distinction between "geometrical" and "mechanical" curves is myopic, since the instrumental origin of a curve need not compromise its intelligibility (see the Cartesian compasses). The Cartesian geometer can intervene with mechanical devices in the traditionally pure theoretical domain of geometry without loss of certainty or exactness. Second, this exactness is owed to the *control* the geometer exercises over the production of the curves utilized in the "construction of problems"; so long as each point of the relevant curve "bear[s] a definite ratio to all the points of a straight line," the curve as a whole can be expressed by a single equation, giving its "precise and exact measure." Epistemic transparency is a function of productive control of motion; or, in other terms, by producing a curve in an exactly regulated manner, one can know with certainty the origin and the determinate measure of each of its (infinite) points. Finally, the outward motions by which admissible curves can be "traced *or conceived*" (my emphasis) are so many allegories or metaphors of inward cognitive movements performed with the same or even greater attention to exactness and orderliness of execution. This inference is supported by the *Rules* in which deduction, passing in a systematic way along a continuous series of determinate ratios (Rule 6), is a *cogitationis motus* (Rule 7); indeed, instantaneous intuition of a whole and continuous deduction from part to part can, in the most desirable instance, appear "to coalesce" into a single operation "through a certain motion of cogitation" (Rule 11). Whatever else needs to be said about the relation of pure *mathesis* to physics, we have to keep in mind this correspondence or homology between mental and extra-mental movements.[140]

(4) When he is not being said to have arithmetized geometry or to have geometrized arithmetic, Descartes is described as having algebraized geometry or, conversely, geometrized algebra. This latter pair of descriptions also fails to reach the core of Descartes' designs. For him algebra is a *tool* for calculating the relations among the lengths and distances figuring in a problem-complex; it has no epistemic (or ontological) standing on its own, even though in its methodical role it borrows such standing from the "nature" of its "objects." Thus, Descartes has to be prodded into releasing to Mersenne et al. an "Introduction à ma Géométrie,"

written by an unnamed Dutchman and later found by Leibniz under the title "Calcul de Monsieur des Cartes." (This short tract sets out more simply and plainly the rules for calculating with algebraic symbols utilized in the *Geometry* and adds several specimen illustrations of elementary problem-solving by calculating the *roots* of equations.) More significantly, he recommends that Desargues employ the algebraic calculus inasmuch as most readers will grasp multiplication more readily than the method of compounding ratios. This recommendation shows us not only the primarily rhetorical or expository function of the algebraic calculus, but also the true root of Descartes' conception of an equation, namely, the composition of ratios.

Compounding ratios is an undefined technique put importantly to use in Euclid despite its being quite illegitimate, since a ratio of magnitudes $(m:n)$ must be treated as though it were itself a magnitude suited for manipulation in (quasi-)arithmetical settings such as multiplication.

Thanks to the principle of homogeneity and the operational freedoms it licenses Descartes is no longer embarrassed by this discrepancy between practice and theory. If the problem at hand is, say, the duplication of the cube (the finding of two mean proportionals between known a and b), we can form the continuous proportion $a/x = x/y = y/b$ and then multiply (compound) the pairs of ratios $a/x, x/y; x/y, y/b$ in turn giving us $ay = x^2$ and $bx = y^2$. Then, also, $x^2 = y^4/b^2$ and $y^4/b^2 = ay$ or, $y^3 = ab^2$, the equation for the duplicate volume of a cube when $a = 2$ and $b = 1$. Moreover, one can always recover the ingredient ratios and proportions from the final equation.

However, a Cartesian equation does not only encapsulate a series of composite ratios; its equational form is decisive in fitting it for a demonstrative role in Cartesian analysis and synthesis (that is, construction). Because an equation, as its name tells us, expresses an equality, and equalities are automatically reversible, the pattern of "upward" inference in analysis ($p \rightarrow q \rightarrow r$, where r is an already-proved theorem or, more precisely, an already-established equation) can be reversed without further ado ($r \rightarrow q \rightarrow p$) to yield a demonstration (another equality). The major "obstacle" in ancient practice to taking analysis as demonstrative in its own right is thus removed with a single stroke. (Still another reason for Descartes' insouciance over failing to give the synthesis of the Pappus-problem in full; that can be done straightforwardly by reversing the analysis.)

The new art of equations also reflects the cognitive stature ascribed to symbolization in the "economy" of Cartesian methodical science. Far from being merely abbreviatory marks, the symbolic letters (Viète's *species*) figuring in an equation embody and announce Descartes' key resolve, first, to treat all magnitudes alike as representable by line-segments (another consequence of the principle of homogeneity) and then, more radically still, to treat the lengths of those line-segments not as "absolute" magnitudes but as functional by-products of the roles their "names" can be seen to play in company with one another and against the backdrop of the *lignes principales,* the Cartesian axes selected *à discrétion* for each locus-problem. In other words, the Cartesian resolve to treat "magnitude in general" by means of algebraic symbolization entails the relativization of magnitude (or measure) as such, or, the same thing seen differently, it is the subordination of magnitude to the rule of ordering. The thematic complicity between the *Geometry* and its own pre-text, the *Rules,* allows us to trace this last implication to an even deeper stratum of Cartesian thinking. Just as human science remains one and the same throughout its diverse fields of attention, inasmuch as pure intellect is uniformly at work in them all (Rule 1), so, too, it is the measures taken by the mind in its deliberately ordered progression from known to unknown (but always potentially knowable, since homogeneous) terms that confer unity on an equation and are, in turn, the genuine referents of the symbolic notation from which equations are forged.

This, however, is not yet the whole story. Equations can and did become the focus of interest for their own sake; indeed, the development of *algebra* after Descartes (and Newton) is largely in the direction of a theory of equations and combinatorial methods of determining their roots (as in the theory of determinants in Leibniz). If what I have suggested in the last four paragraphs and in III. v above is plausible, then this development is quite alien to Descartes' intentions, for its advocates consciously eliminate or at least demote the role of actual constructions. The issue here is by no means narrowly technical; the specifically Cartesian route from *pure mathematics* to a physics of the corporeal world is at stake. Let me try to spell this out as briefly as possible.

The complete mathematical process of which a construction is the terminus consists of a *double transcription.* The algebraic equations expressing the ratios of the line-lengths in a given problem transcribe into symbolic notation the ordered sequence of the steps the mind takes in arranging the terms of that problem according to knowns and unknowns; this, in conformity with the "praecipuum . . . artis secretum" divulged in Rule 6 (*A.-T.,* t. 10, p. 381.7–15). The algebraized equation, in turn, not only encodes and retains that intellectual sequence; it is also the expression of a quantity whose roots are the actual line-segments to be drawn from points at specified distances from the chosen axes. The set of points determined by these measures *is* the locus of a curve. The transcription or inscription of the abstract equation into a visible geometrical configuration *is* the construction being sought.[141]

It is impossible not to observe the crucial significance of this double transcription: both the algebraic notation and the ensuing geometrical curve mark the passage from the order of the intellect into the domain of perception. Furthermore, the direction of the passage is essential, since it informs us that the marks or shapes accessible to perception are caused to be there by the intellect imaging its own activities. Descartes' discussion of the two sources of experience (*experientia*) in Rule 12 is salient here: "We experience . . . whatever arrives at our intellect either from elsewhere [*aliunde*] or from its reflexive contemplation of itself" (*A.-T.*, t. 10, p. 422.25–423.1). The first source, being other than the intellect, enslaves the latter to chance and uncertainty; the second, reflexive source, being the self-identical intellect, enables it to master its own productions even when these are presented in an alien domain. Pre-Cartesian geometry is slavishness; Cartesian geometry, mastery.

Not, however, limitless mastery! The process of construction, mimicking as it does the mind-body problem in technical miniature, shares in its opacity. The limits to the power of intellect to image itself in the domain of perception are given by the very restriction on admissibility (that is, the requirement of kinematic determinateness) that simultaneously ensures the exactness and certainty of constructions. *Only* those equations all of whose exponents are rational correspond to admissible curves (those Descartes calls "geometric" and Leibniz later calls "algebraic" in distinction from "transcendental" curves). Conversely, only those curves constructed in the requisite manner succeed in transcribing algebraic equations of the appropriate form. This means that there are curves occurring in the sensible domain (the logarithmic spiral, say) *or* producible by instruments (the quadratrix, say) as well as equations expressing authentic problems (such as $x^x + x = 1$, the formula for cutting an angle in a given ratio), that permanently elude the grip of Descartes' technique, despite his being thoroughly acquainted with most of these formations. The range of intelligibility, then, does not coincide with the domains either of perceptual accessibility or of algebraic tractability. This limitation comes out in another, striking way in the phenomenon of *imaginary roots*. Descartes' anticipation of the fundamental theorem of algebra— that an equation has as many roots as the unknown has dimensions— will not jibe exactly with his own demand that all the roots be actually constructible in a single, uniform "space." Once again technical limitations haunt Descartes' projected identification of the mathematical with the corporeal *simpliciter*.

Despite these intrinsic limitations, the Cartesian commitment to constructivism remains emphatic. Had the term not been usurped for another purpose, we might have said that Descartes the geometer is an "exhibitionist" since both the procedures of the intellect in prosecuting the links among terms in a problem and the algebraic notation which symbolizes these procedures must be exhibited—must be, so to speak, bodied forth in a forum not of the mind's own making, however much it stands under the authority of mind.

iv Essence and Existence in Cartesian Pura Mathesis

It is now time to reckon up the implications of these principal characteristics of Cartesian geometry for the relationship between essence and existence as that relationship shows itself in the mathematical domain. In light of what I have been suggesting above, the question of existence *sub specie matheseos* seems readily answered: To exist (geometrically) is to have been actually constructed in accordance with determinate regulations. Or, put differently, the existence of a particular geometrical item (such as a root = a line-segment) is evidenced by its mode of coming into being, its methodical genesis. This ready answer should not, however, mask the further complication introduced by Descartes' discretionary selections of his *lignes principales* (the coordinate axes) and the unit-length suited to a given problem: the magnitude of a constructed root does not individuate that root in an absolute way since the measure of magnitude has been relativized by those two selections. Furthermore, the same equation (of degree 1) has more than one positively or negatively valued root; hence, the existence of a single root does not yield uniqueness, but counts as typifying the infinite subclass of roots of the "same" length, relative, that is, to the initial choices. (This same point can be made, *mutatis mutandis,* of the corresponding curves.)

This last aspect of geometrical existence brings us face to face with the issue of mathematical essence. Descartes' formulation in Meditation 5, speaking of the triangle, "determinata quaedam . . . natura, sive essentia, sive forma, immutabilis et aeterna," is bound to suggest that the ontology of Cartesian *pura mathesis* is fundamentally the same as, or continuous with, traditional ontologies, of Platonic or Aristotelian provenience, depending on whether emphasis falls on the notion of an eternal form known prior to sensory experience or on that of an immutable essence separate in thought from sensory corporeal instances. Descartes' *practice* in the *Geometry* (together with his theoretical program in the *Rules*), gives the lie, I think, to both of these historical associations and thereby opens up a novel horizon for thinking about the mode of being peculiar to mathematical entities with respect to their essences and their existence alike.[142]

This can be seen most easily from the following considerations. Cartesian "figures" are not the geometrical shapes exhibited in Euclidean geometry (and here the example cited in Meditation 5 has to be reassigned to the outer *integument* of true mathematics); they are not, that is, figures with their own integrity, shapes to be captured by demonstrated theorems or by constructions always responsive to the particular, pre-given, nature of each relevant shape. Nor are the figures of Cartesian geometry *defined* in advance of the problem-solving in which they are put to use, not simply because Descartes presupposes acquaintance with the antecedent tradition, but more significantly because the shift from theorems to problems also transfers his focus from the pregiven character of certain configurations (squares, circles, and so on) to their *utility* in a constructive procedure.

This transfer of interest from the pre-"given" forms of Greek geometry to constructive utility has repercussions palpable throughout the corpus of Descartes' mathematics. First, the abstract symbolic equation takes precedence, at least in the *ordo cognoscendi,* over the corresponding structure into which it may be transcribed. Leibniz was acutely sensitive to the implications of this precedence when he set about designing his *characteristica geometrica* or *analysis situs:* The symbolic equation of, say, a circle gives us no direct or advance information concerning the geometrical shape resulting from it. "For algebraic characters [as in Descartes] do not express all those matters which ought to be considered in space [such as the similarity and dissimilarity of figures] . . . nor do they directly signify the *situs* of the points."[143] In Cartesian terms, it is the equation which holds us closer to the ordering and measuring activity of pure mind and farther from the figurate extension of the visible.

Accordingly, were we to begin from Euclidean shapes, we would have to dis-figure or de-form them to arrive at the concatenation of the better-known simples of which they intellectually consist (see Rule 12, *A.-T.*, t. 10, p. 422) and, thence, at the algebraic formulation which gives the law of their intelligible and precise genesis via the appropriate ordering and compounding of such simples. In this regard, a Cartesian mathematical essence is a formula, not a (Euclidean) form.[144]

The influence of this transposition is also felt in Descartes' determination of the "nature" of curves in Book 2 of the *Geometry*. Not only are these admissible in geometry just insofar as they result from suitably conjoined motions (performed by hand or with instruments), the nature or essential being of an admissible curve is to be a locus of (infinite) points each of which is, in principle, clearly and distinctly constructible. This determination of "nature" is not as innocent as it might initially sound, and when we compare Descartes' procedures with those of Apollonius, his primary source in matters of conic sections, the salient theoretical differences begin to stand in relief.

First, for Apollonius the locus-property of a conic section is not counted among its *archicha symptōmata,* "those features belonging to it primarily," nowadays called its "planimetric properties," even though the three- and four-line locus-property do follow necessarily from the nature of each section. Descartes, on the other hand, begins from the locus-property (the ratios of known and unknown line-segments) and then reconstructs or regenerates the conic section from this. In doing so, he effects two further changes: The ordinates and abscissas, which were for Apollonius intrinsic to each of the conic sections, become the *lignes principales* or axes instituted prior to the construction of any conic; they belong, that is, to the conditions of the problem, not to the figural nature of the section. Second, he eliminates any trace of what Apollonius named the *eidos* or defining form in the case of the ellipse and the hyperbola, as though to indicate by that name the presence of any invariantly intelligible relation *within* the configuration of curved and straight-lines featured in the section. For Descartes, as already suggested, this invariance occurs at the level of the formula and is only derivatively present in the graphic shape insofar as if the latter has been artfully constructed. We are reminded of Viète's use of the phrase *species sive formae rerum* to denote the algebraic letters referring to indeterminate (or general) magnitude(s), instead of the visible *and* knowable figure somehow recurring whenever a specimen geometrical item is encountered.[145]

Descartes' radical transformation of the ancients' stable geometrical shapes, accessible prior to demonstration and to construction, into the symbolic formulae for the potential generation of such shapes, also alters in a profound way the understanding of the connection between universality and individuality. *Grosso modo,* one could suggest that Descartes abandons an arithmetical or set-theoretical picture of their connection in favor of a geometrical conception. In the former picture, individuals of certain types are seen as falling under or within a general class or sort which itself possesses in full the class-character displayed more or less completely by those individuals; all individual ellipses, for example, are tokens of the type *ellipse* of which true theorems are demonstrated. Each token somehow exhibits the general elliptical character or nature at which those theorems are aimed. For Descartes, individuals or particulars are evolved by progressive delimitation from an indeterminate, general formula or function which does not denote any specific type or nature. Thus, Descartes' general equation for *any* conic section, namely $Ax^2 + Bxy + Cy^2 + Dx + Ey + F = O$, *where A . . . F* are integral constants, is not the equation of any *one* type of conic section, nor does it capture some *one* essence in which all types and their tokens participate in varying degress. Instead, it presents a continuum of abstract possibilities given determinate actuality in particular cases by variations in the values of the coefficients. This extensive continuum is, as it were, punctuated by discrete acts of selection to yield individual sections of a certain type. Hence, to exist individually is, in Cartesian mathematics, to be the result of evaluating the variables bound within the general equation. To have discrete being of a certain type or kind, in the Cartesian perspective, is not to share in or exhibit some *one* "determinate nature or essence or form," as his traditional readers are likely to have interpreted these terms; it *is* to be a limited variation on a formally unlimited theme in which the uniformly pure intellect gives a continuous, virtuoso performance.[146] And yet, to come back full circle, the unlimitedness in question here is itself checked by factors apparently falling outside the ambit of mind and, indeed, of its imaginary auxiliary. Some algebraic formulae do not correspond to any constructible figures since their terms defy the requirement of kinematic continuity and control; other formulae have no real, but only "imaginary" roots, which cannot be inscribed in the space available to Cartesian geometry. Conversely, some figures inscribable in this space prove to be only illusory wholes since they cannot be generated from algebraic equations of the appropriate sort (having only rational exponents). The subsequent history of the metaphysics of geometry, from Leibniz, Wallis, and Newton to Lambert, Kant, and Euler is in large part an effort to eliminate the sources of these extrinsic checks on the constructive powers of the mind.

v   *Morals* Par Provision

Much remains to be filled out in this tentative sketch of *pura mathesis;* nonetheless, I want to end this final chapter by tracing some possible lessons from what I have delineated so far.

I began in section II by pointing to some of the issues raised by the transition from doubt of the senses to doubt of mathematics in Meditation 1, as well as by the reverse path from the truth of mathematics (as a matter of essences) to the existence of *material things* in Meditations 5 and 6. The preceding attempt at excavating the subtext to which the *Meditations* seems to allude should put us in a position to gauge how differently or similarly these issues might look *sub specie metheseos.*

Let me start with the issue of resemblance or nonresemblance between a sensation, an image, or an idea, on the one hand, and its respective "object," on the other. Even in the case of sensation, readers of the *Dioptrics* are likely to have been disconcerted by the invocation of the principle of resemblance and by the doubt occasioned by the possibility of its systematic violation, inasmuch as Descartes had already tried to show that veridical perception is possible without any strong likeness between the sensory datum and the sensed object. By what he calls in that text a "natural geometry," even a blind man is able to see thanks to his ability to make sound inferences from sensations of touch and pressure to the shape, magnitude and position of an external physical object. In the case of the mathematical items considered in the *Geometry,* the claim that resemblance is a necessary condition for veracity is weakened to the point of vanishing altogether.[147] The relations germane here are (1) between a concatenation of purely intellectual simples and the algebraic equation into which it is transcribed; (2) between that equation and the "graph" by which it may be constructed; and (3) between that visible graph and a shape or configuration believed to belong to an independently existing material object. As far as relations (1) and (2) are concerned, resemblance is out of the question, although we could say that the second term in each represents or expresses the first relation; for example, the equation expresses, in the sense of codifying, the operations of the intellect in disposing and ordering the terms of a problem, while the graph expresses one version of the equation, in the sense of constructing it under certain determinations (the "values" of the variables and the choice of coordinates). As for relation (3), it raises the most troublesome problems, at least at the start, only for someone who believes that mathematics proceeds by abstracting such shapes from the perceptual experience of external bodies, a belief Descartes seems at pains to controvert by the evidence of his own mathematical practice.

This mistaken belief works hand in hand with the second prejudice shaping our naive or native view of the origin of our knowledge of the world, namely, the principle of causality, applied in such a way that a nonmental object is assumed to be the cause of which a mental entity is the effect.[148] Descartes' distinction in the *Rules* between experience as it comes *aliunde*—"from elsewhere"—and experience arising from the mind's reflexive knowledge of itself has already paved the way for the more extensive challenge to this application of the principle of causality in the mathematical domain. There Descartes' constructivism is potently at work to show the reader the geometer's mind in the act of producing the symbolic notation and the subsequent graphs which it causes to function as its nonresembling representatives or deputies. The inward private motions of the intellect somehow become public in this vicarious fashion. ("Somehow" strikes the right note of unclarity, since this dual process of publication or embodiment [in symbols and in linear graphs], indispensable as it is from the evidentiary standpoint, nonetheless remains fundamentally enigmatic both with respect to its *modus operandi* and to its public works. I shall return briefly to this twin enigma in a bit.)

At all events, what comes to sight in Descartes' mathematical practice, as well as in his comments on the roots of that practice, is the essentially productive character of *ideas* themselves. A primitive Cartesian *idea,* whatever else it is or seems to be, serves as a "pâtron ou un original," an *instar archetypi,* in the language of Meditation 3, first as generating other *ideas* and, then, as a template for the derivation of other things, including "les choses extérieures," from our thoughts and thus from our formative powers. William Carlo, at the end of his richly detailed study of the differences between the Thomistic and the Cartesian idea of *ideas,* concludes that the item Descartes identifies as an idea "was the principle of artistic production." I want only to add that this shift from the cognitive to the artisanal is not a mistake but comes about by design. In his "Reply to the Third Objections" Descartes justifies the way he has used the term "idea" by reminding his interlocutors that "it was already commonly received by the philosophers for signifying the forms of the conceptions of the divine understanding." Descartes, we might conjecture, takes over for the mind's role model a Christianized Demiurge, not the Aristotelian *nous* whose whole actuality consists in thinking its own thinking.[149]

If the requirements of resemblance and of causality *ab alio* are disarmed when the sources of mathematical achievement are brought retrospectively into view, all that remains as an occasion for diffidence vis-à-vis mathematics is the principle of externality or of independent, extramental existence. Needless to say, this "remnant" is more perplexing and more tractable than any of the preceding occasions for doubt. To make the measure of its power we would need to begin shuttling back and forth from the subtext of *pura mathesis* to the overt argument of the last two Meditations. Since I cannot rise to that intricate challenge in the present setting, let me end this study by calling attention to three of the manifestations of the power intrinsic to this final principle.

First, Cartesian "dualism" is already at issue in mathematics and hence is not parochially tied to the psycho-physical dualism in which my *own* body and, above all, its responses to pleasure and pain, play a selfish role. For the mind to externalize itself through its productive expression in equations and then in constructions accessible to perception it must be both independent of, and connected with, the corporeal. It is in the *Rules,* more extensively than in any other writing, that Descartes faces this conundrum and embodies it in his doctrine of the *imagination.* The latter, said to be "a true body extended and figured," must somehow be capable both of taking inward, incorporeal directions and of giving them outwardly perceptible shape. The exercise of this second capacity is always in jeopardy since the outward shape *qua* outward returns to impress the imagination and to encounter the intellect as something both alien and exempt from its autonomy. Kant will articulate this same conundrum under the rubric "The Schematism of the Pure Concepts of the Understanding."[150]

Second, the externality required by this process of imaginative self-exhibition presents a seemingly insuperable obstacle to the complete intellectualization of phenomena set into motion by the project of Cartesian mathematical physics. As we saw in the details of Descartes' geometry, not all algebraic equations can be constructed or represented phenomenally. Those that can do not require any particular external body or discrete bodily shape to validate them; they do require, however, a receptacle or a phenomenal continuum, not of the mind's own making, in which successful constructions can make their appearances. If the "evil demon" symbolizes the seductiveness of *pretheoretical* nature, the seductiveness Descartes aims to resist by exposing the ways it compromises the project of Cartesian science, it remains indispensable to that very project that the appearances it sets out to master continue to appear, or, more exactly, that phenomenality in general remain as a foil to the dramatic displays of mathematical power. However far Cartesian science goes towards conquering and possessing nature, however successfully it mobilizes techniques for transforming prescientific nature into new appearances conformable to human will, this science cannot sunder its attachment to the givenness (or, perhaps, createdness) of appearance as such.

Third, and finally, the very success anticipated by Cartesian mathematical science brings doubt in its wake—doubt, ultimately, over whether the mathematical laws obeyed by phenomenal nature or imposed upon it to bring about a new ordering of phenomena, are in the last analysis the uniquely necessary laws. Malebranche's compromise—to argue that our knowledge of extension and its quantifiable modes need only be similar to the world of extra-mental experience—would not have satisfied Descartes in his most radical or anhypothetical frame of mind.[151] As he wrote to Mersenne in 1640: "As for Physics, I would believe myself to know nothing of it if I knew only how to say how things can be, without demonstrating that they cannot be otherwise."[152]

This triad of perplexities is, I think, Descartes' main legacy to radically modern philosophy. They can be most helpfully restated in the following way. The ambition animating modern philosophy is, as I have already suggested in chapter 1, the *imitatio Dei*, of the *Deus* who creates *ex nihilo*. What Kant later says of the mathematician's definitions, *sic volo, sic jubeo*—"thus I will, thus I command"—may be more amply applied to the entire field of mathematical knowing and, indeed, to all genuine knowing as these are understood by Descartes.[153] Becoming master and possessor of nature requires first of all and most fundamentally that one become master of the inward resources through which the course of inventive production can be clearly discerned and brought under the reign of methodical order. There is, however, an ineluctable gap dividing our comprehension of these two versions of mastery. We give proof of private ingenuity in a public, outward, and hence corporeally accessible way. The world in which Cartesian mastery is meant to exhibit its triumphs remains *worldly* in a sense not wholly explicable by Cartesian mathematical physics. This *worldliness* exhibits itself, first, as the *externality* of the corporeal or material, even when a body or array of bodies has been deliberately produced. The offspring of the very autarchy and self-referring generosity aimed at the negation of otherness turn out always to have an *alibi*, to be other and elsewhere than their progenitive sources. (Compare Hegel on nature as the self-alienation of the concept.) This inexplicable, but inextirpable worldliness also shows up in the *extendedness* of the corporeal items the mind can either generate or analyze in terms of its own devices of order and measure. In his *Conversations with Burman* Descartes replies to the objection that our thinking is extended and divisible by asserting that it *is* indeed extended as far as its duration is concerned (*quoad durationem*), but not insofar as its nature is concerned (*quoad naturam suam*).[154] Whatever else might eventually be said about this (and related) passage(s), it is clear that neither the successiveness of a sequence of thoughts nor the temporal "stretch" during which a single thought is entertained is strictly equivalent to that occupancy of three-dimensional space definitive, according to Descartes, of the essence of extension and, consequently, of body. Extension could at most be said to figure or schematize duration (as in Kant). Both of these aspects of the world's worldliness—externality and extendedness or corporeality—seem to point towards one central dilemma, namely, that the intelligibility, *but not the being*, of body is coincident with the dianoetic or technical operations of imaginative mind.

The second member of this triad of perplexities is very closely tied to the first. Whether as given (*qua* the presystematic domain to be mastered) or as reconstructed (via ingenious mastery and mathematical physics), worldly items *make an appearance*. The reduction of molar appearances to microphysical entities or events will no more suffice to *obliterate* the experienced phenomenality of the worldly than will the analysis of sensations in terms of the movements of "animal spirits." Descartes clearly held out the prospect that we could, in principle, replace "natural" appearances with artificial appearances of our own devising; his early notes for a projected "Thaumantis Regia," the palaces of Thaumas, god of wonder, bear later fruit in his recollection, at the end of the *Meteorology*, "of an invention for making signs [such as a rainbow] appear in the sky, which could cause great admiration in those who would not know the reasons."[155] The responsibility and the prestige of the covenantal rainbow can pass from divine to human hands, so that, as Nicolas Grimaldi writes: "As soon as 'a science of miracles' [see A.-T., t. 1, p. 21] exists, there is only one miracle any longer: that science exists."[156] Nonetheless, knowing *how* the phenomena (can be made to) appear is not the same as being appeared to by them. It is, then, no wonder that Hobbes, obsessed by optics, will write in *De corpore* (Bk. 4, ch. 25): "Of all the phenomena which exist near by us *to phainesthai* itself is the most admirable."[157]

The third and final member of this triad is the question of the unqualified necessity of the *laws* of worldly motion and causality. The mathematical format of such laws, replacing as it has for the moderns the noetic shapeliness of immobile forms, leaves in suspense the identity of the sources and supposed inviolability of those laws. If the format is originated in and by the mind, do the particular laws filling in that format also have *their* source there or elsewhere? Who legislates for the world? With this query we are returned to the Kantian point from which I began in chapter 1.

Notes

Chapter 1

1.   Among the many contemporary studies of the semantic history of "modern" and "modernity" I have found the following most fruitful: Hans Robert Jauss, "Literarische Tradition und gegenwärtiges Bewusstsein der Modernität"; and Reinhart Koselleck, " 'Neuzeit'. Zur Semantik moderner Bewegungsbegriff." (Heine seems to have coined the term "Modernität" in his *Reisebilder* of 1826.) For an interpretation of modernity not, I think, incompatible with the one set out here, see Hans Blumenberg, *Die Legitimität der Neuzeit*. Also pertinent to my concerns is Anthony J. Cascardi, "Genealogies of Modernism."

2.   B. Pascal, "Preface pour le Traité du vide," in *Oeuvres complètes*, p. 533. Karl Gutzkow, *Werke*, Bd. 11.

3.   For Perrault, see the discussion in Jauss, op. cit., pp. 164–168.

4.   See Friedrich Schlegel, *Über das Studium der griechischen Poesie*, in *Schriften zur Literatur*, pp. 84–192 et passim; and F. W. J. Schelling, *Philosophie der Kunst*, in *Sämtliche Werke*, Abt. 1, Bd. 5, p. 444.

5.   Cited by W. Dilthey (*Gesammelte Werke*, Bd. XIV/1, p. 56, n. 99), from the manuscript of Schleiermacher's lectures on the history of philosophy.

6.   Richard Rorty, *Philosophy and the Mirror of Nature*, ch. 1. As my account of Descartes in ch. 3 implies, I am not in any way endorsing Rorty's presentation of the "mechanics" of this invention.

7.   For a sampling of recent affirmations of the constructivist thesis in a variety of disciplines, including mathematics, see Paul Watzlawick, ed., *The Invented Reality;* and M. A. Arbib and Mary Hesse, *The Construction of Reality.*

Two earlier studies of the history and substantive implications of "construction" are Helga Ende, *Der Konstruktionsbegriff im Umkreis des Deutschen Idealismus;* and Bernhard Taureck, *Das Schicksal der philosophischen Konstruktion.* Both suffer, in my judgment, from failure to pay concentrated attention to the *mathematical* genesis of the leitmotiv of modern philosophy. (Compare, in addition, the very critical review of Taureck by Ingetrud Görland in *Hegel-Studien.*) In contrast, Amos Funkenstein, in his *Theology and the Scientific Imagination From the Middle Ages to the Seventeenth Century,* does see the issue of construction partly in terms of the distinctions between ancient and early-modern mathematics (see ch. 5, B: "Construction and Metabasis, Mathematization and Mechanization," pp. 299–327). For an attempt to capture the implications of human making in a broader net, see Elaine Scarry, *The Body in Pain. The Making and Unmaking of the World,* esp. ch. 4. (Neither of these last two authors, however, calls attention to the lexical and semantic uniqueness of the Hebrew term *bara,* never employed in the *Torah* for human making or production.)

8.   Some words ought to be said about the premathematical uses of *constructio* and about its paths of entry into the mathematical lexicon which (partially) crystallized in the early seventeenth century.

In Latin *constructio* is a stylistic or oratorical term designating a "fit connection" of words or phrases in a speech (e.g., "eratque verborum et dilectus elegans et apta et quasi rotunda constructio" [his words were carefully chosen; his sentences compact and well-turned],

Cicero, *Brutus* 78.272.) Although the verb *construere* can be used for making or fabricating in general (e.g., "ut aedificium idem destruxit facillime qui construxit," Cicero, *De l* 20.72), the preferred trope for putting together a speech is *exaedificatio* (Cicero, *De oratore* 2.15; 2.37).

Constructio comes into its own in later antiquity and throughout the middle ages as a grammatical term, especially in Priscian's *Institutiones*. It refers to the syntactic agreement of the cases of nouns and adjectives, of subjects and verbs, etc. Priscian makes its Greek provenance clear when he writes: "Omnis enim constructio, quam Graeci Syntaxin vocant, ad intellectum vocis est reddenda" (*Grammatici Latini*, Bd. 3, 201.187–188). (Greek *syntaxis* appears in a mathematical context only in the original title of Ptolemy's *Almagest*.)

At all events, it is grammatical "construction" that is the concern of, say, Dante in his *De vulgari eloquentia* ("Est enim sciendum, quod Constructionem vocamus regulatam compaginem dictionum" [the rule-governed joining of words], 2.6.2) and of the medieval speculative grammarians who questioned whether phrases or propositions, as well as words, had determinent syntactic factures. (On these matters, see R. Jakobson and Paolo Valesio, "Vocabulorum constructio in Dante's Sonnet 'Se vedi li occhi miei' "; and Jan Pinborg, "Can Constructions Be Construed? A Problem in Medieval Syntactical Theory.")

On balance, then, *constructio* in the premodern period belongs to the arts of grammar and rhetoric, even where suggestions of the architectural art can be heard. (In French literary criticism of the seventeenth century the same general point holds, even while the parallelism of oratory and architecture is pursued much more conspicuously. *Bâtir* and *bastiment* are the preferred terms for architectural production, while *contruire* still carries clear traces of its origin in oratory and syntax (See, for example, La Mesnardière, *La Poétique* [Paris, 1640], p. 8: "Il faut sur toutes choses que le sujet soit construit avec tant de liaison des ses incidens divers, quo'on n'en puisse tirer aucun sans détruitre tout l'ouvrage," cited in H. Coulet, "La Métaphore de l'architecture dans la critique littéraire au XVII [E] siècle," p. 299.)

On Priscian's evidence, then, Latin *constructio* is the equivalent of Greek *syntaxis*, which is not a technical mathematical term. *Constructio* and its vernacular counterparts nonetheless are domesticated in the mathematical vocabulary from the sixteenth century on, without, however, becoming the dominant or unique translation of the underlying idioms of geometrical activity in the Greek tradition. (See ch. 2, IV, which this note presupposes.)

Three semantic domains deserve attention: (1) the language of activity used *within* the body of proofs and problems; (2) the language used to render Greek *kataskeuē* when it denotes *one* of the six parts of every proof of problem (as it does in the accounts provided by Hero of Alexandria and Proclus); (3) the language used to specify problem-solving in general, or in wider connection with methods of analysis and resolution.

On the first point, we can observe among the sixteenth-century translators and commentators greater faithfulness to the diversity of usage in their Euclidean exemplar than can be found in current versions. So, for instance, Clavius, in his enormously influential translation, turns the Greek *systēsathai* into *constituere*, not *construere* (see, e.g., his versions of Bk. 1, Prop. 1 and Bk. 6, Prop. 25, in the first edition [Rome, 1572]). At about the same time Nicolò Tartaglia respects the same convention in his vernacular version (Venice, 1569); thus, Bk. 1, Prop. 1 becomes "Possiaamo sopra una data retta linea constituir un triangolo equilatero" (20 recto). (Federico Commandino is another case in point; when "operations" are performed within a proof or problem, these are designated by terms peculiar to the figures involved and not by a general term for construction. See Commandino's Latin renditions of Bk. 1, Prop. 1 [*constituere*] and Prop. 46 [*desribere*] in the edition of 1572, printed at Pesaro.)

As regards the second point, it is noteworthy that *two* distinct renditions of the salient term in Hero and Proclus may be found in the sixteenth- and early seventeenth-century tradition. (The ancient texts in question are Heron of Alexandria, *Definitiones*, in *Opera*, vol. 4. 120.21 et seq.; and Proclus, *In Primum Euclidis Elementorum Librum Commentarii*, 203.1–12.) Alongside the expected *constructio* for the term *kataskeuē* (rendered by Heath as "the construction or machinery"), there is an alternate convention in which *delineatio* is the preferred translation. To give just two examples: Conrad Dasypodius in his 1579 version of Heron (= *Hieronis Alexandrini Nomenclaturae Vocabulorum Geometricorum translatio* [Strasbourg, 1579]) gives *delineatio* for *kataskeuē* and adds "Delineatio vero addit ea quae dato desunt ad investigationem quaesiti" (22 verso). In the early seventeenth century J. H. Alsted speaks quite similarly: "Kataskeuē, delineatio, quae praemittitur demonstrationi. Ea nihil est aliud, quam praeparatio subiecti, ad quaesiti investigationem et inventi demonstrationem" (*Methods admirandorum mathematicorum* ch. 5).

When *constructio* is used as the equivalent of that part of every "perfect" theorem or problem called the *kataskeuē*, then further associations come to light and help us to see why *constructio*, not *delineatio*, is on its way to being the canonical term. Thus, to take Clavius again, his *Prolegomena* links problem-solving in geometry with a *problem* as understood in Aristotle's topical dialectic. "Sic etiam quaesitum illud apud Mathematicos, quo aliquid jubent construere [et] cuius contrarium effici etiam potest, problema appellatur" (see 1654 Edition, Frankfurt, Jonae Rosae, Folio 7 recto). While this link does not seem to have been taken up as such by later authors, the association of *problems* (not theorems) with constructing or effecting does become increasingly dominant. Hence, as regards the third semantic domain mention above, *constructio* becomes the title for that whole portion of the solution to a problem which is then followed by a demonstration. Thus, Paolo Bonasoni, in *Algebra Geometric* (ca. 1575) first specifies "what is sought" (the *quaesitum*), then gives its construction (*constructio*), after which he provides a demonstration (*demonstratio*) that the construct is identical with what was sought (see *The Algebra Geometrica of Paolo Bonansoni, circa 1575*, passim). This format is preserved and expanded in the works of Simon Stevin. So, the first problem in his *Problematum Geometricorum . . . Libri V* (Antwerp, 1583) runs on the following lines: *Explicatio dati* (explication of the given); *Explicatio quaesiti* (explication of what is sought); *Constructio: Demonstratio; Conclusio* (see *The Principal Works of Simon Stevin, Volume 2: Mathematics*, pp. 168–70).

This format, together with the singular emphasis given to construction as the hallmark of problems (whereas *kataskeuē* according to Heron and Proclus, is a part of every theorem *and* every problem), suggests quite strongly that the new interpretation of analysis, the method of resolution, determines or directs this semantic development. Zabarella's definition of the resolutive order, cited in the text above, is pertinent here as well: resolution (or analysis) furnishes us with an instrument by which we can "produce and generate some end."

It needs to be added, however, that *construction* does not signify mere or arbitrary production (e.g., of some figure). Its original "grammatical" and "oratorical" uses seem to me to be still audible in this new setting. Solving a problem by construction (or by "problematic analysis") always involves interpreting both the *given* and *what is sought*—that is, finding the best way of fitting together congruously the elements ingredient in these. The ingenious architects of the new geometry still had an ear for grammar and rhetoric. This gives an ironic twist to Renan's famous declaration: "The founders of the modern mind are philologists" (*L'Avenir de la science*, 4th ed. [Paris, 1890], p. 141).

As for the persistence of mathematic construction *sensu stricto*, H. J. M. Bos, in "Arguments on Motivation in the Rise and Decline of a Mathematical Theory: the 'Construction of Equations,' 1637–ca.1750," has tried to show in a very precise way how the original programs of graphical or geometrical construction gave way to the new discipline of solving equations by radicals. To the extent that this gives us an accurate picture, it implies that, as the motif of construction "loses steam" in mathematics, it is gaining momentum in philosophical extrapolations from mathematics (e.g., in Vico and Kant). On the other hand, Bos's fine study tends to overshadow the more persistently constructive character of the purely algebraic solution of equations, namely, that the equations themselves, together with their roots, are constructions. (See above, 3, IV, iii, on the function of equations in Descartes.)

Moreover, the "spirit" of constructivism in its wider sense persists despite these shifts in technical orientation. Two citations from much later mathematicians evince this persistence:

> In all Mathematical Science we consider and compare relations. In Algebra the relations which we first consider and compare, are relations between successive status of some changing thing or thought. . . . Our marks of temporal and local site . . . are at once signs and instruments of that transformation by which thoughts become things, and *spirits put on body;* and the act and passion of mind seem clothed with an outward existence, and we behold ourselves from afar. (Sir William Rowan Hamilton, "Metaphysical Remarks on Algebra" [1831], cited in T. L. Hankins, "Algebra as Pure Time: William Rowan Hamilton and the Foundations of Algebra," p. 336.)

> We are then to think of the result of logical thought as certain spatial configurations of symbols; and our study will then consist in studying the further effects brought about by the processes of symbolization and self-reflection. Now in our subject we are to regard our symbols as without properties except that of permanence, distinguishability and that of being part of certain symbol-complexes. But this latter is essential, i.e. that these symbols enter into certain spatial (not Euclidian or continuous, etc., but spatial as opposed to temporal to be described later), relations. These relations themselves can then be symbolized and the new symbols are again in space and have certain spatial relations, etc. So much for the further effect of symbolization of the spatial properties. But in addition we have this self-reflectiveness. This is a reflection of the process. This process is then itself symbolized and symbolized by a spatial symbol.

> We thus have a continued activity which produces symbols which are spatial. This activity turns on itself and symbolizes its temporal character by a spatial symbol. These spatial symbols have certain spatial relations which are in turn symbolized by a spatial symbol. (Emil Post, "Diary" in *The Undecidable,* ed. M. Davis [Hewlett, N.Y., 1965], p. 420.)

9.   See my studies "Vico, Doria e la geometria sintetica," "Vico and Marx: Notes on a Precursory Reading," and "Vico, Nominalism and Mathematics."

10.   Vico, *De antiquissima Italorum sapientia,* in *Opere filosofiche,* p. 77, and the *Risposta* of 1791, ibid., p. 136. On Vico's distinction between divine generation (*ad intra*) and divine making (*ad extra*), see the very interesting remarks in J. A. Aersten, "Wendingen in waarheid. Anselmus van Canterbury, Thomas von Aquino en Vico."

11.   Heinrich Heine, *Zur Geschichte der Religion und Philosophie in Deutschland* [1835], in *Sämtliche Werke,* Bd. 9, p. 242.

12.   On this motif of the (eventual) unanimity of philosophers, see Kant's essay "Verkündigung des nahen Abschlusses eines Traktats zum ewigen Frieden in der Philosophie," in I. Kant, *Werke,* hrsg. von Wilhelm Weischedel (Frankfurt a.M., 1964), Bd. 3, pp. 405–16.

13.   Kant, *Werke,* Bd. 6, pp. 340–41.

14.   A full defense of this claim would require, among other things, careful examination of the relations between ostensive (geometrical) and symbolic or characteristic (algebraic) construction in Kant's thinking.

For the historical point, see Kant's use of the expression "Konstruktion der Gleichungen" in the Preface to his *Metaphysik der Sitten, Rechtslehre,* in *Werke,* Bd. 4, p. 313. Christian Wolff's definition of the synonyms "Constructio aequationum, effectio geometrica, die Ausführung der Gleichungen" in his *Mathematisches Lexicon* is as follows:

> Wird genennet, wenn man durch Hülfe Geometrischen Figuren den Werth der unbekannten Grosse in einer Gleichung in einer geraden Linie findet; oder sie ist die Erfindung einer geraden Linie, welche die unbekannte Grosse in einer Algebraischen Gleichung andeudet. (1. Abteilung, Bd. 11, coll. 421–23.)

15.   Kant, *Kritik der reinen Vernunft,* A724–727/B752–755. (Hereafter cited as *KdrV.*)

16.   Ibid., A719/B747.

17.   For this distinction, see Kant, *Kritik der Urteilskraft,* par. 77.

18.   Kant, *Kritik der praktischen Vernunft, Werke,* Bd. 4, p. 189 (= A124 of original edition.)

19.   Kant, *Kritik der Urteilskraft,* par. 91: "Von der Art des Fürwahrhaltens durch einen praktischen Glauben" (=*Werke,* Bd. 5, pp. 597–604).

20.   Ibid., p. 599.

21.   Yirmiahu Yovel, *Kant and the Philosophy of History,* pp. 21–22, 271–72. (Emil Lask's 1902 dissertation, "Fichtes Idealismus und die Geschichte," remains the most fruitful starting point for reflection on the issues evoked here; see, especially, pp. 44–56, on "Die Mathematik als Mittelglied zwischen analytischer und emanatistischer Logik").

22.   Kant, *Anthropologie,* par. 35.

23.   The one item conspicuously absent from the foregoing narrative is the motif of "construction" employed and expounded by Kant in his "metaphysics of nature," i.e., the *Metaphysische Anfangsgründe der Naturwissenschaft* of 1789. The presentability of a concept in intuition (via geometrical construction) is here urgently needed to establish the objective validity or real possibility of fundamental physical notions and relations (e.g., composition of motions, the filling of space by matter, the play of attractive and repulsive forces). Unless this condition is fulfilled, no fruitful union of mathematics with physics can be envisioned. And yet, Kant's efforts to meet this condition are only partially successful; where they fail, the reasons for failure are quite intriguing. So, in the case of phoronomy—the science of the velocity and direction of purely *quantitative* mobiles—we can readily effect the geometrical construction of the principle of the composition of two motions into a third (this is in fact the familiar parallelogram construction). In the case of dynamics, however, where the *qualitative* resistance of one body to being moved by another must be taken into account, geometrical constructibility on its own cannot resolve what Kant takes to be the fundamental conflict within the domain of physical explanation: atomism (or mechanism) versus dynamism. Indeed, at one point Kant warns that "one must not raise any objection against the concept [*sc.* the dynamical concept of infinite space continuously filled by matter] from difficulties in the construction of the concept, or rather, from the misinterpretation [*Missdeutung*] of the construction" (*Akademie-Ausgabe,* Bd. 4, p. 522). Elsewhere, at the very end of section on dynamics, Kant's tone is even more hedging: "This is now all that metaphysics can ever achieve towards the construction of the concept of matter . . . namely to regard [*anzusehen*] the properties of matter as dynamical and not as unconditioned primal positions, such as a purely mathematical treatment would postulate" (p. 534). Interpretive regard or recommendation here takes the place of construction and is one more index of the dwindling power of construction to secure objectivity to concepts.

    The question of constructibility in this Kantian text has recently been subjected to careful, and divergent, treatments by Gordon G. Brittain, Jr., "Kant's Two Grand Hypotheses," Robert E. Butts, "The Methodological Structure of Kant's Metaphysics of Science," and

Howard Duncan, "Kant's Methodology: Progress Beyond Newton?" On the later reception of construction in physical science see V. Verra, "La 'construction' dans la philosophie de Schelling."

24.   Fichte, "Über das Verhältnis der Logik zur Philosophie oder transcendentale Logik," in *Nachgelassene Schriften*, Bd. 9, pp. 42–43.

25.   Ibid., p. 34; Kant, *Akademie-Ausgabe*, (Berlin, 1936), Bd. 21, p. 45. See Schelling, *Neue Deduktion des Naturrechts* (1795), in *Schriften von 1794–1798*, in *Sämtliche Werke*, Bd. 1, p. 128; "Ich herrsche über die Welt der Objekte; auch in ihr offenbart sich keine andre, als meine Causalität. Ich kündige Mich an als Herrn der Natur, und fordere, dass sie durch das Gesetz meines Willens schlechthin bestimint sey."

26.   Nietzsche, *Sämtliche Werke, Kritische Studienausgabe*, hrsg. von G. Colli und M. Montinari, (Berlin, 1980), Bd. 12, p. 154.

27.   Ibid. See also Bd. 11, p. 622: "Die Auslegung der Natur: wir legen uns hinein–der furchtbare Charakter."

28.   Nietzsche, *Beyond Good and Evil*, par. 6.

29.   Nietzsche, *Sämtliche Werke*, Bd. 13, p. 332.

30.   Salomon Maimon, *Über die Progressen der Philosophie*, p. 20.

31.   Since I have willfully, and without deference, made a mélange of the authors most often identified with postmodernism, some show of justice and clarification is due to the individual contours of the thinker from whom the term "deconstruction" has taken its prestige and near-ubiquity.

It ought to be noted, first, that Derrida has more than once expressed surprise, even irritation, at the popular fortune of this rubric. (See "The Time of a Thesis: Punctuations," p. 74; or "Response de Jacques Derrida.") His own usage stems, he remarks, from Heidegger's vocabulary of *Destruktion* and *Abbau* (see the fine study by Rudolphe Gasché, *The Train of the Mirror. Derrida and the Philosophy of Reflection*, pp. 109–20). Of interest in this same setting is Derrida's recent allusion to deconstruction as "attack(ing) the systemic (i.e., architectonic) constructionist account of what is brought together, of assembly" (*Memoires for Paul De Man*, p. 730).

These lexical points aside, it is also pertinent that for Derrida the "structures" or "texts," in which, as he says, "deconstruction . . . is always already at work in work" (ibid.), cannot be understood as the deliberate or self-conscious productions of a subject, transcendental ego, or the like. "Structure" belongs to a history or series to which *eidos, essence, form . . . construction, . . . totality, Idea, . . . system, etc.* also belong, and "continues to borrow some implicit signification from them and to be inhabited by them" (*Writing and Difference*, p. 301). This linkage of structure, etc. to the tradition of metaphysics (as construed by Derrida) means, furthermore, that the very notions of constitution, production, and creation—the notions, that is, cardinal to the radically modern accounts of construction— must themselves be somehow purged of the language of metaphysics. Hence, "we shall designate by the term *différance* the movement by which language or any code, any system of reference in general, becomes 'historically' constituted as a fabric of differences" (*Speech and Phenomena*, p. 141). Historical constitution, shorn of its metaphysical implications, is a movement of the fateful play (of signifiers), excluding the autarchy or autonomy inscribed at the core of the modern constructivist conviction; compare, once again, Kant's claim that in virtue of mathematical construction "reason . . . becomes, so to speak, master over nature" (*KdrV*. B753).

It is, finally, this theme of historical constitution that can direct us back to the roots, if not the ultimate origins, of Derrida's preoccupations. It is in his long introduction to the translation of Husserl's *The Origin of Geometry* that these roots become visible for the first time. In lieu of the detailed study this preface demands, let me suggest that the Husserlian motifs of the constitution, production or genesis of sense, evidence, and objectivity enter here into a fatal alliance with Husserl's own adumbrations of a transcendental history (e.g., of geometry). Neither the inscriptionally retained past nor the prophetically envisioned culmination of history (as the history of pure reason) can be brought into line with the primal evidence (*Urevidenz*) of the lived present/presence characteristic of the sense-engendering activity of transcendental subjectivity. For matters to turn out otherwise, for this perfect alignment or absorption to be feasible, man, so Derrida notes, would have to be identical to the Cartesian god, the creator of eternal truths. (See *L'Origine de la géométrie* [Paris, 1962], p. 28, n. 1.) These considerations warrant the following speculation: What Derrida in this initial exposé of (absent) origins has brought under indictment is one pole or one modality of the configuration of radical modernity I have been scrutinizing—the putative mastery of the constructing mind or subject—(the modality still vivid in Husserl's rather Fichtean appropriation of Kant). Once this mastery has been dismantled, its apparent productions or constructions no longer wear their "maker's mark" on their sleeves. Freed from the gravity of this illusory mastery productions, constructions, or "structures" become signs and sources of disenchanted playfulness. Radical modernity, seen through the sometimes surrealist lens of deconstruction, is in all senses a *jeu d'esprit*.

One final remark on the Nietzschean genealogy referred to in the text: While a certain appropriation and simultaneous disavowal of *Hegel* is unmistakably an efficient cause of post-modernism, under its various guises, it is crucial to keep in mind that this is the Hegel of the most "Nietzschean" reading of Hegel conceivable. For an illuminating study of this Hegelian genealogy, see Judith Butler, *Subjects of Desire. Hegelian Reflections in Twentieth Century France*.

32.   Maurice Blanchot, "Le Rire des Dieux," p. 102.

33.   Rudolf Carnap, *The Logical Structure of the World*, p. 285 (= par. 177). On Carnap's constructivism and its dilemmas, see Hiram Caton, "Carnap's 'First Philosophy'."

34.   George D. Romanos, *Quine and Analytical Philosophy. The Language of Language*, p. 8.

35.   On the matrix of Frege's *Begriffsschrift*, see D. R. Lachterman, "Hegel and the Formalization of Logic."

36.   N. Goodman, *Ways of Worldmaking*, p. 100.

37.   Ibid., pp. 100–101.

38.   Ibid., p. 7.

39.   Ibid., p. 107.

40.   See *Philosophy and Literature* 4 (1980): 107–120, at p. 116; Goodman, op. cit., p. 138.

41.   E. Husserl, *Krisis der europäischen Wissenschaften und die transzendentale Phänomenologie*, p. 67.

42.   R. Descartes, *Passions de l'âme*, arts. 94 and 127 (*chatouillement*); arts. 153–161 (*générosité*).
(On the fascinating topic of *chatouillement* in this text, see Alexandre Matheron, "Psychologie et politique: Descartes, la noblesse du chatouillement." On *générosité*, see G. Rodis-Lewis, "Le dernier fruit de la métaphysique cartésienne: La générosité."

43.   Michel Foucault, *L'Archéologie du savoir*, p. 264; Maurice Blanchot, *L'Entretien infini*, pp. 304–305.

44.   Martin Luther, *Disputatio contra scholasticam theologiam* (1517), in Erich Vogelsang, *Der junge Luther*, p. 321, n. 17.

45.   G. Deleuze, *La philosophie critique de Kant*, p. 19.

46.   Kant, *Opus Postumum*, in *Akademie-Ausgabe*, Bd. 21, p. 4 and p. 6.

Chapter 2

1.   Jean-Jacques Rosseau, *Discours sur les sciences et les arts*, in *Oeuvres complètes* T. 3, p. 29.

2.   Much in the way of textual information and some substantive illumination in regard to these sixteenth-century debates may be found in the studies of Giulio Giacobbe and the monograph of Hermann Schüling, *Die Geschichte der axiomatischen Methode im 16, und beginnenden 17, Jahrhundert.*

3.   Johannes Kepler, *Harmonice Mundi*, Liber 1, Prop. 45 (=*Gesammelte Werke*, Bd. 6, p. 55). The Latin text reads: "et pronunciamus recte quod latum Septanguli sit ex *Non Entibus:* puta scientialibus."

4.   Paul Tannery, "Sur l'Arithmétique Pythagoricienne," *Bulletin des sciences mathématiques* (1885), p. 86 (reprinted in his *Mémoires scientifiques*, T. 2, pp. 179–201).

5.   J. Kepler, *Astronomia Nova, Pars Optica*, in *Gesammelte Werke*, Bd. 2, 92; G. Galilei, *Le Opere di Galileo Galilei*, vol. 8, pp. 349–62 (English translation in Stillman Drake, *Galileo at Work. His Scientific Biography*, pp. 422–36). Book 2, ch. 13, of Hobbes's *De corpore* is entitled "De analogismo." For Descartes, see ch. 3, III, v. infra.

6.   See Euclid, *Elementa*, Vol. 5, Pars 2, p. 83.3–4 and Proclus, *In Primum Euclidis Elementorum Commentarii*, pp. 60. 9 (hereafter cited as "Proclus, *In Euclid.*") David Fowler, in his book *The Mathematics of Plato's Academy. A New Reconstruction*, has given a brilliant and intricate reconstruction of the pre-Eudoxian theory of *anthyphairesis*, or reciprocal subtraction (as in Euclid Bk. 10, Props. 2 and 3), and shown it to be immune to the disconcerting effects of the discovery of incommensurable lines (see pp. 294–308). His reconstruction leads him to doubt the historicity of a "Pythagorean crisis." If "crisis" is taken here to mean something akin to the foundational "crises" in nineteenth-century analysis and twentieth-century set-theory, Fowler's doubt is chastening. However, the abandonment of *anthyphairesis* in favor of the "Eudoxean" theory in Euclid Bk. 5, together with the preservation of the restricted theory of numerical ratios in Euclid Bk. 7, indicates an unflagging desire in the Classical period to come to grips with the *aporiai* uncovered by incommensurability. The same desire is powerfully present in Plato's dialogues and "unwritten teachings." (See Paolo Cosenza, *L'Incommensurabile nell' Evoluzione Filosofica di Platone;* for Aristotle, see Silvio Maracchia, "Aristotele e l'incommensurabilità," and the reply by Wilbur Knorr, "Aristotle and Incommensurability: Some Further Reflections." On pre-Euclidean attempts to circumvent the issue of proportions altogether, see Benno Artmann, "Über voreuklidische 'Elemente' deren Autor Proportionen vermied.")

7.   Gilles-Gaston Granger, *Essai d'une philosophie du style*, p. 37.

8.   C. S. Peirce, *The New Elements of Mathematics*, vol. 4, p. 236. Peirce also wrote apropos the reading of Euclid: "In order to understand a book, it is necessary to learn the whole history of the author's thought" (p. 72).

9.   *Meno*, 82d8; 83e1; 85a4. (On the deliberate and provocative blurring of arithmetical and geometrical lines of inquiry here, see Malcolm Brown, "Plato Disapproves of the Slave-Boy's Answer.")

10.   Cf. Friedhelm Beckmann, "Neue Gesichtspunkte zum 5. Buch Euklids," pp. 32–35.

11.   Ps.-Arist., *De lineis insecabilibus*, 968a2; and Arist., *Metaph.* 1021a6–9. The text, the provenance, and the argument of *De lineis* are all tangled matters; see, on the passage cited, M. Timpanaro Cardini, *Pseudo-Aristotele. De lineis insecabilibus*, ad loc.

12.   On this fundamentally important rendition of *arithmos*, see Jacob Klein, *Greek Mathematical Thought and the Origin of Algebra*, pp. 46–60. The relevance of Klein's rendition for issues in contemporary set-theory has recently been shown by John Mayberry, "A New Begriffsschrift (1)," pp. 239–46.

13.   Julius Stenzel, *Zahl und Gestalt bei Platon und Aristoteles*, ch. 10, "Logos Aoristos," pp. 162–69; Granger, op. cit., p. 41.

14.   Ian Mueller, *Philosophy of Mathematics and Deductive Structure in Euclid's Elements*, p. 118, and cf. p. 66.

15.   See J. A. Lohne, "Essays on Thomas Harriot," 296–98, where Harriot's use of *noeticae radices* is illustrated from unpublished manuscripts.

16.   For Hobbes, see *Examinatio et Emendatio Mathematicae Hodiernae, Opera Latina*, vol. 4, pp. 126–31; for Leibniz, see *Mathematische Schriften*, Bd. 5, pp. 377–82 (hereafter cited as "*GM*").

17.   Euclid, *Elementa*, Vol. 5, Pars. 1, p. 216.10–13. For the transformed understanding of homogeneity in early modern times, see J. Klein, op. cit., pp. 172–176; 214–218 and accompanying notes.

18.   See Pascal, "De l'esprit géométrique," in: *Oeuvres complètes*, p. 589; and Jean-Louis Gardies, "Pascal et l'axiome d'Archimède."

19.   Cf. John Murdoch, "Euclides Graeco-Latinus: A Hitherto Unknown Translation," p. 290, n. 37. The passage cited is a fourteenth-century marginal addition to Paris, MS BN 7373.

20.   Mueller, op. cit., p. 132; see also his paper "Homogeneity in Eudoxus' Theory of Proportion."

21.   Compare Stillman Drake's *caveat* against ascribing the full-fledged notions of equation and function to Galileo, in his introduction to Galileo, *Two New Sciences*, pp. xxiii–xxv; and see also Carl B. Boyer, "Proportion, Equation, Function: Three Steps in the Development of a Concept."

22.   See Heron of Alexandria, *Opera*, Vol. 4, p. 78 and *Metrica*, pp. 48–52. Heron's insouciance in this matter appears due to his preference for the Babylonian practice of solving areal problems via "numerical" approximations over the Euclidean theory of natural kinds of figures. See Bruin's remarks in the text, p. 7, and, more generally, Fritz Krafft, "Kunst und Natur. Die Heronische Frage und die Technik in der klassischen Antike."

23.   Eutocius, *Commentarii in Archimedis Libros de Sphaera et Cylindro*, in Archimedes, *Opera Omnia*, vol. 3, pp. 120.16–122.9; Theon of Alexandria, *Commentaires de Pappus et de Theon d'Alexandrie sur l'Almageste*, vol. 3, 532.1–535.9. See also the scholia to Euclid Bk. 6, Def. 5, collected by E. S. Stamatis in *Euclidis Elementa*, vol. 5, Pars 2, pp. 1–9. A quite valuable study by Ken Saito, "Compounded/Ratio in Euclid and Apollonius," shows, among other things, how Euclid circumvents the strategy of compounding ratios (in *Elements*, Bk. 6, Prop. 19 and in the *Data*) by employing the method of "reduction to linear ratio by means of the applications of areas" (p. 37). It is also noteworthy that Pappus, while making more frequent use of compounding, often proves "the same lemma twice, once using compound ratios, once without them" (Pappus of Alexandria, *Book 7 of the Collection*, part 1, p. 74).

24. See, for example, L. Courturat, ed., *Opuscules et fragments inedits de Leibniz*, p. 149: "Rationes et numeri res homogeneae sunt, addi potest ratio numero, etc., quod et ex aequationibus Algebraicis apparet."

25. Descartes, *Oeuvres*, publiés par Charles Adam and Paul Tannery, Nouvelle presentation (Paris, 1975), t. 2, p. 555. [Hereafter cited as "A.-T."]

26. For an understanding of this tradition see J. Murdoch, "The Medieval Language of Proportions"; Andrew G. Molland, "The Denomination of Proportions in the Middle Ages," and "An Examination of Bradwardine's Geometry"; and Edith Sylla, "Compounding Ratios. Bradwardine, Oresme and the First Edition of Newton's *Principia*."

27. Adriaan van Roomen, *Apologia pro Archimede ad clarissimum virum Josephum Scaligerum* (Würzburg, 1597), cited in Giovanni Crapulli, *Mathesis universalis. Genesi di una idea nel XVI secolo*, p. 210. See Crapulli, pp. 104–23 for an extensive discussion of van Roomen's work.

28. Cited in T. L. Heath, *The Thirteen Books of Euclid's Elements*, vol. 2, p. 133. Hereafter cited as "Heath, *Euclid*."

29. I am indebted for this insight to Eva Brann, in her annotated translation "The Cutting of the Canon," *The Collegian*, esp. pp. 23–38. A more accessible translation, with brief notes, has been published by Thomas J. Mathiesen, "An Annotated Translation of Euclid's Division of the Monochord." For a defense of a fourth-century dating of the *Sectio*, see Andrew Barker, "Method and Aims in the Euclidean *Sectio Canonis*."

30. Theon of Smyrna, *Expositio rerum mathematicarum ad legendum Platonem utilium*, pp. 83.24–25, p. 107.15sq.

31. al-Tusi, quoted in A. P. Youschkevith, *Les Mathémetiques arabes (VIIIe-XV^{e}siècles)*, p. 89. See pp. 80–90 ("Nombres irrationels et théorie des proportions") as well as the valuable discussions in Edward B. Plooij, *Euclid's Conception of Ratio and his Definition of Proportional Magnitudes as Criticized by Arabian Commentators*, and Barbara H. Sude, "Ibn Al-Haytham's Commentary on the Premises of Euclid's Elements." (For the Arabic transmission of the *Elements*, see Gregg DeYoung, "The Arabic Textual Tradition of Euclid's Elements.")

32. Richard Dedekind, *Essays on the Theory of Numbers*, p. 40; and "Brief an Lipschitz," cited in Beckmann, op. cit., p. 40.

33. See Beckmann, op. cit., p. 37.

34. R. Dedekind, *Brief an Weber* (1888), in *Gesammelte mathematische Werke*, Bd. 3, p. 489.

35. See Jules Vuillemin, *La Philosophie de l'algèbre*, pp. 527–31; and Henrich Scholz, "Warum haben die Griechen die Irrationalzahlen nicht aufgebaut?" esp. pp. 42–43.

36. Samuel Kutler, "The Source of the Source of the Dedekind Cut." For a contrary view, see Milenko Nikolic, "The Relation between Eudoxus' Theory of Proportions and Dedekind's Theory of Cuts." See also Jean-Louis Gardies, "Eudoxe et Dedekind," for corroboration of Cutler's view.

37. Leibniz, *GM*, Bd. 7, p. 208.

38. For the medieval renditions, see George D. Goldat, "The Early Medieval Traditions of Euclid's Elements," pp. 258–59. Aristotle's definition appears in *EN* V6, 1131a31. For "similarity or sameness," see Theon of Smyrna, op. cit., p. 82.6; and Iamblichus, *In Nichomachi Arithmeticam introductionem Liber*. Compare, as well, the intriguing account given by Nicomachus, *Introductionis Arithmeticae Libri II*, p. 120.2–5.

39. Aristotle, *Metaphysica, Iota* 3.1054a32–66.

40. This discussion of Wolff is based largely on Hans-Jürgen Engfer, "Wolffs Interpretation des Euklidmodells," in *Philosophie als Analyse*, pp. 231–37.

41. Johann Heinrich Lambert, *Abhandlung vom Criterium veritatis,* par. 70 (*contra* Wolff on postulates), par. 79 (on Euclid's starting-point).

42. J. H. Lambert, *Deutscher gelehrter Briefwechwsel,* Bd. 1, p. 337; and *Über die Methode der Metaphysik, Theologie und Moral richtiger zu beweisen,* p. 28.

43. J. H. Lambert, *Anlage zur Architektonik,* in *Philosophische Schriften,* Bd. 4, p. 144; *Abhandlung vom Criterium veritatis,* p. 79.

44. Lambert, *Über die Methode . . . , op. cit.,* p. 62; *Philosophische Schriften* Bd. 2, p. 12; *Deutscher gelehrter Briefwechsel, ed. cit.* [=Brief an Kant 3 Oct. 1770); *Über die Methode . . .* Notanda 77. (On Lambert's theories of mathematics see Wilhelm S. Peters, "Johann Heinrich Lamberts Konzeption einer Geometrie auf einer imaginären Kugel"; Gereon Walters, " 'Theorie' und 'Ausübung' in der Methodologie von Johann Heinrich Lambert"; and H. Ende, op. cit., pp. 17–19.

45. A. G. Kästner, "Was heisst in Euklids Geometrie möglich?" Kant's remarks are printed under the title "Über Kästners Abhandlungen," in *Gesammelte Schriften, Akademie Ausgabe,* Bd. 20, pp. 410–23, with Schulze's published review at the foot of each page.

46. Kästner, op. cit., p. 394.

47. Ibid., p. 392.

48. Ibid., p. 397.

49. Kant, "Über Kästners Abhandlungen," p. 411. On the further significance of Kant's view of the objective reference of mathematical concepts, See Günter Zöller, *Theoretische Gegenstandsbeziehung bei Kant,* pp. 183–255.

50. Kant, *Werke,* Bd. 5, p. 11, Ammerkung.

51. Heath, op. cit.; B. L. van der Waerden, *Science Awakening.*

52. Proclus, *In Euclid.,* p. 203.11–17. It is this occurrence of *kataskeuē* that J. Hintikka has taken to be central to the Euclidean *and* post-Euclidean (including Kantian) understanding of mathematical construction; see his essay, "Kant and the Tradition of Analysis," in *Logic, Language-Games and Information;* and Hintikka and U. Remes, *The Method of Analysis. Its Geometrical Origin and its General Significance* esp. pp. 41–48. My results in this and the subsequent chapter should make it plain that, at a minimum, Hintikka views both Euclidean and modern (e.g., Caretesian) construction from much too narrow a perspective.

53. Cf. Proclus, *In Euclid.,* p. 423.18 sq. (cited by Heath, vol. 1, p. 348). See Maria Timpanaro-Cardini, "Two Questions of Greek Geometrical Terminology," pp. 185–86.

54. Compare Attilio Frajese, "Sur la signification des postulates euclidiens," p. 392: "Si on doit donner une valeur à cette construction [sc. Euclid Bk. 1, Prop. 1], il semble plutôt qu'on affirme là que les deux circonferences se recontrent parce qu'on admet que le triangle équilateral existe, que l'inverse."

55. Proclus, *In Euclid.,* p. 398.13–15.

56. Euclid, *Opera Omnia,* vol. 6, p. 2.

57. See, for example, Leibniz, *GM,* Bd. 5, p. 143, par. 6.

58. Wilbur R. Knorr, *The Evolution of the Euclidean Elements* esp. pp. 211–313. Compare D. H. Fowler, "Ratio in Early Greek Mathematics.

59. Apollonius of Pergae, *Quae graece existant cum commentariis antiquis,* vol. 2, p. 124 (=Fragmentum 35 in *Pappus in Elem. X,* p. 701 [Woepcke]).

60. Euclid, *Opera Omnia,* vol. 5, p. 414.

61. Ibid., vol. 3, p. 370. (This is printed as Theorem 115.)

62. Proclus, *In Eucid.*, p. 78.3–8. See the notes on this passage in Leonardo Taran, *Speusippus of Athens. A Critical Study with a Collection of the Related Texts and Commentary*, pp. 422–31.

63. Ibid., p. 78.8–13.

64. See the study by Alan C. Bowen, "Menaechmus *versus* the Platonists: Two Theories of Science in the Early Academy." (Proclus' own position in this debate deserves very close attention, especially as regards his pivotal concepts of phantasia and *Kinēsis phantastikē;* see the studies of Stanislaus Breton, *Philosophie et mathématique chez Proculus*, esp. pp. 110–23; and Annick Charles-Saget, *L'Architecture du divin, Mathématique et philosophie chez Plotin et Proclus*, esp. pp. 190–205.)

65. See Simplicius, *In De Caelo* [ed. Heiberg], p. 304.24–27 (*Commentaria in Aristotelem Graeca* [Berlin, 1893], vol. 7), where the examples are the triangle composed of three lines and the cube composed of six squares. Alternatively, *tethentōn* here might echo Aristotle's usage in the *Posterior Analytics*, where "premissed" would give the correct meaning. (Guthrie, in his Loeb translation, p. 99, supports Simplicius' interpretation: "when all the constituents have been put together.")

66. Compare Friedrich Otto Sauer, "Euklid und der Operationalismus," pp. 243–49.

67. Malcolm Brown, "Some Debates about Eudoxus' Mathematics, p. 16.

68. H. G. Zeuthen, "Die geometrische Konstruktion als 'Existenzbeweis' in der antiken Geometrie," p. 223. The most concerted defense of Zeuthen's basic thesis appears in Eckhard Niebel, *Untersuchungen über die Bedeutung der geometrischen Konstruktion in der Antike.*

69. Among those dissenting, in whole or in part, from the "Kantian" orthodoxy established by Zeuthen (and Heath), mention might be made of A. Frajese, op. cit., and B. L. van der Waerden, "Die Postulate und Konstruktionen in der frühgriechische Geometrie." A more philosophically inspired opposition is voiced by F. O. Sauer, op. cit., and in his monograph, *Physikalische Begriffsbildung und mathematisches Denken*, esp. pp. 141–58. In addition, the recent study by W. R. Knorr, "Construction as Existence Proof in Ancient Geometry," brings welcome corroboration to my critique of Zeuthen. Knorr goes on to emphasize the independent fascination exercised within Greek mathematics by problems and problem-solving techniques (i.e., constructions); as a consequence he is prepared to give the problematic style a certain priority over the theoretic style, at least in *some* cases (e.g., in Apollonius' *Conics*). In his view the working geometers proceeded in their business without being wedded to any determinate philosophical views (see pp. 140–42). *This* conclusion is at odds with the position I have been defending in the case of Euclid; what is implicit in technique is an understanding of the relation between *technē* and comprehension (either as "pre-understanding" or as *theōria*). Knorr pursues meticulously and illuminatingly the actual or presupposed techniques of problem-solving in his book *The Ancient Tradition of Geometric Problems*, esp. pp. 101–49; 339–81. However, here, too, technical élan is set apart from philosophical meditation on the roots and limits of *technē* itself. Ultimately, Knorr's explanation of this compartmentalization seems to be political or sociological. See his remarks in an earlier paper, "Infinity and Continuity: The Interaction of Mathematics and Philosophy in Antiquity": "This [i.e., Athens in the fifth and early fourth centuries] was a society experiencing a crisis of authority, both intellectually and politically. Theoretically at least, every citizen had independent political standing and had the opportunity, indeed the necessity, of presenting his own cause before the assembly. . . . Philosophers like Plato, of a conservative inclination, were concerned with recovering a consensus on values by describing the absolute bases of true belief and knowledge" (p. 145). Autonomous *technai* appear to be democratic and progressive; philosophy, "conservative." The documented participation of mathematicians such as Theactetus and Eudoxus, Menaechmus and Euclid,

in the affairs or traditions of the Academy seems to go against the grain of this putative division of labor.

70. For initial orientation, one might usefully consult William Sacksteder, "Hobbes: The Art of the Geometricians"; and Ingetrud Pape, *Tradition und Transformation der Modalität*, Bd. 1, pp. 109–73, on Leibniz.

71. I. Newton, *Principia*, vol. 1, p. 15.

72. Ibid. For an exploration of these Newtonian themes see T. K. Simpson, "Newton and the Liberal Arts"; and James D. Liljenwall, "Kepler's Theory of Knowledge," pp. 154–65.

73. On the transport of a ruler to "generate" new irrationals, see J. Vuillemin, op. cit., pp. 537–38; on restrictions on the use of rule and compass, see A. Frajese, op. cit., and esp. Arthur Donald Steele, "Über die Rolle von Zirkel und Lineal in der griechischen Mathematik," for a detailed survey of ancient practices.

74. For Heron, See Proclus, *In Euclid.*, p. 323.5sq. For Pappus, see his discussions of the classes of problems in *Collectio*, vol. 1, Bk. 11.20 and Bk. 4.57; and the account of his divergence from earlier tradition in Wilbur Knorr, *The Ancient Tradition of Geometric Problems*, pp. 341–48 and n. 1–45.

75. B. L. van der Waerden, *Science Awakening*, p. 511. For the various solutions of the "Delian problem" see Eutocius, *Commentarii in libros* [Archimedis] *de sphaera et cylindro*, in Archimedes, *Opera Omnia*, vol. 3, p. 54.26–106.24. These are critically analyzed by W. Knorr, *The Ancient Tradition*, pp. 50–66. See also Malcolm Brown, "Plato on Doubling the Cube. *Politicus* 266 AB."

76. See *Alpharabii . . . Opera Omnia* (Paris, 1638), p. 26 (where "Agelea" should be corrected to "Algebra"). As Muhsin Mahdi (in "Science, Philosophy and Religion in Alfarabi's *Enumeration of the Sciences*") points out, this division comprises "the applicability of mathematical knowledge to natural bodies, the production of instruments, and in general the 'principles of practical, political arts' " (p. 125). On the topic of classification in the medieval Arabic tradition, see Ahmad A. al-Rabe, "Muslim Philosophers' Classifications of the Sciences: al-Kindi, al-Farabi, al-Ghazali, Ibn Khaldun."

77. Plutarch, *Vita Marcelli* 305E–F; Ps.-Eratosthenes, apud Eutocius, op. cit., pp. 88.4–96.27. For a contrary reading of the so-called "Tadelstelle" in Plutarch, see E. Niebel, op. cit., pp. 112–22. (I do not mean to imply that testimonies such as the above give a complete sense of the techniques of construction in Greek mathematics or of their possible significances. For instance, one would have to consider carefully the constructions of the conic sections and the use of *neusis* ("verging" or "inclination") in Archimedes. On these matters see Wilbur R. Knorr, "The Hyperbola-Construction in the *Conics*, Book 11: Ancient Variations on a Theorem of Apollonius," "Archimedes' Neusis-Constructions in Spiral Lines."

78. For example, in *The Statesman*, the "kingly art," having earlier been set parallel to the *productive* art of weaving (281D), is subsequently set apart from those subordinate arts that are responsible for *genesis* as an art aimed at preserving their productive handiwork (287D–E). This suggests that the preservative or directive arts stand in need of the productive arts even while being irreducible to the latter.

79. Compare the account of an authentically "Platonic" approach to problem-solving in Alexander Mourelatos, "Plato's 'Real Astronomy': *Republic* 527D–531D," esp. pp. 60–62, "The Meaning of *Problema*."

80. Carpus apud Pappus, *Collectio*, Bd. 8, Praefatio, sect. 3.

81. Compare the interpretation of Archimedes' mechanical heuristics in Jean-Louis Gardies, "La Méthode mécanique et le Platonisme d'Archimède." See also Tohru Sato, "A

Reconstruction of *The Method* Proposition 17, and the Development of Archimedes' Thought on Quadrature," where *The Method* is plausibly assigned to the latest period of Archimedes' career.

82. Archimedes, *Opera Omnia*, vol. 2, pp. 428.29–429.1

83. Ibid., vol. 1, p. 2.19–20.

84. On this concept of normative idealization, see Paul Lorenzen and Wilhelm Kamlah, *Logische Propädeutik*, esp. pp. 228–29. A more exotic, and exhilarating, account of the "manual" beginnings of geometry may be found in the essays of Michel Serres, "Mathematics and Philosophy: What Thales Saw . . . ," in *Hermes, Literature, Science, Philosophy*, pp. 84–97, and "Origins de la géométrie 3,4,5," in *Hermès V: Le Passage du Nord-Ouest*, pp. 165–95.

85. Edited by H. Menge in vol. 8 of *Euclidis Opera Omnia*.

86. *Metaph.* 83.998a3–5.

87. See *Meno* 85b; and M. Brown, "Plato Disapproves of the Slave-Boy's Answer," n. 11.

88. See, for example, Def. 4 in the *Data* (= *Opera Omnia*, vol. 6, p. 2): "Tē thesei dedosthai legontai sēmeia te kai grammai kai gnōniai ha ton auton aei topon epechei." On *chōrion* see Charles Mugler, *Dictionnaire historique de la terminologie géometrique des Grecs*, s.v.; and the important comments by J. Cook Wilson, "On the Geometrical Problem in Plato's *Meno* 68E."

The most explicit early argument in behalf of a three-dimensional, corporeal space seems to be due to the Christian commentator on Aristotle, John Philoponus. See his 'Corollarium de Loco' in *Commentaria in Aristotlelem Graeca*, ed. H. Vitelli, vol. 17, pp. 557–85; and the studies by Wolfgang Wieland, "Zur Raumtheorie des Johannes Philoponus"; and David Sedley, "Philoponus' Conception of Space." Of great value are Sorabji's remarks in his editorial introduction (pp. 18–22) of *Philoponus and the Rejection of Aristotelean Science*. He shows that Philoponus' three-dimensional corporeal extension *cannot* be identified with geometrical extension (p. 20).

The frequently asserted identity of *chora* in the *Timaeus* and Cartesian extension, maintained, for example, by Heidegger, has been illuminatingly criticized by Alain Boutot, *Heidegger et Platon. Le problème du nihilisme*, pp. 216–30.

89. The disparity between the *Optics* and the *Elements* should be taken into account in any attempt to show that Euclid embraced without reserve the now-contemporary (metrical) notion of "Euclidean space." This is not done in the otherwise highly instructive studies by Imre Toth (e.g., "Das Parallelenproblem in Corpus Aristotelicum") and by Vittorio Hösle, "Platons Grundlegung der Euklidizität der Geometrie." For general and helpful discussions of Euclidean and later Greek optics, especially in connection with matters of perspective and anomalous appearances, see A. Mark Smith, "Saving the Appearances of Appearances. The Foundations of Classical Geometrical Optics"; and C. D. Brownson, "Euclid's Optics and it Compatibility with Linear Perspective."

90. Couturat, *Opuscules et Fragments inédits de Leibniz*, pp. 98–99.

91. Ibid., p. 99; and *GM*, Bd. 7, p. 60.

92. *De an.* 3.3, 429a7 and *De mem.* 451a6–10. (Compare 453a 19–20 on melancholics: *Toutous gar phantasmata kinei malista*)

93. *De mem.* 450b21–26.

94. *De an.* 429b13–17. See G. Rodier, *Aristote, Traité de l'âme* t.2, ad loc. (pp. 444–450) for a survey of ancient and modern opinions. Mine comes closest to that of Trendelenburg, op. cit. p. 450.

95.   See the discussion of intellectual versus imaginative perfection in Maimonides, *Guide of the Perplexed;* Leo Strauss, "Maimunis Lehre von der Prophetie und ihre Quellen"; Shlomo Pines, "The Arabic Recension of *Parva Naturalia* and the Philosophical Doctrine Concerning Veridical Dreams according to *al-Risala al-Manamiyya* and Other Sources"; and Laurence Kaplan, "Maimonides on the Miraculous Element in Prophecy."

96.   See Leibniz, *Leibnitiana,* p. 6.

97.   Oskar Becker, *Mathematische Existenz,* p. 198.

98.   Ibid., p. 199. (See n. 67, above.) Proclus, *In Platonis Rem Publicam Commentarii,* vol. 1, 235.18 and 121.2–3; cf. also vol. 2, 107.18–20 for the assertion by "some of the ancients" that *phantasia* and *nous* are identical.

99.   H. G. Zeuthen, "Notes zur l'histoire des mathématiques, IX: Sur les connaissances géométriques des Grecs avant la reforme Platonicienne," *Oversigt o.d. Kgl. Danske Vidensk. Selsk. Forh.* (913), p. 434; cited in E. Niebel, op. cit., p. 23.

100.   Charles Kahn, *The Verb "Be" in Ancient Greek,* and "Why Existence Does Not Emerge as a Distinct Concept in Greek Philosophy?" (For an interesting critique of Kahn's main thesis, see Mohan Matten, "Greek Ontology and the 'Is' of Truth.")

Jaakko Hintikka has recently challenged the claims of "The Frege Trichotomy" (i.e., the "is" of existence/predication/identity) and its bearing on Aristotelian ontology. See his " 'Is,' Semantical Games and Semantical Relativity," "Semantical Games, The Alleged Ambiguity of 'Is' and the Aristotelian Categories," and "The Varieties of Being in Aristotle."

101.   See, for example, Heath, *Euclid,* vol. 1, pp. 119–20. Two "classical" accounts of the alleged structural resemblance (or dependence) between *Posterior Analytics* and the *Elements* are H. D. P. Lee, "Geometrical Method and Aristotle's Account of First Principles"; and Benedict Einarson, "On Certain Mathematical Terms in Aristotle's Logic." A valuable counterstatement may be found in Ian Mueller, "Greek Logic and Greek Mathematics."

102.   Alfonso Gomez-Lobo, "Aristotle's Hypotheses and the Euclidean Postulates," and "The So-Called Question of Existence in Aristotle, *An. Post.,* 2.1–2." Mario Mignucci, "In margine al concetto aristotelica di esistenza," and *L'argomentazione dimostrativa in Aristotele. Commento agli Analitici Secondi,* vol. 1, ad loc.

For a recent restatement of the hitherto conventional view, see Russell Dancy, "Aristotle and Existence."

103.   J. A. Smith, "*TODE TI* in Aristotle."

104.   Alcman, Frag. 58, 1 [Page], cited in Seth Benardete, "The Grammar of Being," p. 491; Benardete's article is an illuminating critique of Kahn's book, op. cit., n. 105.

105.   *Post. An.* A10.77a1–3.

106.   See Pierre Hadot, "Zur Vorgeschichte des Begriffs 'Existenz': *HYPARXEIN* bei den Stoikern," p. 126. For dissent from Hadot's interpretation, see Victor Goldschmidt, "'*HYPARXEIN* et *HYPHISTANAI* dans la philosophie stoïcienne."

107.   See Pierre Hadot, "L'Etre et l'étant dans le Néoplatonisme"; and C. Kahn, "On the Terminology for Copula and Existence." On the background to Victorinus' construal of *on/einai,* see Gerhard Huber, *Das Sein und das Absolute. Studien zur Geschichte der ontologischen Problematik in der spätantiken Philosophie,* esp. pp. 89–116.

108.   Cf. Plutarch, *Moralia* 1073e (= *Stoicorum Veterum Fragmenta,* ed. J. von. Arnim [Stuttgart, 1864], vol. 2, p. 167 (hereafter cited as "*SVF*"). On the members of the class of "incorporeals" see Emile Brehier, *Le Théorie des incorporels dan l'ancien Stoïcisme.* On the identity and the status of the *lekta,* see, among many mutually discrepant studies, Guido

Cortassa, "Pensiero linguaggio nella teoria stoica del LEKTON"; Andreas Graeser, "The Stoic Theory of Meaning"; and M. Mignucci, *Il Significato della logica stoica*, esp. pp. 88–103.

109.   See A. A. Long, "Language and Thought in Stoicism," p. 110, n. 59.

110.   See the brief comments in my "Review of J. Ritter, hrsg., *Historisches Wörterbuch der Philosophie*, Bde. 1–4," p. 200 and the bibliographical references cited there. Part of the semantic promiscuity of *ma 'na* as it enters into the Latin tradition can now be seen by consulting *Avicenna Latinus: Liber de prima philosophia sive scientia divina.* s.v. *ma 'na.* On the impact of the Arabo-Latin fashioning of the notion of concept/intention, see Johannes Lohmann, "Saint Thomas et les Arabes (Structures linguistiques et formes de pensée)."

111.   Pierre Pachet, "La Deixis selon Zénon et Chrysippe," 20, p. 245. (But see also Michael Frede, *Die stoische Logik*, pp. 53. 53 sq., for a different interpretation of *deixis* and its indirect or anaphoric employment.)

112.   On Chrysippus' proposal and its context, see Jean-Paul Dumont, "Mos geometricus, mos physicus."

113.   Jaap Mansfield, "Intuitionism and Formalism: Zeno's Definition of Geometry in a Fragment of L. Calvenus Taurus," p. 70. (Mansfield's reconstruction of the fragment, together with his thesis that it yields Zeno's definition of geometry, has quite recently been challenged by Harold Tarrant, "Zeno on Knowledge or on Geometry? The Evidence of anon. *In Theaetetum.*"

114.   Jean-Francois Courtine, "Note Complémentaire pour l'histoire du vocabulaire de l'être."

115.   See Christopher Stead, *Divine Substance*, ch. 6: "The Word *OUSIA* in Late Antiquity."

116.   On the theme of the distance or intimacy between the divine and the (visible) cosmos, see Hans Blumenberg, *Die Genesis der kopernikanischen Welt*, I. Teil, Kap. 1: "Der Kosmos und die Tragödie"; and Alexandre Kojève, "The Christian Origin of Modern Science."

117.   This is quite noticeably true in two otherwise helpful and important exegetical studies, viz. F. Rahman, "Essence and Existence in Avicenna"; Parviz Morewedge, "Philosophical Analysis and Ibn Sina's 'Essence-Existence' Distinction." On the other hand, Emil Fackenheim has brought clearly to light the politico-theological setting of this and cognate issues in Islamic and Judaic thinking; see "The Possibility of the Universe in Al-Farabi, Ibn Sina and Maimonides."

118.   Cf. F. Rahman, op. cit., p. 8; and Ghassan Finianos, *Les grandes divisions de l'être "mawjud" selon Ibn Sina*, pp. 123–31.

119.   Avicenna, *La métaphysique du Shifa'*, *Livres I a V*, p. 108. See the discussion in Ibn Sina, *Livre des Directives et remarques (Kitāb al-'Ishārāt wa l-tanbīhāt)*, pp. 87–88; and the significantly different translation of these same passages concerning 'the intention of existence' in G. Finianos, op. cit., pp. 132–33.

120.   Ibn Sina, *al-Najāt, Metaphysique*, p. 213, cited in A.-M. Goichon, *La Distinction de l'essence et de l'existence d'après Ibn-Sina (Avicenne)*, p. 141.

121.   Ibid., p. 14 (from Ibn Sina's *Mantiq al-mashriqiyyin* [Logic of the Orientals]).

122.   *Livre des directives et remarques*, op. cit. pp. 353–54.

123.   See Ibn Sina's remark in his commentary on *Metaph.* Lambda, cited in Edward Booth, *Aristotelian Aporetic Ontology in Islamic and Christian Thinkers*, pp. 109–10 (quoted from A. Badawi, ed., *Aristū 'inda l-'Arab*, pp. 23.21–24.4). It ought to be noted that Ibn Sina claims that Aristotle and his Greek commentators also alluded to a first cause of existence

as well as a first cause of motion (see A. Badawi, ibid., p. 180, and compare Simplicius, *In Phys.* (=*CAG*, X), pp. 1362–63. Kant's statement "This question [viz. "How is Nature itself possible?"] is the highest point on which transcendental philosophy may ever touch" (*Prolegomena*, par. 36 ad init.) can be seen as the successor to the questions raised in these Avicennian texts. This dimension of the question is *not* brought to light in Allan Bäck, "Avicenna on Existence."

124.   See E. Fackenheim, op. cit., passim.

125.   See Maimonides, *Guide of the Perplexed*, 2.13; and Warren Zev Harvey, "A Third Approach to Maimonides' Cosmogony-Prophetology Puzzle," p. 289, n. 9. Alfred Ivry's recent study, "Maimonides on Creation," takes the decisive phrase *ba'da l-'adama l-'mutlaq* to signify "after absolute privation," *not* "after absolute non-being."

126.   Cited from the *Kitab al-Najāt, Metaphysics,* by E. Fackenheim, op. cit., p. 40.

127.   In addition to the works cited in the notes above, I have found much of value in Ernst Behler, *Die Ewigkeit der Welt, Erster Teil: Die Problemstellung in der arabischen und jüdischen Philosophie des Mittelalters,* pp. 88–114; and Herbert A. Davidson, "Avicenna's Proof of the Existence of God as a Necessarily Existent Being."

For further light on the syntactic and semantic underpinnings of the Arabic term for existence considered here (viz. *wujud*), see Farid Jabre, "EINAI et ses dérivés dans la traduction, en arabe, des catégories d'Aristote"; On the explication provided by al-Farabi, see Amine Rochid, "Dieu et l'être selon al-Farabi: Le chapitre de 'l'être' dans le Livre des Lettres"; and Georges C. Anwati, "La Notion d'*al-Wujud* (existence) dans le *Kitab al-Hudud* d'al Farabi." The importance of rendering *wujud* with precision is brought out by Wilbur Knorr's remark (in *The Ancient Tradition,* p. 376) that in Heron of Alexandria's scholium to Euclid, Bk. 2, preserved only in the Arabic translation by al-Naírízí, the expression *innahu mawjud* points to the underlying Greek *hōs zētoumenon,* "as if found," *not* to *hōs on,* "as being." "Finding," "being found," *se trouver* are close to the etymology of the Arabic root.

128.   Cited above, n. 99.

129.   Samuel Sambursky, *The Physics of the Stoics,* p. 85.

130.   See *Gesamtausgabe,* Abt. 2, Bd. 24, p. 165. (See also Heidegger's discussion of the relations among *physis, technē,* and *thaumazein* in *Grundfragen der Philosophie. Ausgewählte "Probleme" der "Logik,"* pp. 177–81.)

131.   Ibid., p. 168. Whatever may be the case in Christianized versions of the creation-motif, it is important to note that in, e.g., Ibn Sina, a salient distinction is drawn between the cause of generation/production in time (*huduth*) and the cause of maintenance over time (*thabat*); cf. H. Davidson, art. cit., pp. 177–78. The latter holds ontological priority over the former when it is a question of the essential dependence of any being on the one being which is necessary *per se.* Moreover, Ibn Sina distinguishes al-huduth al-dhati (essential coming into existence) from al-huduth al-zamani (temporal coming into existence), where, once more, temporal considerations are subsidiary. See Michael E. Marmura, "Avicenna on Causal Priority," p. 79, n. 2.

132.   Ibid., p. 153.

133.   Besides the references cited in n. 69, above, see A. Frenkian, *Le postulat chez Euclide et chez les Modernes;* and Kurt von Fritz, "Die ARXA1 in der griechischen Mathematik."

134.   In addition to the references cited in n. 89, above, see also E. M. Bruins, *La géométrie non-Euclidienne dans l'antiquité;* and C. S. Peirce, *New Elements of Mathematics,* vol. 4, pp. 704–705, esp. Peirce's claims: "I maintain that Euclid was himself a non-Euclidean geometer" (p. 704), and "In this connection let me ask: is it possible that Euclid, the author of the *Data* and *Porisms,* was unaware that metrical geometry is not the real basis of geometry" (p. 705).

135.   Here I find myself in partial agreement with F. O. Sauer, op. cit. (see n. 69), and in disagreement with the importantly provocative studies of Árpád Szabó, "Greek Dialectic and Euclid's Axiomatics," and *Anfänge der griechischen Mathematik* (Munich, 1969), pp. 361–78, the main thesis of which is that the appeal to motion or moveability implicit in the first three postulates represents an attempt to circumvent or disarm the Eleatic paradoxes.

The entire question of the axiomatic or nonaxiomatic (where "axiomatic" embraces the "postulates" as well) character of the *Elements* is treated in *Proceedings of the Pisa Conference on the History and Philosophy of Science*, 1978, vol. 1: *Theory Change, Ancient Axiomatics and Galileo's Methodology* (Dordrecht, 1981), pp. 113–225.

136.   Heath, *Euclid*, vol. 1, p. 146, citing A. Trendelenburg's *Erläuterungen zu den Elementen der aristotelischen Logik*, p. 197.

137.   See Heath, *Euclid*, vol. 1, p. 195, and *The Works of Archimedes*, p. clxxv.

138.   *Metaph.* A6.987b.15ff. On what might be the genuinely Platonic intention motivating this thesis see Jacob Klein, "The World of Physics and the 'Natural' World," pp. 16–17.

139.   Heath, *Euclid*, vol. 1, p. 200.

140.   Robert J. Wagner, "Euclid's Intended Interpretation of Superposition."

141.   Still other issues in Greek mathematics seem to implicate questions of "existence" in the modern, Kantian, sense. These include: (1) the assumption that a fourth-proportional can be found, especially in connection with the so-called "method of exhaustion" (as in Euclid XII, 2; see, on the whole affair of quadrature and implicit continuity assumptions, O. Becker, "Warum haben die Griechen die Existenz der vierten Proportionale angenommen?" *Quellen und Studien zur Geschichte der Mathematik, Astronomie und Physik* 2, Abt. B [1933]:369–87); (2) the use of "neusis-construction"—i.e., verging-lines—by Archimedes and others for determining certain properties of tangents to a spiral (see W. R. Knorr, "Archimedes' Neusis-Constructions in Spiral Lines"); (3) the Euclidean demonstration, in Book 8, that only five regular solids "exist." Let me comment briefly on the last of these issues. In line with the case I have been making throughout this chapter, Euclid's proof should be interpretable as demonstrating *not* that the five solids *exist* (are constructible) but that only these *five* meet the specified conditions of geometrical intelligibility. The emphasis falls, accordingly, on the unalterable *number* of the solids, not on the warrants for asserting their existence.

Apart from the discussion in Proclus our most important ancient source of enlightenment concerning the question of the "existence" of geometrical beings is the Fifth-Century A.D. Neo-Platonist Marinus. His short commentary on Euclid's *Data* (*Ta Dedomena*) consists in the main of a review and a critical filtering of the various explanations given in the tradition of what Euclid means by a *datum*. The list of candidates is already noteworthy: *tetagmenon* (ordered and fixed, as in the case of a straight line passing through two fixed points); *gnōrimon* (recognizable to the intellect); *Rhēton* ("sayable," in the exact sense of commensurable with a unit-measure, either *per se* or in square); *Porimon* (what *can* be provided or furnished because and only because it has already been furnished in the past); and the various combinations of these simple definitions taken two-by-two.

Three things stand out from Marinus' account. First, no mention is made of *hyparchon* (or *on*) as a possible rendition of *dedomenon*. Second, the full and precise definition Marinus endorses makes *given* equivalent to *gnorimon kai porimon*, where *gnorimon*, recognizable and thus admissible, is the notion of wider scope. Third, Marinus takes pains to distinguish between *Porimon* and *Poriston:* "We are treating at this time of what is already furnished or realized, that which we properly call *Porimon*. For what has not yet been furnished, but is capable of being so, we properly call *Poriston* (see *Euclidis Opera*, ed. Menge, vol. 6, p.

240.22–26). What immediately strikes a chord in memory is this third point: As Euclid did in the statement of the postulates and in the vocabulary of construction generally, so Marinus here stresses the already achieved character of the *datum*. Its mode of being and hence of accessibility to the teacher or student of geometry is anterior to the work to which it may be put, for example in the solution of problems, In Maurice Michaux's words "il s'agît des choses réalisable parce que déjà réalisées" (*Le Commentaire de Marinus aux Data d'Euclide. Étude critique* [Louvain, 1947], p. 26; see this somewhat neglected work for further analyses of Marinus' definition, which is also of great value for the understanding of Euclid's *Data*). Also of great value for understanding the differences between *ta dedomena* in Ancient analysis and "the known" in modern analysis is Robert H. Schmidt, "The Analysis of the Ancients and The Algebra of the Moderns" (Fairfield, Conn., 1987: A Golden Hind Editorial). Schmidt argues that *to dedomenon* ought to be translated "the recipient," not "the given," and explains: "But the Greek art of analysis was centered around the *dedomenon*, which was the indirect object of an act of giving for which the other geometrical figures were responsible" (p. 12). See also Schmidt's remarks in his edition of Apollonius *On Cutting Off a Ratio*, trans. from the Arabic manuscript by E. M. Macierowski (Fairfield, Conn., 1987), pp. 157–59.

The early-modern understanding of "poristic analysis" and of "porism" stems from Pappus, not from Marinus (see Pappus, *Collectio*, ed. Hultsch, vol. 2, pp. 634.3–636.30, as well as p. 648ff. on Euclid's lost *Porisms*). Indeed, Marinus' distinction between *Porimon* and *Poriston* appears to be aimed against Pappus (see Michaux, op. cit., pp. 20–21). Nevertheless, Pappus' apparent use of existential language in describing *theoretical* analysis, in contrast to *problematic* analysis, is open to a quite different reading. When he writes that in the former we first "lay down what is being sought as being and as true" (*hōs on hypothemenos kai hōs alēthes* [636.1–2]), the last member of the phrase, *kai hōs alēthes*, may be taken exepegetically; that is, to lay something down as "being" is to assume or hypothesize the truth of the relevant geometrical description. (Compare Hultsch's translation: *primum id quod quaeritur re vera ita se habere statuimus*.) This reading is buttressed by the conclusion of Pappus' description: *alēthes estai kai to zētoumenon* (the thing sought will be true), where the omission of any term answering *to on* suggests that *hōs on* at the beginning was used veridically, *not* "existentially."

For the vastly more complicated issue of the poristic art in Euclid, Pappus, and in their modern exegete, Viète, see Richard D. Ferrier, "Two Exegetical Treatises of Francois Viète, Translated, Annotated and Explained," ch. 8, "A Restoration of Viète's Poristic Art."

142.   Compare Stanley Rosen, *Plato's Sophist: The Drama of Original and Image*, pp. 93–95.

Chapter 3

1.   See Michel Serres, "Knowledge in the Classical Age: La Fontaine and Descartes," p. 27. (On Descartes' sense of self-disguise, see further the famous *larvatus prodeo* in A.-T., t. 10, p. 213; and the conjectures of Jean-Luc Nancy, "Larvatus Pro Deo."

2.   Much of the material included in this first section has already appeared in my article, "Descartes and the Philosophy of History." For another view of Descartes' attitude toward history, see Pierre Guenancia, "Remarques sur le rejet cartésien de l'histoire." Guenancia emphasizes the discontinuity of Cartesian time as a succession of instants, in virtue of which "the idea of becoming loses all meaning" (p. 569). This association of history with becoming runs counter to the traditional conception of exemplary history discussed in this section.

3.   The argument, and much of the text, of this third section appear in my essay "Objectum Purae Matheseos."

4. See Paul Valéry, "Descartes," p. 29.

5. *A.T.*, t. 10, p. 204.

6. For Bruno, see *La cena delle ceneri, in Dialoghi Italiana*, p. 39; for Bacon, see *Novum Organum* 1.84; and cf. F. E. Guyer, " 'C'est nous qui sommes les anciens.' "

7. Joachim Du Bellay, *La Défense et illustration de la langue Francoyse*, ed. H. Chamard (Paris, 1948), p. 65.

8. *A.-T.*, t. 10, p. 204. See St. Jerome's gloss on the verse of *Ecclesiastes*—"Et non est recens sub sole"—in *Patrologia Latina*, ed. Migne, vol. 23, col. 1071. Jerome is reminded of his teacher Donatus' similar adage: "Pereant . . . qui ante nos nostra dixerunt."

9. See *A.-T.*, t. 1, p. 158, and t. 2, p. 436.

10. Ibid., t. 10, p. 214, and cf. t. 10, 156–57 for Descartes' repudiation of the Lullian "art."

11. Ibid., t. 10, pp. 331–32.

12. Ibid., t. 2, p. 339.

13. Ibid., t. 10, p. 204.

14. See *Nouvelles lettres et opsucucles inédits de Leibniz*, ed. Foucher de Careil, p. 289; and compare G. Kahl-Furthmann, "Descartes' Betonung seiner Unabhängigkeit von der Tradition and Leibnizens Kritik."

15. J. A. Worp, *De gedichten van Constantijn Huygens*, vol. 4, p. 143; cited in E. J. A. Dijkterhuis et al., *Descartes et le cartésianisme hollandais*, p. 226.

16. Henri Gouhier, *Les premiéres pensées de Descartes*, p. 149. (After completing this section [see n. 2, above], I encountered the following studies of the rhetoric of the *Discourse*, which valuably supplement or complement my remarks on these pages: J.-L. Nancy, "Mundus est fabula"; John D. Lyons, "Subjectivity and Imitation in the *Discours de la Methode*"; Stephen H. Daniel, "Descartes on Myth and Ingenuity/Ingenium" and Dalia Judovits, *Subjectivity and Representation in Descartes*, ch. 3.)

17. *A.-T.*, t. 10, p. 214.

18. Ibid., t. 10, p. 367. cf. Jean-Luc Marion, *L'Ontologie grise de Descartes*, p. 44: "science sans généalogie, où le savant se produit sans père."

19. *A.-T.*, t. 10, pp. 502–503; cf. ibid., t. 10, pp. 497–98.

20. *A.-T.*, t. 1, p. 570. See also G. F. A. Gadoffre, "Sur la chronologie du *Discours de la Methode*."

21. *A.-T.*, t. 6, p. 7.

22: Cicero, *De oratore* 11.9.36. On the later destiny of this slogan see Reinhart Koselleck, " 'Historia magistra vitae.' Über die Auflösung des Topos im Horizont neuzeitlich bewegter Geschichte."

23. Polybius, *Histories*, 12.28.1–5. On Polybius' politics of history, see Robert D. Cumming, *Human Nature and History*, vol. 1, pp. 135–79.

24. Jean Bodin, *Methodum ad facilem historiarum cognitionem*, ch. 1, par. 1.

25. *A.-T.*, t. 6, pp. 7–8.

26. *A.-T.*, t. 1, p. 570.

27. Cicero, *De inventione*, 1.27.

28.   Quintilian, *Institutio Oratoria*, 11.4.2. On this and other relevant ancient texts, see Wesley Trimpi, "The Quality of Fiction: The Rhetorical Transmission of Literary Theory," pp. 43–51, or *Muses of One Mind. The Literary Analysis of Experience and Its Continuity*, ch. 3.

29.   Aristotle, *Topics* 1.2.101a25–64. On the links between *ta phainomena* and *ta legomena*, see G. E. L. Owen, "Tithenai ta phainomena"; Martha Nussbaum, "Saving Aristotle's Appearances"; and my unpublished lectures "The Methods of Aristotelian Discourse" and "*Physis, Phainomena* and Dialectic."

30.   *A.-T.*, t. 6, p. 8. On the refusal of otherness in Descartes' thinking, see Michel Guerin, "Le Malin Génie et l'instauration de la pensée comme philosophie."

31.   Arist. [?], *Magna Moralia* 11.15.1213a15–26.

32.   *EN* 9.12.1172a2–8.

33.   *A.-T.*, t. 6, p. 6. See the remarks of Luis Noussan-Lettry, "Die Anerkennung des Historischen in der Lebenserfahrung und der Weg des Denkens bei Descartes."

34.   *A.-T.*, t. 6, p. 11. On the workshop motif in Descartes, see Lucien Laberthonnière, *Etudes sur Descartes*, in *Oeuvres*, t. 2, pp. 287–315: "De la contemplation de la nature par l'esthète à son exploitation par l'ingénieur."

35.   *A.-T.*, t. 9-1, p. 141.

36.   William E. Carlo, "Idea and Concept: A Key to Epistemology."

37.   *A.-T.*, t. 9-1, p. 33. Cf. ibid., t. 7, p. 42, where *Instar archetypi* is at the base of the French phrase "comme un patron ou un original." See, as well, Richard B. Carter, "Volitional Anticipation and Popular Wisdom in Descartes"; and Andre Doz, "Sur la signification de 'instar archetypi.' Descartes, Troisième Méditation."

38.   Francis Bacon, *Novum Organum*, 1, Aphorism 3.

39.   *A.-T.*, t. 3, pp. 722–23.

40.   Machiavelli, *Il Principe*, ch. 15, ad init.

41.   *A.-T.*, t. 6, p. 374.

42.   See Adrien Baillet, *La Vie de Monsieur Descartes*, t. 1, p. 163.

43.   B. Pascal, "De l'esprit géométrique," *Oeuvres complètes*, p. 601.

44.   On the matrix and the significance of Descartes' formulations of these rules, see Jules Sirven, *Les Anneés d'apprentissage de Descartes* (1596–1628), pp. 180–225.

45.   See Johannes Clauberg, *Defensio Cartesiana*, in *Opera Omnia philosophica*, vol. 1, p. 4, where the *Discourse* is called "exotericum, popularem minusque accuratum"; and Daniel Lipstorp, *Specimina philosophiae Cartesianae*, Pars prima, p. 8: "clavis . . . omnium liberalium artium et scientiarum."

46.   J. S. Mill, *An Examination of Sir William Hamilton's Philosophy*, in Collected Works of John Stuart Mill, vol. 9, p. 478.

47.   *A.-T.*, t. 2, p. 268. Descartes' further avowal, "toute ma physique ne soit autre chose que mécanique" (to Debeaune, 30 April 1639 = *A.-T.*, t. 2, p. 542), provides another link in the chain connecting "abstract geometry" with a mechanized mathematical physics. (On the conceptual armature of Cartesian physics, see John A. Shuster, "Descartes and the Scientific Revolution, 1618–1634: An Interpretation"; Shuster, however, argues that Descartes came to jettison the scheme of a "Universal mathematics" before turning his head to the "merely discursive mechanism" of *Le Monde* and later works. See n. 120, below.)

48.   See Descartes, *Oeuvres et lettres*, ed. A. Bridoux, introduction, p. 16.

49. To Mersenne, December 1637 [?], *A.-T.*, t. 1, p. 478.

50. See the "Preliminary Discourse," in *The Encyclopedia. Selections,* trans. Stephen J. Gendzier (New York, 1967), p. 21.

51. To give a few representative samples: (1) "arithemetization of geometry"—Julian L. Coolidge, *A History of Geometrical Methods,* p. 126; (2) geometrization of arithmetic—Yvon Belaval, *Leibniz, critique de Descartes,* ch. 4; (3) "geometrization of algebra"—Carl B. Boyer, "Descartes and the Geometrization of Algebra"; (4) "algebraization of geometry"–Léon Brunschvicg, *Les étapes de la philosophie mathématique,* chs. 7–8.

52. *A.-T.,* t. 2, p. 490.

53. See Paul Tannery's "Note sur le probléme de Pappus," in *A.T.,* t. 6, pp. 721–25, for the germane references.

54. *In Euclid,* p. 394.17–19.

55. See *GM, Bd.* 7, p. 21: "Punctum (spatii scilicet) est locus simplicissimus, seu locus nullius alteri loci."

56. See ch. 2, n. 117, above.

57. Apollonius, *Quae greace exstant cum commentariis antiquis,* 1, p. 4.3. On the pre-Apollonian methods for generating the come sections and the accompanying terminology see Diocles, *On Burning Mirrors. The Arabic Translation of the Lost Greek Original,* pp. 3–17.

58. Apollonius, op. cit., 1, p. 42.22; p. 48.18–19.

59. But, *only* "analogous." See Richard B. Carter, *Descartes' Medical Philosophy,* pp. 253–66. Much earlier, Vico's colleague Paolo Mattia Doria challenged the modern (*sc.* Cartesian) identification of a planimetric with essential or definitory properties; see my "Vico, Doria e la geometria sintetica," p. 28, n. 53.

60. Pappus, *Collectio,* p. 680.2–30.

61. The most lucid account of these correlations may be found in Jules Vuillemin, *Mathématique et métaphysique chez Descartes,* pp. 108–12. In addition to Vuillemin's fundamental work, I am also indebted in this section to A. Boyce Gibson, "La 'Géométrie' de Descartes au point de vue de sa methode"; and Jean Hyppolite, "Du Sens de la géometrie de Descartes dans son oeuvre."

62. *A.-T.,* t. 6, p. 471. On the earlier history of Descartes' reckoning with the problem of finding mean proportionals, see ibid., vol. 10, pp. 651–59; and Pierre Costabel, "La solution par Descartes du probleme des moyennes proportionelles (Regle VI)."

63. *A.-T.,* t. 6, p. 485.

64. Ibid., t. 6, p. 475.

65. Ibid., t. 6, p. 369.

66. Although this emphasis is, I think, novel, very good arguments for Descartes' basically "constructivist" position may be found in J. Vuillemin, op. cit., pp. 165–166; and Timothy J. Lenoir, "Descartes and the Geometrization of Thought: The Methodological Background of Descartes' *Géométrie.*" Despite its other, egregious virtues, Yvon Belaval's work, *Leibniz, critique de Descartes,* esp. ch. 4, does not take into account the central place given to actual or potential constructions, since Belaval prefers to stress the contrast between Cartesian "intuitionism" and Leibnizian "formalism."

67. See, e.g., Leibniz, *GM, Bd.* 7, pp. 203–16. For Chasles' sobriquet, see Julian L. Coolidge, "The Origin of Analytic Geometry," p. 231.

68.   See David Brewster, *Memoirs of the Life, Writings and Discoveries of Sir Isaac Newton,* vol. 1, p. 22, n. 1.

69.   There are, I would argue, salient indications in all these authors that the "problematic" approach is *not* preoccupied with ascertaining or constructing the "existence" of geometrical entities but with discovering the conditions required for the felicitous execution of a proposed task. On Oenopides, see A. Mourelatos, "Plato's 'Real Astronomy': Republic 527D–531D." Marinus' *Commentary on Euclid's Data,* in vol. 6 of the *Opera omnia* is highly relevant in this context. Among the many glosses on ("given") discussed by Marinus (e.g., *tetagmenon*) "existent" is not to be found. (See Maurice Michaux, *Le Commentaire de Marinus aux Data d'Euclide. Etude critique.*)

Finally, neither Pappus' account of the meaning of a "problem" (*Collectio,* 11.650.17–19) nor Proclus' more extended discussion of the same matter (*In Euclid.,* p. 80.15–81.4) refers to the question of "existence" *tout court.* (In the latter text *el estin* seems clearly to mean "if it is possible or not.")

70.   *A.-T.,* t. 4, p. 38.

71.   Ibid., t. 10, p. 376. On the sense in which Descartes and Fermat understood themselves as continuing or restoring the Greek geometrical tradition, see Jacob Klein, "The World of Physics and the 'Natural' World." However, one should also note Descartes' own comment in his letter of January 1638 to Mersenne: "mon dessein n 'a point été . . . de reparer les livres perdus d'Apollonius, comme Viète, Snellius, Marinus Ghetaldus, etc., mais seulement de passer au-dela de tous côtés" [*A.-T.,* t. 1, p. 491].

72.   The first phrase quoted is the legend on the title page of Stevin's work on statics (*The Principal Works of Simon Stevin,* vol. 1, p. 47); the second is from his *Dialecticke ofte Beweysconstz,* p. 5 recto.

73.   On the sense of *objectum* in Descartes, see Johannes Lohmann, "Descartes' 'Compendium Musicae' und die Entstehung des neuzeitlichen Bewusstsein." For a more general semantic history, see Lawrence Dewan, " 'Objectum.' Notes on the Invention of a Word."

74.   *A.-T.,* t. 11, p. 31.

75.   Ibid., t. 7, pp. 155–56.

76.   H.-J. Engfer, *Philosophie als Analysis,* pp. 122–67.

77.   On "magnitude in general" as the theme for modern algebra, see J. Klein, *Greek Mathematical Thought and the Origin of Algebra,* part 2, passim.

78.   *A.-T.,* t. 10, p. 229.

79.   See *Journal tenu par Isaac Beeckman de 1604 à 1634,* t. 2, p. 170; and Lüder Gäbe, *Descartes Selbstkritik. Untersuchungen zur Philosophie des jungen Descartes,* pp. 129–32.

80.   *A.-T.,* t. 10, p. 460.

81.   Ibid., t. 6, p. 550.

82.   Ibid., t. 10, p. 476.

83.   H.-J. Engfer, op. cit., p. 142.

84.   *Collectio,* pp. 634.24–636.14. (Cf. Engfer's full analysis in op. cit., pp. 78–89, where he follows very closely the interpretation of Hintikka and Remes, in *The Method of Analysis.* The most important objections to their reading of the Pappus passage can be found in Erkka Maula, "An End of Invention"; and Arpad Szabó, "Analysis and Synthesis."

85.   This pairing of "construction" and "demonstration" as near synonyms, as attested in the passages I go on to cite, does not, of course, exhaust the range of meanings Descartes

gives to the traditional term "demonstration." See Desmond Clarke, "Descartes' Use of "Demonstration" and "Deduction."

86. *A.-T.*, t. 2, p. 83.

87. Ibid., t. 2, p. 511.

88. Ibid., t. 10, pp. 342–44.

89. Archimedes, *Opera Omnia*, Bd. 2, p. 439; Geminus apud M. Ammonius, *In Anal. Pr.*, ed. M. Wallies (Berlin, 1899 = *CAG*, vol. 4, Pars Vi, p. 5.27–28); André Robert, "Descartes et l'analyse des anciens," p. 242.

90. *A.-T.*, t. 6, p. 453.

91. H. Hankel, *Theorie des complexen Zahlensystems*, 1. Teil, pp. 71–72.

92. *A.-T.*, t. 10, p. 335; and see I. Gäbe, op. cit., p. 119.

93. For the Renaissance tradition of the relevant Galenic texts, see John H. Randall, "The Development of Scientific Method in the School of Padua"; and Neal W. Gilbert, *Renaissance Concepts of method*, s.n. "Galen." For Zabarella, see *Opera Logica*, p. 230 (= *De methodo*, Bd. 3, p. 4). The debates alluded to are illuminated in Wilhelm Risse, "Zur Vorgeschichte der cartesischen Methodenlehre"; and Cesare Vasoli, *La dialettica e la retorica dell' Umanesimo*, esp. pp. 333–601.

94. *A.-T.*, t. 7, pp. 155–56, and ibid., t. 9-1, p. 122.

95. Zabarella, op. cit., col. 159.

96. *A.-T.*, t. 6, p. 442.

97. Ibid., t. 6, p. 442.

98. Ibid., t. 6, p. 457.

99. Ibid., t. 2, pp. 327–28. See, too, ibid., t. 1, p. 490, for one of the many occurrences of *par hazard*, and t. 2, p. 149 and p. 327, where Fermat, once again, is accused of proceeding à tâtons.

100. *An. Post.* 1.34.89b10ff.

101. *A.-T.*, t. 6, p. 372.

102. N. Tartaglia, *General trattato di numeri et misure* (Venice, 1556), vol. 2, parte 6, F.A5 recto, as cited in Angelo Crescini, *Le origini del metodo analitico*, p. 311, n. 23.

103. Guillaume Gosselin, *De arte magna*, p. 3. [Cited in J. Klein, lib. cit., p. 263.]

104. See *The Principal Works of Simon Stevin*, vol. 2B: *Mathematics*, p. 582 [= *L'Arithmetica*, (Leiden, 1585), p. 265].

105. Franciscus Vieta, *Opera Mathematica*, p. 2 (end of *In Artem Analyticen Isagoge*, Caput 11). (See the T. Richard Witmer's new English translation, *François Viéte, The Analytic Art*, p. 15 and n. 12.

106. *GM*, Bd. 7, p. 208.

107. Thomas Harriot, *Artis analyticae praxis*, preface.

108. S. Maimon, *Über die Progresen der Philosophie*.

109. Pappus, *Collectio*, vol. 2, p. 680, quoted by Descartes (*A.-T.*, t. 6, pp. 378–79), from the Latin translation of Fredericus Commandinus (Pesaro, 1588).

110. See Michael S. Mahoney, "The Royal Road. The Development of Algebraic Analysis from 1550 to 1650, with Special Reference to the Work of Pierre de Fermat," p. 57.

111. *A.-T.*, t. 6, p. 20.

112.   Compare J. Klein on the symbolic status of straight lines in Descarte, lib. cit., pp. 203–11. See also the extremely interesting remarks on the relation between the magnitudes of bodies or surfaces and the ratios of line-lengths in L. Gäbe, "La régle XIV. Lien entre géométrie et algèbre."

113.   A.-T., t. 6, p. 555.

114.   For a contrary view aimed at mitigating the discrepancies between Ancient and Cartesian intentionality, see A. G. Molland, "Shifting the Foundations. Descartes' Transformation of Ancient Geometry." For a discussion of this and other recent studies, see Massimo Galuzzi, "Recent interpretationi della Géométrie di Descartes."

115.   A.-T., t. 1, p. 71. See the discussion of logarithmic and other "non-geometric" curves in J. Vuillemin, Mathématique et métaphysique chez Descartes, pp. 9–55; on the debate between Descartes and Leibniz in these matters, compare Emile Turrière, "La notion de transcendence géométrique chez Descartes et Leibniz."

116.   A.-T., t. 6, p. 392.

117.   See ch. 2, n. 74. The assertion in the text needs to be somewhat qualified by Descartes' own remarks on the difference between geometry and mechanics (as these are traditionally understood) in the second book of La Géométrie (A.-T., t. 6, pp. 389–90).

118.   A.-T., t. 6, p. 386.

119.   Ibid., t. 2, p. 384. Compare his much earlier letter to Mersenne (November 13, 1629), giving his reasons for rejecting non-algebraic curves [=ibid., t. 1, pp. 69–71].

120.   Ibid., t. 10, p. 332. The controversial "patch-work" reading of the Rules proposed by J. M. Weber (La Constitution du texte des Regulae) has, in my judgement, been effectively countered by Lüder Gäbe, in his introduction to Regulae ad directionem ingenii/Regeln zur Ausrichtung der Erkenntniskraft, pp. xxi–xxxvii. However controversy over the dating and editional integrity of the Regulae is still unabated, especially in regard to the two components of Rule 4, labelled by modern critics "4-A" and "4-B," respectively. Rule 4-B contains the unique mention of mathesis universalis and is added as an appendix to the Hanover manuscript copied by Leibniz, but not to the Dutch translation of 1684 by Glazemaker or to the Latin Edition of 1701. Marion (op. cit.) and Van de Pitte (in "Descartes Mathesis universalis") have defended the unity of Rules 4-A and 4-B, while John Schuster ("Cartesian Method as Mythic Speech: A Diachronic and Structural Analysis") and Pamela A. Kraus (in "From Universal Mathematics to Universal Method: Descartes 'Turn' in Rule IV of the Regulae") have argued for their incompatibility and for the much earlier composition of 4-B (early 1619).

Three brief remarks must suffice here.

(1) Even if the original text of 4-B was added on to the manuscript of the Regulae, this does not, by itself, establish that it was composed at an earlier date. I could be a record of "second thoughts." In particular, the criticisms it contains of the "vast array of numbers and inexplicable figures by which it (sc. algebra) is overwhelmed" squares with Descartes' own decision to abandon the variegated symbolism of this so-called "cossists" and to adopt, in the Geometry, line-segments alone to represent relations of magnitude.

(2) The phrase mathesis universalis, although a hapax legomenon in Descartes, was quite clearly transmitted to his disciples (van Schooten; Bartholinus 1 prior to the posthumous publication of the Reguae.) van Schooten first met Descartes in Leiden in 1637!

(3) The key issue is whether ordo et mensura, the themes of mathesis universalis according to 4-B, characterize mathematics in the restricted sense of the received disciplines (cf. Kraus, op. cit., p. 168). In the pages that follow, I am intent on showing how order and measure name the much more general concepts of the regulated succession of mental

motion and the fundamental homogeneity of the domain (i.e., extension) over which that motion ranges.

121.   Their separability is argued at length by L. J. Beck, *The Method of Descartes;* in contrast, see L. Gäbe's remarks in his "Einleitung," op. cit. p. xxxix, and the works cited below, nn. 123 and 124.

122.   Proclus, *In Euclid.,* p. 45.18–21.

123.   See Frederick Van de Pitte, "Descartes' *Mathesis Universalis,*" at p. 157, n. 11, where he corrects Giovanni Crapulli's ascription of the phrase to a slightly later work by van Roomen, *Apologia pro Archimede* (1597).

124.   In addition to the works of G. Crapulli and F. P. Van de Pitte, op. cit., see Jürgen Mittelstrass, "Die Idee einer mathesis universalis bei Descartes."

125.   *A.-T.,* t. 10, p. 381. On Rule 6 and its context, see J.-L. Marion, op. cit., pp. 85–98; and Pamela A. Kraus, "The Structure and Method of the Regulae ad directionem ingenii," pp. 113–33.

126.   *Categories,* ch. 5.2b22; *Metaph. Nu* 1.1088a27–29. (Note that in the *Categories* text Aristotle adds the qualification *hosa mē esti genē.*)

127.   See Pierre Costabel in Descartes, *Reglès utiles et claires pour la direction de l'esprit en la recherche de la vérité,* p. 273, and cf. p. 275.

128.   *A.-T.,* t. 10, p. 441 (Rule 14). On the imagination, in addition the funadmental reflections in J. Klein, lib. cit., pp. 197–211, see Emile Boutroux, *L'Imagination et les mathématiques selon Descartes* (with the review by Bertrand Russell in *Mind*), and P. A. Kraus, "The Structure and Method of the *Regulae ad directionem ingenii,*" pp. 52–181.

129.   Etienne Gilson, *Descartes, 'Discours de la méthode,' texte et commentaire,* p. 184.

130.   *GM,* Bd. 7, p. 205.

131.   On this increasingly acrimonious debate in which Descartes' authority was solicited, see P. Costabel, "Descartes et la racine cubique des nombres binômes" (repr. in P. Costabel, *Demarches originales de Descartes savant,* op. cit., pp. 121–40).

132.   The text of this manuscript [=*A.-T.,* t. 10, pp. 265–76] has now been superlatively edited by P. T. Federico, *Descartes on Polyhedra,* where, however, no mention is made of Ciermans. On the theorem itself, see P. Costabel, "Le theoréme de Descartes-Euler," in *Demarches originales de Descartes Savant,* pp. 15–25.

133.   See, e.g., *A.-T.,* t. 1, p. 501. A new version of Descartes' "Calculus," more extensive than the two previously published texts (viz. "Calcul de M. Descartes," *A.-T.,* t. 10, pp. 659–80, and "Recueil de calcul qui sert à la géométrie," *Correspondance,* ed. Adam-Milhaud, t. 3, pp. 323–52), has been reported by Pierre Costabel, "Découverte d'un nouveau manuscrit de l'Introduction à la Géométrie."

134.   *A.-T.,* t. 9-1, p. 212; cf. ibid., t. 11, pp. 688–90 (Correspondence between Roberval and Des Noyer).

135.   See n. 50, above.

136.   Fifth Meditation, *A.-T.,* t. 7, p. 71; Sixth Meditation, ibid., p. 71 and p. 80.

137.   See n. 134, above, and consult Descartes' correspondence with Henry More (*A.-T.,* t. 5, p. 378; pp. 402–403 et alibi) on the issue of whether, e.g., impenetrability belongs to the essence of bodies. (On further aspects of this theme, see Jean Laporte, "La connaissance de l'étendue chez Descartes."

138. *GM*, t. 7, p. 205. The differences between the Cartesian and Leibnizian versions of *mathesis universalis* have been usefully studied by Roswitha Engelbrecht, "Der Begriff 'Mathesis Universalis' bei Descartes und Leibniz."

139. *A.-T.*, t. 6, p. 390. On the techniques involved, see H. J. M. Bos, "On the Representation of Curves in Descartes' *Géométrie*."

140. The prevalence of this general theme—motions of the mind—in Descartes' thinking has been convincingly exhibited in J.-M. Beyssade, *La philosophie premiére de Descartes*, pp. 129–76.

141. The shift from the constructive to the non-constructive or formal approach in algebra is traced lucidly by Carl B. Boyer, "Cartesian and Newtonian Algebra in the Mid-Eighteenth Century." (See also ch. 1, n. 8.)

142. On the analogous treatment of number (i.e., as identical to the "things" numbered), see Helen Lauer, "Descartes' Concept of Number."

143. *GM*, Bd. 5, p. 142. On the philosophical underpinnings of Leibniz' *analysis situs*, see G.-G. Granger, "Philosophie et mathématique leibniziennes," esp. pp. 13–16.

144. I have borrowed the motif of "dis-figuration" from J.-L. Marion, *Sur la théologie blanche de Descartes*, pp. 231–63, whose analysis is detailed and exemplary.

145. See J. Klein, lib. cit., p. 150–85.

146. Compare Pascal's even more explicit proposal for this new manner of conceiving "the universal" in mathematics; for example, a point, two straight lines, and a right-angle all become *cases* of the motion "conic section" when the mode of generation *and* the position of the geometer's eye are taken into account. See R. Taton, "L'Oeuvre de Pascal en géométrie projective," esp. pp. 55–58.

147. On the "similarity-thesis" see Richard Kennington, "The 'Teaching of Nature' in Descartes' Soul Doctrine"; and the very helpful study by Gerard Lebrun, "La Notion de 'ressemblànce' de Descartes à Leibniz."

148. See K. C. Clatterbaugh, "Descartes' Causal Likeness Principle." *Philosophical Review* 89 (1980):379–402.

149. Compare *A.-T.*, t. 9-1, p. 33, with ibid., t. 7, p. 43.

150. See my discussion of Kant in ch. 1, II.

151. N. Malebranche, *De la Recherche de la vérité*, t. 2, p. 118.

152. *A.-T.*, t. 3, p. 39.

153. *Kants handschriftlicher Nachlass, Gesammelte Schriften (Akademie Ausgabe)*, Bd. 16, p. 579.

154. *A.-T.*, t. 5, p. 148.

155. Ibid., t. 6, p. 343. Compare the catalogue of proposed marvels in Descartes' early manuscript "Experiementa" (ibid., t. 10, pp. 215–16) and the very helpful study by G. Rodis-Lewis, "Machineries et perspectives curieuses dans leurs rapports avec le Cartésianisme."

156. Nicolas Grimaldi, *L'Experience de la pensée dans la philosophie de Descartes*, p. 40, n. 68.

157. Hobbes, *Opera Philosophica quae latine scripsit omnia*, vol. 1, p. 316.

Bibliography

(This listing of primary and secondary sources includes *only* those works cited or referred to in the text and footnotes.)

I. Primary Sources

Alfarabi. *Opera omnia*, Ed. Guillelmus Camerarius. Paris, 1638 (repr. Frankfurt, 1969).

Alsted, Johann Henrich. *Methodus admirandorum mathematicorum*. Herbornae Nassivorum, 1613.

Apollonius of Perga. *Quae graece exstant cum commentariis antiquis*. Ed. J. L. Heiberg. Stuttgart, 1891–93. Vols. 1–2.

——. *On Cutting Off a Ratio*. Trans. by E. M. Macierowski. Ed. by R. H. Schmidt. Fairfield, Conn., 1987.

Archimedes. *Opera Omnia*. Ed. J. L. Heiberg and E. S. Stamatis. Stuttgart, 1972. Vols. 1–3.

——. *The Works of Archimedes*. Ed. by T. L. Heath. Cambridge, 1912.

Aristotle. *Opera Omnia*. Ed. I. Bekker. Berlin, 1831–1870. Vols. 1–3.

——. *Traité de l'âme*. Ed. avec commentaire par Georges Rodier. Paris, 1900.

Avicenna. *Avicenna Latinus: Liber de prima philosophia sive scientia divina, I–X: Lexiques*. Par S. van Riet. Louvain-La Neuve, 1983.

——. *La Métaphysique du Shifa', Livres 1 à V*. Trad. par Georges C. Anwati. Paris, 1978.

——. *Livre des Directives et remarques (Kitāb al'ishārāt wa l-tanbīhāt)*. Trad. par A.-M. Goichon. Paris, 1951.

Bacon, Francis. *Novum Organum*. Ed. Thomas Fowler. Oxford, 1878.

Badawi, A. R., (Ed.,). *Aristū ʿind al-ʿArab*. Cairo, 1947.

Baillet, André. *La Vie de Monsieur Descartes*. Paris, 1691.

Beeckman, Isaac. *Journal tenu par Isaac Beeckman de 1604 à 1634*. Ed. C. de Waard. The Hague, 1942. T. 2.

Bodin, Jean. *Methodum ad facilem historiarum cognitionem*. Paris, 1566.

Bonansoni, Paolo. *The Algebra Geometrica of Paolo Bonansoni, Circa 1575*. Ed. and trans. by R. Schmidt. Annapolis, Md., 1985.

Bruno, Giordano. *Dialoghi Italiani*. Ed. G. Gentile. Florence, 1958.

Cicero. *De l'invention*. Ed. et trad. par Henri Bornecque. Paris, 1932.

——. *Cato Maior de Senectute*. Ed. L. Huxley. Oxford, 1887.

——. *Rhetorica*. Ed. A. S. Wilkins. Oxford, 1901–1903.

Clauberg, Johannes. *Opera omnia philosophica*. Amsterdam, 1691.

Clavius, Christopher. *In Disciplinas Mathematicas Prolegomena*. In *Euclidis Elementorum Libri XV . . . auctore Christophoro Clavio*. Frankfurt, 1654.

Commentaria in Aristotelem Graeca. Edita consilio et auctoritate Academiae Litterarum Regiae Borussicae. Berlin, 1882 sq. Vol. 4.vi: Ammonius, *In Analytica Priora*. Ed. M. Wallies.

233

1899; Vol. 7. Simplicius, *In De Caelo*. Ed. J. L. Heiberg. 1893; Vol. 17. Joannes Philoponus. *In Aristoteli Physicorum Libros Quinque Posteriores*. Ed. H. Vitelli 1888.

D'Alembert, Jean. "Preliminary Discourse." *The Encyclopedia. Selections*. Trans. Stephen J. Gendzier. New York, 1967.

Dante. *Opere*. A cura di M. Porena e M. Pazzaglia. Bologna, 1966.

Dasypodius, Conrad. *Oratio . . . De Disciplinis Mathematicis . . . Eiusdem Lexicon Mathematicum ex diversis collectum antiquis scriptis*. Strasbourg, 1579.

Dedekind, Richard. *Essays on the Theory of Numbers*. Trans. W. W. Beman. New York, 1901.

―――. *Gesammelte mathematische Werke*. Braunschweig, 1932. Bd. 3.

Descartes, René. *Correspondance publiée avec une introduction et des notes par Charles Adam et Girard Milhaud*. Paris, 1936–63, Tomes 1–8.

―――. *Discours de la méthode*. Texte et commentaire Par Etienne Gilson. Paris, 1925.

―――. *Oeuvres*. Publiés par Charles Adam et Paul Tannery. Nouvelle présentation. Paris, 1975. T. 1–11.

―――. *Oeuvres et lettres*. Textes presentés par André Bridoux. Paris, 1953.

―――. *Reglès utiles et claires pour la direction de l'esprit en la recherche de la vérité*. Trad. J.-L. Marion, avec des notes mathématiques de Pierre Costabel. The Hague, 1977.

―――. *Regulae ad directionem ingenii/Regeln zur Ausrichtung der Erkenntniskraft*. Kritisch rediviert, übersetzt und herausgegeben von Heinrich Springmeyer, Lüder Gäbe, Hans Günter Zekl. Hamburg, 1973.

Dilthey, Wilhelm. *Leben Schleiermachers. Zweiter Band: Schleiermachers System als Philosophie und Theologie. Aus dem Nachlass*. Hrsg. von Martin Redeker. Gottingen, 1966. (= *Gesammelte Schriften*, Bd. 14/1).

Diocles. *On Burning Mirrors. The Arabic Translation of the Lost Greek Original*. Ed. with English Trans. and Commentary by G. J. Toomer. Berlin, Heidelberg, New York, 1976.

Diogenes Laertius. *Vitae Philosophorum*. Ed. H. S. Long. Oxford, 1964. Vols. 1–2.

Du Bellay, Joachim. *La Défense et illustration de la langue Francoyse*. Ed. H. Chamard. Paris, 1948.

Euclid. *Elementa*. post. J. L. Heiberg, ed. K. S. Stamatis. Leipzig, 1969–74. Vols. 1–5.

―――. *Opera Omnia*. Ed. J. L. Heiberg and H. Menge. Leipzig, 1883–1916. Vols. 1–9.

―――. *The Thirteen Books of Euclid's Elements*. trans. by Sir Thomas L. Heath. New York, 1956. Vols. 1–3.

―――. *Sectio Canonis*. In *Musici Scriptores Graeci*. Ed. K. von Jan. Leipzig, 1895. Vol. 1.

Eutocius. (See Archimedes, *Opera Omnia*, cit. supra, Vol. 3.)

Fichte, Johann Gottlieb. *Nachgelassene Werke*. Hrsg. von. I. H. Fichte. Bonn, 1834–35. Bde. 1–3.

―――. *Sämtliche Werke*. Hrsg. von I. H. Fichte. Berlin, 1845–46. Bde. 1–8.

Galilei, Galileo. *Le Opere di Galileo Galilei*. Nuova Ristampa della Edizione Nazionale. Florence, 1965 sq. Vols. 1–2.

―――. *Two New Sciences*. Trans. Stillman Drake. Madison, Wisc., 1974.

Grammatici Latini. Ed. H. Keil. Leipzig, 1855–78.

Harriot, Thomas. *Artis analyticae praxis*. London, 1631.

Hegel, G. W. F. *Jenaer Kritische Schriften.* Hrsg. von H. Buchner and O. Pöggelier. Hamburg, 1968 (= *Gesammelte Werke,* Bd. 4).

Heidegger, Martin. *Die Grundprobleme der Phänomenologie* (= *Gesamtausgabe,* 2. Abteilung, Bd. 24). Frankfurt, a.M., 1975.

------. *Grundfragen der Philosophie. Ausgewählte "Probleme" der "Logik"* (= ibid., Bd. 45). Frankfurt a.M. 1984.

Heine, Heinrich. *Zur Geschichte der Religion und Philosophie in Deutschland* [1835] (= *Sämtliche Werke,* Bd. 10. hrsg. von Hans Kaufmann [Munich, 1964]).

Heron of Alexandria. *Metrica.* Ed. E. M. Bruins. Leiden, 1964 (= *Textus Minores,* Vol. 34).

------. *Opera.* Ed. J. L. Heiberg. Leipzig, 1912. Vol. 4.

Hobbes, Thomas. *The English Works.* Ed. W. Molesworth. London, 1839–45. Vols. 1–11.

------. *Opera philosophica quae latine scripsit omnia.* Ed. W. Molesworth. London, 1839–45. Vols. 1–5.

Husserl, Edmund. *Die Krisis der europäischen Wissenschaften und die transzendentale Phänomenologie.* The Hague, 1954 (= *Husserliana,* Bd. 6).

Iamblichus. *In Nicomachi Arithmeticam Introductionem Liber.* Ed. H. Pistelli. Leipzig, 1894.

Kant, Immanuel. *Gesammelte Schriften.* Hrsg. von der Königlich Preussischen Akademie der Wissenschaften. Berlin, 1900 sq.

------. *Werke.* Hrsg. von Wilhelm Weischedel. Wiesbaden, 1958; repr. Frankfurt a.M., 1964.

Kästner, A. G. "Was heisst in Euklids Geometrie möglich?" *Philosophisches Magazin,* Bd. 2, Stück 4 (1790), pp. 391–402.

Kepler, Johannes. *Gesammelte Werke.* Hrsg. von Max Caspar. Munich, 1938 sq. Vols. 2 and 6.

Lambert, Johann Heinrich. *Abhandlung vom Criterium Veritatis.* Hrsg. von K. Bopp. Berlin, 1915 (= *Kant-Studien,* Ergänzungsheft 36).

------. *Deutscher gelehrter Briefwechsel.* Hrsg. von J. Bernoulli. Berlin, 1781. Bd. 1.

------. *Philosophische Schriften.* Hrsg. von H.-W. Arndt. Hildesheim, 1965ff. Bds. 1–7.

------. *Über die Methode die Metaphysik, Theologie und Moral richtiger zu beweisen.* Hrsg. von K. Bopp. Berlin, 1918 (= *Kant-Studien, Ergänzungsheft* 42).

Leibniz, G. W. *Leibnitiana. Elementa philosophiae arcanae de summa rerum.* Ed. Ivan Jagodinsky. Kazan, 1913.

------. *Mathematische Schriften.* Hrsg. von C. I. Gerhardt. Halle, 1849–63. Bde. 1–8.

------. *Nouvelles lettres et opuscules inédits de Leibniz.* Ed. A. Foucher de Careil. Paris, 1857.

------. *Opuscules et Fragments inédits de Leibniz.* Ed. Par Louis Couturat. Paris, 1903.

------. *Philosophische Schriften.* Hrsg. C. I. Gerhardt. Berlin, 1875–90. Bde. 1–8.

Lipstorp, Daniel. *Specimina philosophiae Cartesianae.* Lyon, 1653.

Machiavelli, Nicolò. *Il Principe.* A cura di Luigi Firpo. Turin, 1977.

Maimon, Salomon. *Über die Progressen der Philosophie.* Berlin, 1793.

Maimonides, Moses. *Guide of the Perplexed.* Trans. Shlomo Pines. Chicago, 1963.

Malebranche, Nicolas. *De la Recherche de la vérité.* Ed. M. Francisque Bouiller. Paris, 1880. T. 2.

Marinus. [See Euclid. *Opera Omnia*, cit. supra, Vol. 8.]

Marx, Karl. *Early Writings*. Trans. R. Livingstone and Gregor Benton. New York, 1975.

————. *Grundrisse. Introduction to the Critique of Political Economy*. Trans. Martin Nicolaus. New York, 1973.

————. and Engels, Friedrich. *Werke*. Berlin 1956 sq.

Newton, Isaac. *Philosophiae naturalis principia mathematica*. The Third Edition (1726) with Variant Readings. Assembled and edited by Alexandre Koyré and I. Bernard Cohen, with the assistance of Anne Whitman. Cambridge, Mass., 1972. Vols. 1–2.

Nicomachus of Gerasa. *Introductionis Arithmeticae Libri II*. Ed. R. Hoche. Leipzig, 1866.

Nietzsche, Friedrich. *Sämtliche Werke. Kritische Studienausgabe in 15 Bänden*. Hrsg. von Giorgio Colli und Mazzino Montinari. Berlin, 1980.

————. *Werke*. Hrsg. von F. Naumann and R. Kröner. Leipzig, 1894 sq. (= *Grossoktavausgabe*). Bd. 10.

Pappus of Alexandria. *Collectionis quae supersunt e libris manuscriptis*. Ed. F. Hultsch. Berlin, 1876–78. Vols. 1–3.

————. *Book 7 of the Collection*. Parts 1–2. Ed., with Trans. and Commentary by Alexander Jones. New York, Berlin, Heidelberg, 1986.

————. *The Commentary of Pappus on Book X of Euclid's Elements*. Ed. G. Junge and W. Thomas. Cambridge, Mass., 1930 (= *Harvard Semitic Series*, Vol. 8).

Pascal, Blaise. *Oeuvres complètes*. Ed. par Jacques Chevalier. Paris, 1954.

Patrizi, Francesco. *Della nuova geometria*. Ferrara, 1587.

Patrologiae cursus completus, series latina. Ed. J.-P. Migne. Paris, 1844–64. Vol. 23.

Plato. *Opera Omnia*. Ed. J. Burnet. Oxford, 1900–1907. Vols. 1–5.

Plutarch. *Moralia*. Ed. G. N. Bernardakis. Leipzig, 1889–96. Vols. 1–7.

————. *Vitae Parallelae*. Ed. K. Ziegler et al. Leipzig, 1914–35.

Polybius. *Historiae*. Ed. Th. Büttner-Wobst. Leipzig, 1889–1904.

Proclus, Diadochus. *In Platonis Rem Publicam Commentarius*. Ed. W. Kroll. Leipzig, 1899–1901. Vols. 1–2.

————. *In Primum Euclidis Elementorum librum commentarii*. Ed. G. Friedlein. Leipzig, 1873.

Pseudo-Aristotele. *De lineis insecabilibus*. Introduzione, traduzione e commento a cura di M. Timpanaro Cardini. Milan, 1970 (= Testi e documenti per lo studio dell' Antichità, 32).

Quintilian. *Institutio oratoria*. Ed. M. Winterbottom. Oxford, 1970. Vols. 1–2.

Rousseau, Jean-Jacques. *Discours sur les sciences et les arts*. (= *Oeuvres complètes*. Pléiade ed. Paris, 1964. T. 3).

Schelling, F. W. J. *Sämtliche Werke*. Hrsg. von K. F. A. Schelling. Stuttgart, Augsburg, 1856–61. Abt. 1, bde. 1 and 5.

Schlegel, Friedrich. *Schriften zur Literatur*. Hrsg. von W. Rasch. Munich, 1972.

Stevin, Simon. *Dialecticke ofte Beweysconst*. Leiden, 1585.

————. *The Principal Works of Simon Stevin*. Ed. D. J. Struik et al. Amsterdam, 1955 sq. Vols. 1–3.

Stoicorum Veterum Fragmenta. Ed. Jacob van Arnim. Stuttgart, 1864. Vols. 1–4. [=*SVF*.]

Theon of Alexandria. *Commentaires de Pappus et de Théon d'Alexandrie sur l'Almagèste.* Texte établi et annoté par A. Rome. Vatican City, 1936. Vols. 1–3.

Theon of Smyrna. *Expositio rerum mathematicarum ad legendum Platonem utilium.* Ed. E. Hiller. Leipzig, 1878.

Vico, Giambattista. *Opere Filosofiche.* Ed. P. Cristofolini. Florence, 1971.

Vieta, Franciscus. *Opera Mathematica.* Ed F. van Schooten. Leiden, 1646 (repr. Hildescheim, 1970).

———. *The Analytic Art.* Trans. T. Richard Witmer. Kent, Ohio, 1983.

Wolff, Christian. *Mathematisches Lexicon* (= *Gesammelte Werke,* hrsg. von J. Ecole u.a. Hildesheim, 1965. I. Abteilung, Bd. 11.) (Originally published 1716).

Zabarella, Jacopo. *Opera Logica.* Köln, 1597; Venice, 1600.

II. Secondary Sources

Aersten, J. A. "Wendingen in Waarheid. Anselmus van Canterbury, Thomas von Aquino en Vico." *Tijdschrift voor Filosofie* 49(1987), pp. 187–228.

al-Rabe, Ahmad A. "Muslim Philosophers Classifications of the Sciences: al-Kindi, al-Farabi, al-Ghazzali,Ibn Khaldun," (Ph.D. Diss., Harvard Univ., 1984).

Anwati, Georges C. "La Notion d'*al-Wujud* (existence) dans le *Kitāb al-Hudud* d'al-Farabi." *Congreso Internacional de Filosofia Medieval* (Madrid, 1979), 1, pp. 505–19.

Arbib, Michael A., and Mary Hesse. *The Construction of Reality.* Cambridge, 1986.

Artmann, Benno. "Über voreuklidische 'Elemente' deren Autor Proportionen vermied." *Archive for History of Exact Sciences* 33 (1985), pp. 291–306.

Bäck, Allan. "Avicenna on Existence." *Journal of the History of Philosophy* 25 (1987), pp. 351–67.

Barker, Andrew. "Method and Aims in the Euclidean *Sectio Canonis.*" *Journal of Hellenic Studies* 101 (1981), pp. 1–16.

Beck, L. J. *The Method of Descartes. A Study of the Regulae.* Oxford, 1952.

Becker, Oskar. *Mathematische Existenz.* 2.Auflage. Tübingen, 1973. (Originally published in *Jahrbuch für Philosophie and phänomenologische Forschung* 8 [1927], pp. 441–809).

———. "Warum haben die Griechen die Existenz der vierten Proportionale angenommen?" *Ouellen und Studien zur Geschichte der Mathematik, Astronomie und Physik* 2, Abt. B (1933), pp. 369–87.

Beckmann, Friedhelm. "Neue Geschichtspunkte zum 5. Buch Euklids." *Archive for History of Exact Sciences* 4 (1967), pp. 1–145.

Behler, Ernst. *Die Ewigkeit der Welt.* Erster Teil. Munich, 1965.

Belaval, Yvon. *Leibniz, critique de Descartes.* Paris, 1960.

Benardete, Seth. "The Grammar of Being." *Review of Metaphysics* 20 (1977), pp. 486–96.

Beyssade, Jean-Marie. *La Philosophie première de Descartes.* Paris, 1979.

Blanchôt, Maurice. *L'Entretien infini* (Paris, 1969).

———. "Le Rire des Dieux." *La Nouvelle Revue Francaise.* Juillet, 1965.

Blumenberg, Hans. *Die Genesis der kopernikanischen Welt.* Frankfurt a.M., 1975.

———. *Die Legitimität der Neuzeit.* Frankfurt a.M., 1966. Eng. trans. R. M. Wallace. Cambridge, Mass., 1983.

238   Bibliography

Booth, Edward. *Aristotelian Aporetic Ontology in Islamic and Christian Thinkers.* Cambridge, 1983 (= *Cambridge Studies in Medieval Life and Thought,* 3rd series, Vol. 20).

Bos, H. J. M. "On the Representation of Curves in Descartes' *Géométrie.*" *Archive for History of Exact Sciences* 24 (1981), pp. 295–38.

———. "Arguments on Motivation in the Rise and Decline of a Mathematical Theory: the 'Construction of Equations,' 1637–ca. 1750." *Archive for the History of Exact Sciences* 30 (1984), pp. 731–80.

Boutot, Alain. *Heidegger et Platon. Le Problème du nihilisme.* Paris, 1987.

Boutroux, Emile. *L'Imagination et les mathématiques selon Descartes* Paris, 1900.

Bowen, Alan C. "Menaechmus *versus* the Platonists: Two Theories of Science in the Early Academy." *Ancient Philosophy* 3 (1983), pp. 12–29.

Boyer, Carl B. "Proportion, Equation, Function: Three Steps in the Development of a Concept." *Scripta Mathematica* 12 (1946), pp. 5–13.

———. "Descartes and the Geometrization of Algebra." *American Mathematical Monthly* 66 (1959), pp. 390–93.

———. "Cartesian and Newtonian Algebra in the Mid-Eighteenth Century." *Actes du XIième Congrès International d'Histoire des Sciences* 3 (Wroclaw, 1968), pp. 195–202.

Brann, Eva. "The Cutting of the Canon." *The Collegian* (St. John's College, Annapolis, Md.), Supplement (November, 1962), pp. 1–63.

Bréhier, Emile. *La Théorie des incorporels dans l'ancien Stoïcisme.* Cinquieme ed. Paris, 1980.

Breton, Stanislas. *Philosophie et mathématique chez Proclus.* Paris, 1969.

Brewster, David. *Memoirs of the Life. Writings and Discoveries of Sir Isaac Newton.* Edinburgh, 1855.

Brown, Malcolm. "Plato Disapproves of the Slave-Boy's Answer." *Review of Metaphysics* 20 (1967), pp. 57–93.

———. "Some Debates about Eudoxus' Mathematics." Unpublished lecture, Summer Institute in Ancient Philosophy, Colorado Springs, 1970.

———. "Plato on Doubling the Cube. *Politicus* 266AB." In *Plato, Time and Education. Essays in Honor of Robert S. Brumbaugh.* Ed. B. P. Hendley. Albany, 1987. pp. 43–60.

Brownson, C. D., "Euclid's Optics and its Compatibility with Linear Perspective." *Archive for History of Exact Sciences* 24 (1981), pp. 165–94.

Bruins, E. M. *Le Gémétrie non-euclidienne dans l'Antiquité.* Paris, 1968.

Brunschvicq, Léon. *Les Étapes de la philosophie mathématique.* Paris, 1912.

Butler, Judith. *Subjects of Desire. Hegelian Reflections in Twentieth-Century France.* New York, 1987.

Butts, Robert, ed., *Kant's Philosophy of Physical Science.* Dordrecht and Boston, 1986.

Carlo, William E. "Idea and Concept: A Key to Epistemology." In *The Quest for the Absolute.* Ed. F. J. Adelmann. Chestnut Hill, Mass., 1966. Pp. 47–66.

Carnap. Rudolf. *The Logical Structure of the World. Pseudoproblems in Philosophy.* Trans. Rolf A. George. Berkeley, Calif., 1969.

Carter, Richard B. "Volitional Anticipation and Popular Wisdom in Descartes." *Interpretation* 7 (1978), pp. 75–98.

———. *Descartes' Medical Philosophy.* Baltimore, 1983.

Cascardi, Anthony J. "Genealogies of Modernism." *Philosophy and Literature* 11 (1987), pp. 207–25.

Caton, Hiram. "Carnap's 'First Philosophy.' " *Review of Metaphysics* 28 (1975), pp. 623–59.

Charles-Saget, Annick. *L'Architecture du divin. Mathématique et philosophie chez Plotin et Proclus.* Paris, 1982.

Clarke, Desmond P. "Descartes' Use of 'Demonstration' and 'Deduction.' " *The Modern Schoolman* 54 (1977), pp. 333–44.

Clatterbaugh, K. C. "Descartes' Causal Likeness Principle." *Review of Metaphysics* 89 (1980), pp. 379–402.

Coolidge, Julian L. "The Origin of Analytic Geometry." *Osiris* 1 (1936), pp. 231–50.

————. *A History of Geometrical Methods.* Oxford, 1940.

Cortassa, Guido. "Pensiero e linguaggio nella teoria stoica del *LEKTON*." *Rivista di filologia e d'istruzione classica* 106 (1978), pp. 385–94.

Cosenza, Paolo. *L'Incommensurabile nell' evoluzione filosofica di Platone.* Naples, 1977.

Costabel, Pierre. *Démarches originales de Descartes savant.* Paris, 1982.

————. "Découverte d'un nouveau manuscrit de l'"Introduction à la Géométrie"" *Archives de Philosophie* 47 (1984), p. 74.

Coulet, H. "La Métaphore de l'architecture dans la critique littéraire au XVIIᵉ siècle." In *Critique et création littéraires en France au XVIIᵉ Siècle.* Paris, 1977.

Courtine, Jean-François. "Note complémentaire pour l'histoire du vocabulaire de l'être." In *Concepts et catégories dans la pensée antique.* Ed. Pierre Aubenque. Paris, 1980. Pp. 33–87.

Crapulli, Giovanni. *Mathesis Universalis. Genesi di una idea nel XVI secolo.* Rome, 1969.

Crescini, Angelo. *Le Origini del metodo analitico. Il Cinquecento.* Udine, 1965.

Cumming, Robert D. *Human Nature and History.* Chicago, 1968. Vols. 1–2.

Dancy, Russell. "Aristotle and Existence." *Synthese* 54 (1983), pp. 409–42.

Daniel, Stephen H. "Descartes on Myth and Ingenuity/Ingenium." *Southern Journal of Philosophy* 23 (1985), pp. 157–70.

Davidson, Herbert A. "Avicenna's Proof of the Existence of God as a Necessarily Existent Being." In *Islamic Philosophical Theology.* Ed. P. Morewedge. Albany, N.Y., 1979. Pp. 165–87.

Davis, Martin. Ed., *The Undecidable.* Hewlett, N.Y., 1965.

Delatte, André. *La Catoptromancie grecque et ses derivés.* Bibliothèque de la Faculté de Philosophie et Lettres de l'Université de Liège, Fasc. 48, 1932.

Deleuze, Gilles. *La Philosophie critique de Kant.* Paris, 1963.

Derrida, Jacques. "Introduction" à *Edmund Husserl, L'Origine de la géométrie.* Paris, 1962. Eng. trans. J. P. Leavey. Stony Brook, N.Y., 1978.

————. *De la grammatologie.* Paris, 1967.

————. *L'Ecriture et la différence.* Paris, 1967.

————. *La Voix et le Phénomène.* Paris, 1967. Eng. trans. D. Allison. Evanston, Ill., 1973.

————. "Reponse de Jacques Derrida." In *L'Oreille de l'autre.* Sous la direction de Claude Levesque et C. V. McDonald. Montreal, 1982. Pp. 116–89.

———. "The Time of a Thesis: Punctuations." In *Philosophy in France Today*. Ed. A. Montefiore. Cambridge, 1983. Pp. 34–50.

———. *Memoires For Paul de Man*. New York, 1986.

Dewan, Lawrence. " 'Objectum.' Notes on the Invention of a Word." *Archives d'histoire doctrinale et littéraire du moyen âge* 48 (1981), pp. 37–96.

DeYoung, Gregg. "The Arabic Textual Traditions of Euclid's Elements." *Historia Mathematica* 11 (1984), pp. 147–60.

Dijkterhuis, E. J. A., et al. *Descartes et le cartésianisme hollandais*. Paris and Amsterdam, 1950.

Doz, André. "Sur la signification de 'instar archetypi.' Descartes, Troisième Meditation." *Revue philosophique de la France et de l'Etranger* 93 (1968), pp. 380–87.

Drake, Stillman. *Galileo at Work. His Scientific Biography*. Chicago, 1978.

Dumont, Jean-Paul. "Mos geometricus, mos physicus." In *Les Stoiciens et leur logique*. Actes du Colloque de Chantilly, 18–22 Septembre 1976 (Paris, 1978), pp. 121–34.

Einarson, Benedict. "On Certain Mathematical Terms in Aristotle's Logic." *American Journal of Philology* 57 (1936), pp. 33–54, 150–72.

Ende, Helga. *Der Konstruktionsbegriff im Umkreis des Deutschen Idealismus*. Meisenheim a.G., 1973 (= *Studien zur Wissenschaftstheorie*,Bd. 7).

Engelbrecht, Roswitha. "Der Begriff 'Mathesis Universalis' bei Descartes und Leibniz." Phil. Diss. Universität Wien, 1970.

Engfer, Hans-Jurgen. *Philosophie als Analyse. Studien zur Entwicklung philosophischer Analysiskonzeptionen unter dem Einfluss mathematischer Methodenmodelle im 17. und frühen 18, Jahrhundert*. Stuttgart-Bad Constatt, 1982.

Fackenheim, Emil A. "The Possibility of the Universe in al-Farabi, Ibn Sina and Maimonides." *Proceedings of the American Academy for Jewish Research* 16 (1947), pp. 39–70.

Federico, P. J. *Descartes on Polyhedra. A Study of the De Solidorum Elementis*. New York, Heidelberg, Berlin, 1982 (= *Sources in the History of Mathematics and Physical Sciences* 4).

Ferrier, Richard D. "Two Exegetical Treatises of Francois Viète, Translated, Annotated and Explained." Ph.D. Diss. Indiana University, 1980.

Fiebig, Hans. *Erkenntnis und technische Erzeugung. Hobbes' operationale Philosophie der Wissenschaft*. Meisenheim a.G., 1973.

Finianos, Ghassan. *Les Grandes divisions de l'être "mawjud" selon Ibn Sina*. Fribourg, 1976.

Foucault, Michel. *L'Archéologie du savoir*. Paris, 1969.

Fowler, D. H. "Ratio in Early Greek Mathematics." *Bulletin (New Series) of the American Mathematical Society* 1 (1979), pp. 807–46.

———. *The Mathematics of Plato's Academy. A New Reconstruction*. Oxford, 1987.

Frajèse, Attilio. "Sur la signification des postulats euclidiens." *Archives internationales d'histoire des sciences* 4 (1951), pp. 383–92.

Frede, Michael. *Die stoische Logik*. Göttingen, 1974.

Frenkian, Aram. *Le Postulat chez Euclide et chez les Modernes*. Paris, 1940.

Fritz, Kurt von. "Die *APXAI* in der griechischen Mathematik." *Archiv für Begriffsgeschichte* 1 (1955), pp. 13–103.

Funkenstein, Amos. *Theology and the Scientific Imagination From the Middle Ages to the Seventeenth Century*. Princeton, 1966.

Gadoffre, G. F. A. "Sur la chronologie du *Discours de la méthode.*" *Revue d'histoire de la philosophie* 11 (1943), pp. 45–70.

Gäbe, Lüder. *Descartes Selbstkritik. Untersuchungen zur Philosophie des jungen Descartes.* Hamburg, 1972.

———. "La Règle XIV. Lien entre géométrie et algèbre," *Archives de philosophie* 46 (1983), pp. 654–60.

Galuzzi, Massimo. "Recenti interpretazioni della *Géométrie* di Descartes." In *Scienza e Filosofia. Saggi in onore di Ludovico Geymonat.* A curadi Corrado Mangione. Milan, 1985, Pp. 643–663.

Gardies, Jean-Louis. "La Méthode mécanique et la Platonisme d'Archimède." *Revue philosophique de la France et l'Etranger* 170 (1980), pp. 39–43.

———. "Pascal et l'axiom d'Archimede." *Revue philosophique de la France et l'Etranger* 171 (1981), pp. 425–40.

———. "Eudoxe et Dedekind." *Revue d'histoire des sciences* 37 (1984), pp. 119–25.

Gasché, Rudolphe. *The Tain of the Mirror. Derrida and the Philosophy of Reflection.* Cambridge, Mass., 1986.

Giacobbe, Giulio Cesare. "Il *Commentarium de certitudine mathematicarum disciplinarum* di Alessandro Piccolomini." *Physis* 14 (1972).

———. "Francesco Barozzi e la *Quaestio de certitudine mathematicarum. Physis* 14 (1972), pp. 357–74.

———. "La Riflessione metamatematica di Pietro Catena." *Physis* 15 (1973), pp. 178–96.

Gibson, A. Boyce. "La 'Géométrie' de Descartes au point de vue de sa méthode." *Revue de métaphysique et de morale* 4 (1896), pp. 386–98.

Gilbert, Neal W. *Renaissance Concepts of Method.* 2nd ed. New York, 1963.

Görland, Ingetrud. "Rez. von B. Taureck, *Das Schicksal der philosophischen Konstruktion.*" *Hegel-Studien* 13 (1978), pp. 325–32.

Goichon, Anne-Marie. *La distinction de l'essence et de l'existence d' après Ibn Sina (Avicenne).* Paris, 1937.

Goldat, George D. "The Early Medieval Traditions of Euclid's Elements." Ph.D. Diss. Univ. of Wisconsin, 1956.

Goldschmidt, Victor. "HYPARXEIN et HYPHISTANAI dans la philosophie stoicienne." *Revue des études grecques* 85 (1972), pp. 331–44.

Gomez-Lobo, Alfonso. "Aristotle's Hypotheses and the Euclidean Postulates." *Review of Metaphysics* 30 (1977), pp. 430–39.

———. "The So-Called Question of Existence in Aristotle. *An Post*, 2.1–2. *Review of Metaphysics* 34 (1980), pp. 71–89.

Goodman, Nelson. *The Structure of Appearance.* 2nd ed. Indianapolis, 1966.

———. *The Ways of Worldmaking.* Indianapolis, 1978.

Gouhier, Henri. *Les Premières pensées de Descartes.* Paris, 1958.

Graeser, Andreas. "The Stoic Theory of Meaning." In *The Stoics.* Ed. John Rist. Berkeley, 1978. Pp. 77–100.

Granger, Gilles-Gaston. *Essai d'une philosophie du style.* Paris, 1968.

———. "Philosophie et mathématique leibniziennes." *Revue de métaphysique et de morale* 86 (1981), pp. 231–63.

Grimaldi, Nicolas. *L'Experience de la pensée dans la philosophie de Descartes*. Paris, 1978.

———. "Sur la volonté de l'homme chez Descartes et notre ressemblance avec Dieu." *Archives de philosophie* 50 (1987), pp. 95–107.

Guenancia, Pierre. "Remarques sur le rejet cartésien de l'histoire.".*Archives de philosophie* 49 (1986), pp. 561–70.

Guerin, Michel. "Le malin génie et l'instauration de la pensée comme philosophie." *Revue de métaphysique et de morale* 79 (1974), pp. 145–76.

Gutzkow, Karl. *Die Mode und das Moderne* [1836]. In *Werke* Hrsg. von Reinhold Genzel. Berlin, 1910. Bd. 2.

Guyer, F. E. " 'C'est nous qui sommes les anciens.' " *Modern Language Notes* 36 (1921), pp. 257–64.

Hadot, Pierre. "L'Etre et l'étant dans le Néoplatonisme." In *Etudes Néoplatoniciennes*. Neuchatel, 1973. Pp. 27–39.

———. "Zur Vorgeschicte des Begriffs 'Existenz': *HYPARXEIN* bei den Stoikern." *Archiv für Begriffsgeschichte* 13 (1969), pp. 115–27.

Hankel, H. *Theorie des complexen Zahlen-systems*. Leipzig, 1867. I. Teil.

Hankins, Thomas L. "Algebra as Pure Time: William Rowan Hamilton and the Foundations of Algebra." In *Motion and Time, Space and Matter*. Ed. R. Turnbull and P. Machamer. Columbus, Ohio, 1976. Pp. 327–59.

Harvey, Warren Zev. "A Third Approach to Maimonides' Cosmogony-Prophetology Puzzle." *Harvard Theological Review* 74 (1981), pp. 287–301.

Henry, John. "Francesco Patrizi da Cherso's Concept of Space and its Later Influence." *Annals of Science* 36 (1979), pp. 549–73.

Hintikka, Jaakko. "Kant and the Tradition of Analysis." In *Logic, Language-Games and Information*. Oxford, 1973. Ch. 9.

———. " 'Is,' Semantical Games and Semantical Relativity." *Journal of Philosophical Logic* 8 (1979), pp. 433–68.

———. "Semantical Games, the Alleged Ambiguity of 'Is' and the Aristotelian Categories." *Synthese* 54 (1983), pp. 443–68.

———. "The Varieties of Being in Aristotle." In *The Logic of Being*, Ed. S. Knuuttila and J. Hintikka. Dordrecht and Boston, 1986. Pp. 81–114.

———, and Remes, Unto. *The Method of Analysis. Its Geometrical Origin and its General Significance*. Dordrecht, 1974 (= *Boston Studies in the Philosophy of Sciences*, vol. 25).

Hösle, Vittorio. "Platons Grundlegung der Euklidizität der Geometrie." *Philologus* 126 (1982), pp. 184–97.

Hüber, Gerhard. *Das Sein und das Absolute. Studien zur Geschichte der ontologischen Problematik in der spätantiken Philosophie*. Basel, 1955 (= *Studia Philosophica*, Supplementum 5).

Hyppolite, Jean. "Du Sens de la géométrie de Descartes dans son oeuvre." In *Figures de la pensée philosophique*. Paris, 1971, T. 1, pp. 7–19. (Originally published in *Descartes. Cahiers de Royaumont*. Philosophie, 2. Paris, 1957).

Itard, Jean. "L'Introduction à la géométrie de Pascal." In *L'Oeuvre scientifigue de Pascal*. Preface de Rene Taton. Paris, 1964. Pp. 102–19.

Ivry, Alfred, "Maimonides on Creation." In *Creation and the End of Days. Judaism and Scientific Cosmology*. Ed. D. Novak and N. Samuelson Lonhom, Md., 1986. Pp. 185–213.

Jabre, Farid. "EINAI et ses derivés dan la traduction, en arabe, des Catégories d'Aristote." *Mélanges de la Faculté orientale de l'université de Saint Joseph (Beyrouth)* 48 (1973–74), pp. 243–68.

Jakobson, Roman and Paolo Valesio. "Vocabulorum constructio in Dante's Sonnet 'Se vedi li occhi miei.' " *Studi Danteschi* 43 (1966), pp. 7–33.

Jauss, Hans-Robert. "Literarische Tradition und gegenwärtiges Bewusstsein der Modernität." In *Aspekte der Modernität.* Hrsg. von Hans Steffen. Göttingen, 1965. Pp. 150–97.

Judovitz, Dalia. "Autobiographical Discourse and Critical Praxis in Descartes." *Philosophy and Literature* 5 (1981), pp. 91–107.

———. *Subjectivity and Representation in Descartes. The Origins of Modernity.* Cambridge, 1988.

Kahl-Fuhrmann, G. 'Descartes Betonung seiner Unabhängigkeit von der Tradition und Leibnizens Kritik." *Zeitschrift für philosophische Forschung* 4 (1949/50), pp. 377–84.

Kahn, Charles. "On the Terminology for Copula and Existence." In: *Islamic Philosophy and the Classical Tradition.* Ed. S. M. Stern, et al. Oxford, 1972, pp. 141–58 (= *Oriental Studies* 5).

———. *The Verb "Be" in Ancient Greek.* Dordrecht, 1973.

———. "Why Existence Does Not Emerge as a Distinct Concept in Greek Philosophy?" *Archiv für Geschichte der Philosophie* 58 (1976), pp. 323–34.

Kaplan, Laurence. "Maimonides on the Miraculous Element in Prophecy." *Harvard Theological Review* 70 (1977), pp. 233–56.

Kennington, Richard. "The 'Teaching of Nature' in Descartes' Soul Doctrine." *Review of Metaphysics* 26 (1972), pp. 86–117.

Klein, Jacob. *Greek Mathematical Thought and the Origin of Algebra.* Trans. Eva Brann. Cambridge, Mass., 1968.

———. "The World of Physics and the 'Natural' World." Trans. D. R. Lachterman. In *Lectures and Essays,* J. Klein. Ed. R. Williamson and E. Zuckerman. Annapolis, 1985. Pp. 1–34.

Knorr, Wilbur R. *The Evolution of the Euclidean Elements* Dordrecht, 1975 (= *Synthese Historical Library,* 15).

———. "Archimedes' Neusis-Constructions in Spiral Lines." *Centaurus* 22 (1978), pp. 77–98.

———. "Aristotle and Incommensurability. Some Further Reflections," *Archive for History of Exact Sciences* 24 (1981), pp. 1–8.

———. "The Hyperbola-Construction in the *Conics,* Book II: Ancient Variations on a Theorem of Apollonius." *Archive for History of Exact Sciences* 25 (1982), pp. 253–91.

———. "Infinity and Continuity: The Interaction of Mathematics and Philosophy in Antiquity." In: *Infinity and Continuity in Ancient and Medieval Thought.* Ed. Norman Kretzmann. Ithaca, N.Y., 1982. Pp. 112–45.

———. "Construction as Existence Proof in Ancient Geometry." *Ancient Philosophy* 3 (1983), pp. 125–48.

———. *The Ancient Tradition of Geometric Problems.* Boston, Basel, Stuttgart, 1986.

Kojève, Alexandre. "The Christian Origin of Modern Science." Trans. D. R. Lachterman. *The St. John's Review* 35 (Winter, 1984), pp. 22–26. (French original in *Mélanges Alexandre Koyré, II: L'Aventure de l'esprit.* Paris, 1964. Pp. 295–306.

244   Bibliography

Koselleck, Reinhart. "'Historia magistra vitae.' Uber die Auflösung des Topos im Horizont neuzeitlich bewegter Geschichte." In *Natur und Geschichte. Karl Löwith zum 70. Geburtstag.* Hrsg. von H. Brann und M. Riedel. Stuttgart, 1967. Pp. 196–216.

————. " 'Neuzeit.' Zur Semantik moderner Bewegungsbegriff." In *Studien zum Beginn der modernen Welt.* Hrsg. von R. Koselleck. Stuttgart, 1980. Pp. 264–99.

Krafft, Fritz. "Kunst and Natur. Die Heronische Frage und die Technik in der klassischen Antike." *Antike und Abendland* 19 (1973), pp. 1–19.

Kraus, Pamela A. "The Structure and Method of the *Regulae ad directionem ingenii.*" Ph.D. Diss. Catholic University of America, 1980.

————. "From Universal Mathematics to Universal Method: Descartes' 'Turn' in Rule IV of the *Regulae.*" *Journal of the History of Philosophy* 21 (1983), pp. 159–74.

Kutler, Samuel. "The Source of the Source of the Dedekind Cut." In *Essays in Honor of Jacob Klein.* Annapolis, Md., 1976. Pp. 87–94.

Laberthonnière, Lucien. *Etudes sur Descartes.* In *Oeuvres.* Ed. L. Canet. Paris, 1935. T. 2.

Lachterman, David R. "Review of J. Ritter, hrsg. *Historisches Wörterbuch der Philosophie,* Bde. I–IV." *Archiv für Geschichte der Philosophie* 61 (1979), pp. 196–205.

————. "Hegel and the Formalization of Logic." *Graduate Faculty Philosophy Journal* 12 (1987), pp. 153–236.

————. "Vico, Doria e la geometria sintetica." *Bollettino del Centro di studi vichiani* 10 (1981), pp. 10–35.

————. "Descartes and the Philosophy of History." *Independent Journal of Philosophy* 4 (1983), pp. 31–46.

————. "Vico and Marx: Notes on a Precursory Reading." In *Vico and Marx. Affinities and Contrasts.* Ed. Giorgio Tagliacozzo. Atlantic Highlands, N.J., 1983). Pp. 38–61.

————. "Vico, Nominalism and Mathematics." In *Sachkommentar zu Vicos Liber Metaphysicus.* Hrsg. von S. Otto und H. Viechtbauer. Munich, 1985. Pp. 47–85.

————. "*Objectum Purae Matheseos.* Mathematical Construction and the Passage from Essence to Existence." In *Essays on Descartes' Meditations.* Ed. A. Rorty. Berkeley, 1985. Pp. 435–58.

Laporte, Jean. "La Connaissance de l'étendue chez Descartes." *Revue philosophique de la France et de l'Etranger* 123 (1937), pp. 257–89. (Reprinted in *Etudes d'histoire de la philosophie française au XVII᷎ siècle.* Paris, 1951. Pp. 11–36.)

Lask, Emil. *Fichtes Idealismus und die Geschichte* (= *Gesammelte Schriften,* Ed. Eugen Herrigel, Tübingen, 1923. Bd. 1, pp. 3–273).

Lauer, Helen. "Descartes' Concept of Number." *Studia Cartesiana* 2 (1981), pp. 137–42.

Lebrun, Gerard. "La Notion de 'ressemblance' de Descartes à Leibniz." In *Sinnlichkeit und Verstand in der deutschen und französischen Philosophie von Descartes bis Hegel.* Hrsg. von Hans Wagner. Bonn, 1979. Pp. 39–57.

Lee, H. D. P. "Geometrical Method and Aristotle's Account of First Principles." *Classical Quarterly* 29 (1935), pp. 113–24.

Lenoir, Timothy J. "Descartes and the Geometrization of Thought: The Methodological Background of Descartes' *Géométrie.*" *Historia Mathematica* 6 (1979), pp. 355–79.

Liljenwall, James D. "Kepler's Theory of Knowledge. An Inquiry into Book 1 of Johannes Kepler's *Harmonice Mundi* in Light of its Greek Roots and in Relation to the Development of Renaissance Algebra." Ph.D. Diss. Univ. of California, San Diego, 1976.

Lohmann, Johannes. "St. Thomas et les Arabes. (Structures linguistiques et formes de pensée." *Revue philosophique de Louvain* 74 (1976), pp. 30–44.

———. "Descartes' *Compendium Musicae* und die Entstehung des neuzeitlichen Bewusstsein." *Archiv für Musikwissenschaft* 36 (1979), pp. 81–104.

Lohne, J. A. "Essays on Thomas Harriot." *Archive for History of Exact Sciences* 20 (1979), pp. 189ff.

Long, A. A. "Language and Thought in Stoicism." In *Problems in Stoicism.* Ed. A. A. Long. London, 1971. Pp. 75–113.

Lorenzen, Paul, and Wilhelm. *Logische Propädeutik.* Mannheim, 1967.

Lyons, John D. "Subjectivity and Imitation in the *Discours de la methods.*" *Neophilologus* 66 (1982), pp. 508–24.

Mahdi, Muhsin. "Science, Philosophy and Religion in Alfarabi's *Enumeration of the Sciences.*" In *The Cultural Context of Medieval Learning.* Ed. John Murdoch and Edith Sylla. Dordrecht, 1975. Pp. 113–47.

Mahoney, Michael A. "The Royal Road. The Development of Algebraic Analysis from 1550 to 1650, with Special Reference to the Work of Pierre de Fermat." Ph.D. Diss. Princeton Univ., 1967.

Mansfeld, Jaap. "Intuitionism and Formalism: Zeno's Definition of Geometry in a Fragment of L. Calvenus Taurus." *Phronesis* 28 (1983), pp. 59–74.

Maracchia, Silvio. "Aristotele e l'incommensurabilità." *Archive for History of Exact Sciences* 21 (1980), pp. 201–28.

Marion, Jean-Luc. *L'Ontologie grise de Descartes. Savoir aristotélicien et science cartésienne.* Paris, 1975.

———. *Sur la théologie blanche de Descartes.* Paris, 1981.

Marmura, Michael E. "Avicenna on Causal Priority." In *Islamic Philosophy and Mysticism.* Ed. P. Morewedge. Delmar, N.Y., 1981. Pp. 65–83.

Martini, F. "Artikel: 'Modern, Die Moderne.' " In *Reallexicon der deutschen Literaturgeschichte.* Berlin, 1961. Bd. 2, coll. 391–415.

Matheron, Alexandre. "Psychologie et politique: Descartes, la noblesse du chatouillement." *Dialectiques* 6 (1974), pp. 79–98.

Mathiesen, Thomas J. "An Annotated Translation of Euclid's Division of the Monochord." *Journal of Music Theory* 19 (1975), pp. 236–58.

Matten, Mohan. "Greek Ontology and the 'Is' of Truth." *Phronesis* 28 (1983), pp. 113–35.

Maula, Erkka. "An End of Invention." *Annals of Science* 38 (1981), pp. 109–22.

Mayberry, John. "A New Begriffsschrift (1)." *British Journal for the Philosophy of Science* 31 (1980), pp. 231–54.

Michaux, Maurice. *La Commentaire de Marinus aux Data d'Euclide. Etude Critique.* Louvain, 1947.

Mignucci, Mario. *Il Significato della logica stoica.* Bologna, 1965.

———. *L'Argomentazione dimostrativa in Aristotele. Commento agli Analitici Secondi.* Padua, 1975. Vol. 1.

———. "In margine el concetto aristotelico di esistenza." In *Scritti in onore di Carlo Giacon.* Bologna, 1975. Pp. 227–61.

Mill, John Stuart. *An Examination of Sir William Hamilton's Philosophy.* In *The Collected Works of John Stuart Mill.* Toronto, 1979. Vol. 9.

Mitrovitch, Radicha. "La Théorie des sciences chez Descartes d'après sa Géométrie." (Thèse du Doctorat. Paris, 1932.

Mittelstrass, Jürgen. "Die Idee einer *mathesis universalis* bei Descartes." *Perspektiven der Philosophie* 4 (1978), pp. 177–92.

Molland, Andrew G. "The Denomination of Proportions in the Middle Ages." *Actes du Xi^{ème} Congrès International d'Histoire des Sciences* (Wroclaw, 1968), T. 3, pp. 167–70.

――――. "Shifting the Foundations. Descartes' Transformation of Ancient Geometry." *Historia Mathematica* 3 (1976), pp. 21–49.

――――. "An Examination of Bradwardine's Geometry." *Archive for History of Exact Sciences* 19 (1978), pp. 114–75.

Morewedge, Parviz. "Philosophical Analysis and Ibn Sina's 'Essence-Existence' Distinction." *Journal of the American Oriental Society* 92 (1972), pp. 425–35.

Mourelatos, Alexander. "Plato's 'Real Astronomy': Republic 527D–531D." In *Science and the Sciences in Plato.* Ed. John P. Anton. New York, 1980. Pp. 33–73.

Mueller, Ian. "Greek Logic and Greek Mathematics." In *Ancient Logic and its Modern Interpretations.* Ed. John Corcoran Dordrecht, 1974. Pp. 35–70.

――――. "Homogeneity in Eudoxus' Theory of Proportion." *Archive for History of Exact Sciences* 7 (1970), pp. 1–6.

――――. *Philosophy of Mathematics and Deductive Structure in Euclid's Elements.* Cambridge, Mass., 1981.

Mugler, Charles. *Dictionnaire historique de la terminologie géométrique des Grecs.* Paris, 1959.

Murdoch, John. "The Medieval Language of Proportions." In *Scientific Change.* Ed. A. C. Crombie. New York, 1963. Pp. 237–71, 334–43.

――――. "Euclides Graeco-Latinus: A Hitherto Unknown Translation." *Harvard Studies in Classical Philology* 71 (1966), pp. 249–302.

Nancy, Jean-Luc. "Larvatus Pro Deo." *Glyph* 2 (1977), pp. 14–37.

――――. "Mundus est Fabula." *Modern Language Notes* 93 (1978), pp. 635–653.

Niebel, Eckhard. *Untersuchungen über die Bedeutung der geometrischen Konstruktion in der Antike.* Köln, 1959 (= *Kant-Studien, Ergänzungsheft* 76).

Nikolic, Milenko. "The Relation between Eudoxus' Theory of Proportions and Dedekind's Theory of Cuts." In *For Dirk Struik.* Ed. Robert S. Cohen et al. Dordrecht, 1974. Pp. 225–43.

Noussan-Lettry, Luis. "Die Anerkennung des Historischen in der Lebenserfahrung und der Weg des Denkens bei Descartes." *Philosophisches Jahrbuch* 80 (1973), pp. 15–37.

Nussbaum, Martha. "Saving Aristotle's Appearances." In *Language and Logos. Studies in Ancient Greek Philosophy presented to G. E. L. Owen.* Ed. M. Schofield and M. C. Nussbaum. Cambridge, 1982. Pp. 267–93.

Owen, G. E. L. "*Tithénai ta phainomena.*" In *Aristote et les problèmes de méthode.* Louvain, 1961. Pp. 83–103.

Pachet, Pierre. "La deixis selon Zenon et Chrysippe." *Phronesis* 20 (1975), pp. 241–46.

Pape, Ingetrud. *Tradition und Transformation der Modalität* (Hamburg, 1966), Bd. 1: *Möglichkeit-Unmöglichkeit.*

Pedersen, Olaf. "Logistics and the Theory of Functions. An Essay in the History of Greek Mathematics." *Archives internationales d'histoire des sciences* 24 (1974), pp. 29–50.

Peirce, Charles Sanders. *The New Elements of Mathematics.* Ed. Carolyn Eisele. The Hague, 1976. Vols. 1–4.

Peters, Wilhem S. "Johann Heinrich Lamberts Konzeption einer Geometrie auf einer imaginären Kugel." *Kant-Studien* 53 (1961), pp. 51–67.

Pinborg, Jan. "Can Constructions Be Construed? A Problem in Medieval Syntactical Theory." *Historiographia Linguistica* 7 (1980), pp. 201–10.

Pines, Shlomo. "The Arabic Recension of *Parva Naturalia* and the Philosophical Doctrine concerning Veridical Dreams according to *al-Risala al-Manamiyya* and Other Sources." *Israel Oriental Studies* 4 (1974), pp. 104–153. (Reprinted in S. Pines. *Collected Works.* Jerusalem, 1986. Vol. 2, pp. 96–145.)

Plooij, Edward B. *Euclid's Conception of Ratio and His Definition of Proportional Magnitude as Criticized by Arabian Commentators.* Rotterdam, 1950.

Proceedings of the Pisa Conference on the History and Philosophy of Science, 1978. Vol. 1: *Theory Change, Ancient Axiomatics and Galileo's Methodology.* Dordrect, 1981. Pp. 113–225.

Rahman, F. "Essence and Existence in Avicenna." *Medieval and Renaissance Studies* 4 (1958), pp. 1–16.

Randall, John Hermann, Jr. "The Development of Scientific Method in the School of Padua." In *Renaissance Essays.* Ed. P. O. Kristeller and Philip P. Wiener. New York, 1968. Pp. 217–51.

Renan, Ernest. *L'Avenir de la science.* 4th ed. Paris, 1890.

Ricoeur, Paul. "Review of N. Goodman, *The Ways of Worldmaking.*" *Philosophy and Literature* 4 (1980), pp. 107–20.

Risse, Wilhelm. "Zur Vorgeschichte der cartesischen Methodenlehre." *Archiv für Geschichte der Philosophie* 45 (1963), pp. 269–91.

Robert, André. "Descartes et l'analyse des anciens." *Archives de philosophie* 13 (1937), pp. 221–45.

Rochid, Amine. "Dieu et l'être selon al'Farabi: le chapitre de 'l'être' dans le Livre des Lettres." In *Dieu et l'Être. Exégèses d'Exode 3, 14 et Coran 20, 11–24.* Paris, 1978. Pp. 179–90.

Rodis-Lewis, Geneviève. "Machineries et perspectives curieuses dans leurs rapports avec le Cartésianisme." *Dix-Septième Siècle* 5 (1956), pp. 461–74.

———. "Le Dernier fruit de la métaphysique cartésienne: La Générosité." *Les Etudes philosophiques* (1987), pp. 43–54.

Romanos, George D. *Quine and Analytical Philosophy. The Language of Language.* Cambridge, Mass., 1983.

Rorty, Richard. *Philosophy and the Mirror of Nature.* Princeton, 1979.

Rose, Paul Lawrence. *The Italian Renaissance of Mathematics. Studies on Humanists and Mathematicians from Petrarch to Galileo.* Geneva, 1975.

———. "A Venetian Patron and Mathematician of the Sixteenth Century: Francesco Barozzi (1537–1604)." *Studi Veneziani*, n. s. 1 (1976), pp. 119–80.

Rosen, Stanley. *Plato's Sophist: The Drama of Original and Image.* New Haven, 1983.

Russell, Bertrand. "Review of E. Boutroux, *L'Imagination et les mathématiques selon Descartes.*" *Mind*, n. s. 11 (1902), pp. 108–109.

Sacksteder, William. "Hobbes: The Art of the Geometricians." *Journal of the History of Philosophy* 18 (1980), pp. 131–46.

Saito, Ken. "Compounded Ratio in Euclid and Apollonius." *Historia Scientiarum* 31 (1986), pp. 25–59.

Sambursky, Samuel. *The Physics of the Stoics*. London, 1959.

Sato, Tohru, "A Reconstruction of *The Method* Proposition 17, and the Development of Archimedes' Thought on Quadrature," *Historia Scientiarum* 31 (1986), pp. 61–86; 32 (1987), pp. 75–142.

Sauer, Friedrich Otto. "Euklid und der Operationalismus." *Philosophische Perspektiven* 3 (1971), pp. 217–49.

———. *Physikalische Begriffsbildung und mathematisches Denken*. Amsterdam, 1977.

Scarry, Elaine. *The Body in Pain. The Making and Unmaking of the World*. Oxford, 1985.

Schmidt, Robert H. "The Analysis of the Ancients and the Algebra of the Moderns." *A Golden Hind Editorial*. Fairfield, Conn., 1987. Pp. 1–15.

Scholz, Heinrich. "Warum haben die Griechen die Irrationalzahlen nicht aufgebaut?" *Kant-Studien* 33 (1928), pp. 33–72.

Schüling, Hermann. *Die Geschichte der axiomatischen Methode im 16. und beginnenden 17. Jahrhundert*. Hildesheim, 1969 (= *Studien und Materialien zur Geschichte der Philosophie*, Bd. 13).

Schuster, John A. "Descartes and the Scientific Revolution, 1618–1634: An Interpretation." Ph.D. Diss. Princeton Univ., 1977.

———. "Cartesian Method as Mythic Speech: A Diachronic and Structural Analysis." In *The Politics and Rhetoric of Scientific Method*. Ed. J. A. Schuster and R. R. Yeo (Dordrecht and Boston, 1986. Pp. 33–95.

Sedley, David. "Philoponus' Conception of Space." In *Philoponus and the Rejection of Aristotelian Science*. Ed. Richard Sorabji. Ithaca, N.Y., 1987. Pp. 140–53.

Serres, Michel. *Hermes. Literature, Science, Philosophy*. Ed. Josué V. Harari and David F. Bell. Baltimore, 1982.

———. *Hermès V: Le Passage du Nord-Ouest*. Paris, 1980.

Simpson, Thomas K. "Newton and the Liberal Arts." *The St. John's Review* 27 (January 1976), pp. 1–11.

Sirven, Jules. *Les Années d'apprentissage de Descartes, 1596–1628*. Albi, 1928.

Smith, A. Mark. "Saving the Appearances of Appearances. The Foundation of Classical Geometrical Optics." *Archive For History of Exact Sciences* 24 (1981), pp. 73–99.

Smith, J. A. "*TODE TI* in Aristotle." *Classical Review* 35 (1921), p. 19.

Stead, Christopher. *Divine Substance*. Oxford, 1977.

Steele, Arthur Donald. "Uber die Rolle von Zirkel und Lineal in der griechischen Mathematik." *Ouellen und Studien zur Geschichte der Mathematik Astronomie und Physik*, Abt. B. Studien 3 (1934), pp. 288–369.

Stenzel, Julius. *Zahl und Gestalt bei Platon und Aristoteles*. 3 Auflage. Bad Homburg, 1959.

Strauss, Leo. "Maimuni's Lehre von der Prophetie und ihre Quellen." *Le Monde oriental* 28 (1934), pp. 99–139.

Sude, Barbara H. "Ibn Al-Haytham's Commentary on the Premises of Euclid's Elements." Ph.D. Diss. Princeton Univ., 1974.

Sylla, Edith. "Compounding Ratios. Bradwardine, Oresme and the First Edition of Newton's *Principia*." In *Transformation and Tradition in the Sciences. Essays in Honor of I. Bernard Cohen*. Ed. E. Mendelssohn. Cambridge, 1984. Pp. 11–43.

Szabó, Árpád. "Greek Dialectic and Euclid's Axiomatics." In *Problems in the Philosophy of Mathematics*. Ed. I. Lakatos. Amsterdam, 1967. Pp. 1–27.

————. *Anfänge der griechischen Mathematik*. Munich, 1969.

————. "Analysis und Synthesis. (Pappus II. p.634ff. Hultsch)." *Acta Classica Universitatis Scientiarum Debrecensis* [Hungary], 10–11 (1974–75), pp. 155–164. [Not identical with the reply by Szabo published in *Hintikka/Remes*, q.v., pp. 118–30.]

Tannery, Paul. *Mémoires scientifiques*. Toulouse and Paris, 1912. T. 2.

Tarán, Leonardo. *Speusippus of Athens. A Critical Study with a Collection of the Related Texts and Commentary*. Leiden, 1981 (= *Philosophia Antique*, vol. 39).

Tarrant, Harold. "Zeno on Knowledge or on Geometry? The Evidence of Anon. *In Theaetetum*." *Phronesis* 29 (1984), pp. 96–99.

Taton, René. "L'Oeuvre de Pascal en géométrie projective." In *L'Oeuvre scientifique de Pascal*. Paris, 1964. Pp. 17–72.

Taureck, Bernhard. *Das Schicksal der philosophischen Konstruktion*. Munich, 1975 (= *Uberlieferung und Ausgabe*, Bd. 14).

Timpanaro-Cardini, Maria. "Two Questions of Greek Geometrical Terminology." In *Kephalaion. Studies in Greek Philosophy and its Continuation Offered to Professor C. J. de Vogel*. Ed. J. Mansfield and L. M. de Rijk. Assen, 1975. Pp. 183–88.

Toth, Imre. "Das Parallelenproblem im Corpus Aristotelicum." *Archive for History of Exact Sciences* 3 (1967), pp. 249–422.

Trendelenburg, Adolf. *Erläuterungen zu den Elementen der aristotelischen Logik*. Berlin, 1876.

Trimpi, Wesley. "The Quality of Fiction: The Rhetorical Transmission of Literary Theory." *Traditio* 30 (1974), pp. 1–118.

————. *Muses of One Mind. The Literary Analysis of Experience and Its Continuity*. Princeton, 1983.

Turrière, Emile. "La notion de transcendence géométrique chez Descartes et Leibniz." *Isis* 2 (1914), pp. 106–124.

Valéry, Paul. "Descartes." In *Masters and Friends*. Princeton, 1968. Pp. 13–35 (= *The Collected Works of Paul Valéry*, vol. 9).

Van de Pitte, Frederick. "Descartes' *Mathesis Universalis*." *Archiv für Geschichte der Philosophie* 61 (1979), pp. 154–74.

van der Waerden, B. L. *Science Awakening*. 2nd ed. New York, 1961.

————. "Die Postulate und Konstruktionen in der frühgriechische Geometrie." *Archive for History of Exact Sciences* 18 (1978), pp. 343–57.

Vasoli, Cesare. *La dialettica e la retorica dell' Umanesimo*. Milan, 1968.

————. "Fondamento e metodo logico della geometria nell' *Eucides Restitutus* del Borelli." *Physis* 11 (1969), pp. 571–98.

Verra, V. "La 'Construction' dans la philosophie de Schelling." In *Actualité de Schelling*. Paris, 1979. Pp. 27–47.

Vogelsang, Erich. *Der junge Luther*. Berlin, 1933.

Vuillemin, Jules. *Mathématique et métaphysique chez Descartes*. Paris, 1960.

————. *La philosophie de l'algèbre*. Paris, 1962. T. 1.

Wagner, Robert J., "Euclid's Intended Interpretation of Superposition." *Historia Mathematica* 10 (1983), pp. 63–89.

Watzlawick, Paul, ed. *The Invented Reality*. New York, 1984.

Weber, J.-M. *La Constitution du texte des "Regulae."* Paris, 1964.

Weiss, Ulrich. "Wissenschaft als menschliches Handeln. Zu Thomas Hobbes' anthropologischer Fundierung von Wissenschaft." *Zeitschrift für philosophische Forschung* 37 (1983), pp. 37–55.

Wieland, Wolfgang. "Zur Raumtheorie des Johannes Philoponus." In *Festschrift für Joseph Klein zum 70. Geburtstag*. Hrsg. von Erich Fries. Göttingen, 1967. Pp. 114–35.

Wilson, J. Cook. "On the Geometrical Problem in Plato's *Meno* 86E sqq." *Journal of Philology* 28 (1903), pp. 114–35.

Wolters, Gereon. "'Theorie' und 'Ausübung' in der Methodologie von Johann Heinrich Lambert." In *Studia Leibnitiana*, Supplementa, vol. 19: *Theoria cum Praxi. Zum Verhältnis von Theorie und Praxis im 17. und 18. Jahrhundert*. Wiesbaden, 1980. Bd. 1, pp. 109–14.

Youschkevitch, A. P. *Les Mathématiques arabes (VIIIᵉ-XVᵉ siècles)*. Trad. par M. Cazenave et K. Jaoviche. Paris, 1976.

Yovel, Yirmiahu. *Kant and the Philosophy of History*. Princeton, 1980.

Zeuthen, H. G. "Notes sur l'histoire des mathématiques, IX: Sur les connaissances géométrique des Grecs avant la reforme platonicienne." *Oversigt over det Kongelige Danske Videnskabernes Selskabs Forhandlinger* (1913), pp. 431–73.

————. "Die geometrische Konstruktion als 'Existenzbeweis' in der antiken Geometrie." *Mathematische Annalen* 47 (1896), pp. 222–228.

Zöller, Günter. *Theoretische Gegenstandsbeziehung bei Kant. Kant-Studien*, Ergänzungsheft 117. Berlin and New York, 1984.

Index